# Advanced Computational and Mathematical Approaches in Applied Differential Equations

# Advanced Computational and Mathematical Approaches in Applied Differential Equations

Edited by

Snehashish Chakraverty
Department of Mathematics
Applied Mathematics Group
National Institute of Technology Rourkela
Rourkela, Odisha, India

Srinivas Suripeddi
Department of Mathematics
VIT-AP University
Amaravati, Andhra Pradesh, India

Karunakar Perumandla
Department of Mathematics
School of Advanced Sciences
VIT-AP University
Amaravati, Andhra Pradesh, India

MK
MORGAN KAUFMANN PUBLISHERS
ELSEVIER AN IMPRINT OF ELSEVIER

Morgan Kaufmann is an imprint of Elsevier
50 Hampshire Street, 5th Floor, Cambridge, MA 02139, United States



ISBN: 978-0-443-45743-2

For information on all Morgan Kaufmann publications
visit our website at https://www.elsevier.com/books-and-journals

*Publisher:* Mara Conner
*Acquisitions Editor:* Chris Katsaropoulos
*Editorial Project Manager:* Vicky Sharma
*Production Project Manager:* M. Balakrishnan
*Cover Designer:* Raman Kumar

Typeset by VTeX

*This book is dedicated to the researchers, educators, and students whose commitment to the pursuit of knowledge continues to expand the frontiers of computational mathematics and differential equations. We further dedicate this volume to the global scientific community, whose interdisciplinary contributions inspire the innovative mathematical modeling and analytical techniques essential to addressing complex real-world phenomena.*

# Contents

# Contributors

**Sana Abdulla**
Department of Mathematics, VIT-AP University, Amaravati, Andhra Pradesh, India

**B. Bala Sai Sankar**
Department of Mathematics, School of Advanced Sciences, VIT-AP University, Amaravati, India

**Sreenivasulu Ballem**
Department of Mathematics, Central University of Karnataka, Kalaburagi, India

**Thameem Basha**
Ulsan National Institute of Science and Technology, Department of Mathematical Sciences, Ulsan, South Korea

**Veeresh C**
Sri Venkateswara University, Tirupati, India

**S. Chakraverty**
National Institute of Technology Rourkela, Department of Mathematics, Rourkela, India

**T. Dilleshwar Rao**
School of Advanced Sciences, Vellore Institute of Technology, Chennai, India

**Sreedhara Rao Gunakala**
The University of the West Indies St Augustine Campus, Department of Mathematics and Statistics, St. Augustine, Trinidad and Tobago

**Rajarama Mohan Jena**
ITER, Siksha 'O' Anusandhan University, Department of Mathematics, Bhubaneswar, Odisha, India

**Subrat Kumar Jena**
University of Cagliari, Department of Civil, Environmental and Architecture, Cagliari, Italy

**Perumandla Karunakar**
Department of Mathematics, School of Advanced Sciences, VIT-AP University, Amaravati, India

**Abeka Khare**
Shaheed Bhagat Singh Govt. Degree College, Ashta, India

**Venkateswara Konduru**
Sri Venkateswara College of Engineering, Tirupati, India

**Nihar Ranjan Mallick**
National Institute of Technology Rourkela, Department of Mathematics, Rourkela, India

**Kohilavani Naganthran**
Institute of Mathematical Sciences, Faculty of Science, Universiti Malaya, Kuala Lumpur, Malaysia
Center for Data Analytics, Consultancy and Services, Faculty of Science, Universiti Malaya, Kuala Lumpur, Malaysia

**Anasuya Nath**
PG Department of Mathematics, Utkal University, Vani Vihar, Bhubaneswar, India

**Triloki Nath**
Deen Dayal Upadhyay Gorakhpur University, Gorakhpur, India

**Ioan Pop**
Department of Mathematics, Babes-Bolyai University, Cluj-Napoca, Romania

**V. Pujitha**
Department of Mathematics, School of Advanced Sciences, VIT-AP University, Amaravati, Andhra Pradesh, India

**V. Raja**
Department of Mathematics, School of Advanced Sciences, VIT-AP University, Inavolu, Beside AP Secretariat, Amaravati, AP, India

**Vikash Ramcharitar**
The University of the West Indies St Augustine Campus, Department of Mathematics and Statistics, St. Augustine, Trinidad and Tobago

**K. Mahipal Reddy**
Department of Mathematics, VIT-AP University, Amaravati, Andhra Pradesh, India

**Jashobanta Sa**
PG Department of Mathematics, Government (Autonomous) College, Angul, Odisha, India

**Deepak Kumar Sah**
Department of Mathematics, Central University of Karnataka, Kalaburagi, India

**Arup Kumar Sahoo**
Hatter Department of Marine Technologies, University of Haifa, Haifa, Israel

**Basanta Kumar Sahoo**

PG Department of Physics, Government (Autonomous) College, Angul, Odisha, India

**Shivani Saini**

Department of Mathematics and Scientific Computing, National Institute of Technology, Hamirpur, Himachal Pradesh, India

**Zailan Siri**

Institute of Mathematical Sciences, Faculty of Science, Universiti Malaya, Kuala Lumpur, Malaysia

Center for Data Analytics, Consultancy and Services, Faculty of Science, Universiti Malaya, Kuala Lumpur, Malaysia

**Estemsetti Siva Sai**

Department of Mathematics, School of Advanced Sciences, VIT-AP University, Amaravati, Andhra Pradesh, India

**P.T. Sowndarrajan**

School of Advanced Sciences, VIT-AP University, Amaravati, India

School of Advanced Sciences, Vellore Institute of Technology, Chennai, India

**S. Srinivas**

Department of Mathematics, School of Advanced Sciences, VIT-AP University, Amaravati, India

**Smita Tapaswini**

PG Department of Mathematics, Government (Autonomous) College, Angul, Odisha, India

**Shobha V**

Government First Grade College, Kengeri, Mathematics, Bengaluru, India

**Kuppalapalle Vajravelu**

Department of Mathematics, Department of Mechanical, Materials and Aerospace Engineering, University of Central Florida, Orlando, FL, United States

**Rambabu Vana**

Department of Mathematics, School of Computer Science and Artificial Intelligence, SR University, Warangal, India

**Sibyala Vijayakumar Varma**

REVA University, Department of Mathematics, Bengaluru, India

**M. Varunkumar**

Department of Mathematics, School of Advanced Sciences, VIT-AP University, Amaravati, Andhra Pradesh, India

**Ramesh Kumar Vats**
Department of Mathematics and Scientific Computing, National Institute of Technology, Hamirpur, Himachal Pradesh, India

**M. VijayaLakshmi**
Department of Mathematics, School of Advanced Sciences, VIT-AP University, Inavolu, Beside AP Secretariat, Amaravati, AP, India

**Cheong Kai Zen**
Institute of Mathematical Sciences, Faculty of Science, Universiti Malaya, Kuala Lumpur, Malaysia

# About the editors

**Prof. Snehashish Chakraverty**

Prof. Snehashish Chakraverty is a Senior Professor in the Department of Mathematics (Applied Mathematics Group) at NIT Rourkela, India, with more than 30 years of teaching and research experience. He obtained his PhD from IIT Roorkee and completed postdoctoral research at the University of Southampton, UK, and Concordia University, Canada. He has also served as a visiting professor at universities in Canada and South Africa. His research interests encompass differential equations, computational mathematics, fluid and structural dynamics, uncertainty modeling, fuzzy systems, and soft computing techniques. He has authored and edited 33 books and published over 475 research articles in reputed journals. Prof. Chakraverty has supervised 25 PhD scholars and successfully completed numerous sponsored research projects. He serves on the editorial boards of several international journals and is the Chief Editor of the *International Journal of Fuzzy Computation and Modelling*. Recognized among the top 2% of scientists worldwide in the Stanford–Elsevier rankings, he has received several prestigious national and international awards for his contributions to applied mathematics and computational science.

**Prof. Srinivas Suripeddi**

Prof. Srinivas Suripeddi is Professor (higher academic grade) in the Department of Mathematics, School of Advanced Sciences, VIT-AP University, Amaravati. He earned his PhD in fluid dynamics from NIT Warangal and has over three decades of teaching and research experience. His research areas include physiological fluid dynamics, magnetohydrodynamics, nanofluids, and heat and mass transfer. He has published more than 110 research papers in reputed national and international journals and has guided several PhD and M.Phil. scholars. Prof. Srinivas has successfully completed research projects funded by organizations such as DRDO, DST, NBHM, and CSIR. He holds patents and has delivered invited lectures at numerous national and international conferences. His significant contributions to applied mathematics have earned him widespread recognition and over 3,300 citations.

**Dr. Karunakar Perumandla**

Dr. Karunakar Perumandla is an Assistant Professor of Mathematics at VIT-AP University, Amaravati. He received his PhD in Mathematics from NIT Rourkela, with research focused on shallow water wave equations, fuzzy and interval analysis, and numerical methods. He has more than 12 years of teaching experience and has served at several reputed institutions, including Amrita Vishwa Vidyapeetham, Anurag University, and IIIT Vadodara. His research interests include fuzzy modeling, fractional differential equations, wave dynamics, and computational fluid mechanics. Dr. Karunakar has published extensively in SCI/SCOPUS-indexed journals, has contributed book chapters and written scholarly books. He is a recipient of the BRNS-DAE Scholarship and the Prathibha Puraskar Award for academic excellence. His current research focuses on mathematical modeling under uncertainty and advanced computational techniques for applied sciences.

# Preface

Differential equations serve as the cornerstone of contemporary scientific inquiry, functioning as some of the most robust and versatile mathematical instruments available for accurately modeling, comprehending, and forecasting the intricate behaviors inherent in natural, engineering, and biological systems. In light of the rapid and transformative advancements currently unfolding within the fields of computational mathematics, artificial intelligence, fuzzy logic systems, fractional calculus, and high-precision numerical analysis, our collective capacity to rigorously analyze and solve increasingly sophisticated and nonlinear differential equation models has been expanded to unprecedented levels. These interdisciplinary developments not only enhance our theoretical understanding of dynamic systems but also provide the essential computational framework required to address the most pressing technical challenges in modern research and applied science.

This edited volume, *Advanced Computational and Mathematical Approaches in Applied Differential Equations*, brings together contemporary research contributions that demonstrate the breadth and depth of modern mathematical methodologies applied to differential equations. The chapters explore analytical, semianalytical, numerical, computational, and data-driven approaches for solving a wide range of problems arising in fluid dynamics, heat transfer, uncertainty quantification, biological systems, optimization, graph theory, lubrication theory, and structural mechanics.

This volume serves to bridge the gap between theoretical developments and practical applications by presenting innovative solution techniques and computational frameworks. It emphasizes key areas, including fuzzy differential equations, fractional-order systems, homotopy-based methods, spline approximations, transform techniques, optimal control strategies, and machine learning approaches, such as physics-informed neural networks.

The diverse range of topics presented in this volume highlights the rapid evolution of applied mathematics and computational science. We anticipate that this work will serve as a valuable reference for researchers, graduate students, and practitioners seeking advanced methodologies for analyzing and solving differential equation models.

# Acknowledgments

The editors sincerely thank all chapter authors for their valuable research and expertise. Your dedication has made this volume a comprehensive resource on applied differential equations and computational mathematics. We also gratefully acknowledge the reviewers, colleagues, and academic institutions for their constructive feedback. Special thanks to the publisher and editorial team for their professional guidance throughout this project.

# Overview of the book

The book is organized into fifteen chapters covering diverse theoretical and applied topics:

**Chapter 1** investigates thermal diffusion and nonlinear radiation effects in magnetohydrodynamic chemically reactive nanofluid flow over a stretching sheet, highlighting applications in thermal engineering and material processing.

**Chapter 2** presents fuzzy analytical techniques for obtaining solitary wave solutions of generalized Boussinesq equations under uncertainty.

**Chapter 3** develops implicit cubic spline methods for solving intuitionistic fuzzy parabolic partial differential equations with improved numerical accuracy.

**Chapter 4** analyzes a three-species food chain model governed by fractional differential equations involving nonsingular kernels, providing insights into ecological dynamics.

**Chapter 5** addresses uncertain dynamic responses of fractionally damped beam systems through a double-parametric homotopy analysis framework.

**Chapter 6** examines solute transfer through permeable channels while incorporating viscosity effects into the governing mathematical model.

**Chapter 7** investigates lubrication characteristics of exponential slider bearings under variable viscosity conditions.

**Chapter 8** studies velocity-slip effects in channel flows with permeable walls and their influence on fluid transport mechanisms.

**Chapter 9** explores stagnation-point flow of trihybrid nanofluids over shrinking surfaces under induced magnetic field effects.

**Chapter 10** introduces a semianalytical methodology combining the Sawi transform and Adomian decomposition method for solving Burgers' equation.

**Chapter 11** further expands the study of advanced differential equation models through analytical and computational investigations of nonlinear systems.

**Chapter 12** focuses on optimization of the iterated function system parameters for rational fractal linear splines using memory-efficient computational strategies.

**Chapter 13** presents theoretical developments in graph theory through the study of the structure and arboricity properties of $(k-r,k)$-trees.

**Chapter 14** develops an optimal control framework for tumor-immune interactions, incorporating targeted chemotherapy strategies.

**Chapter 15** concludes the volume with modern machine-learning-based approaches, employing modified physics-informed neural networks for structural vibration analysis.

Collectively, these chapters demonstrate the broad applicability of advanced computational and mathematical techniques in addressing contemporary challenges governed by differential equations.

# Introduction

The increasing complexity of contemporary scientific and engineering challenges necessitates the implementation of sophisticated mathematical models and efficient computational methodologies. Differential equations continue to serve as the fundamental framework for such models, offering a rigorous basis for characterizing dynamic processes across diverse academic and technical disciplines. This volume presents fifteen chapters that showcase recent advances in the theory, computation, and application of differential equations. The contributions encompass both classical and emerging areas of research, emphasizing innovative methodologies capable of addressing uncertainty, nonlinearity, fractional dynamics, and high-dimensional systems.

Several chapters investigate uncertainty and imprecision through fuzzy differential equations and intuitionistic fuzzy systems. These contributions introduce modern spline-based methods, fuzzy wave solutions, and uncertainty quantification frameworks that enhance the reliability of mathematical predictions in uncertain environments. The volume also highlights the growing significance of fractional calculus. Fractional-order food chain models and fractionally damped structural systems demonstrate how nonlocal operators provide realistic descriptions of memory-dependent processes observed in physical and biological systems. Innovative computational methodologies form another major theme of the book. Readers will find discussions on homotopy analysis methods, transform-based decomposition techniques, rational fractal interpolation, and advanced numerical algorithms designed to improve solution accuracy and computational efficiency.

Beyond its traditional applications, the book explores various interdisciplinary areas, including graph-theoretical structures, optimal control strategies for tumor-immune interactions, and machine learning frameworks for structural vibration analysis. The incorporation of modified physics-informed neural networks highlights the growing integration of artificial intelligence with mathematical modeling. Collectively, these contributions illustrate the expanding role of computational mathematics in addressing complex, real-world problems while offering significant insights into contemporary research trends within the field of applied differential equations.

1

# Thermal diffusion and nonlinear radiation analysis of magnetohydrodynamics chemically reactive nanofluid flow on a stretching sheet

Shobha V[a], Veeresh C[b], Sreedhara Rao Gunakala[c], Vikash Ramcharitar[c], Thameem Basha[d], Venkateswara Konduru[e], and Sibyala Vijayakumar Varma[f]

[a]*Government First Grade College, Kengeri, Mathematics, Bengaluru, India* [b]*Sri Venkateswara University, Tirupati, India* [c]*The University of the West Indies St Augustine Campus, Department of Mathematics and Statistics, St. Augustine, Trinidad and Tobago* [d]*Ulsan National Institute of Science and Technology, Department of Mathematical Sciences, Ulsan, South Korea* [e]*Sri Venkateswara College of Engineering, Tirupati, India* [f]*REVA University, Department of Mathematics, Bengaluru, India*

## 1.1 Introduction

Hybrid nanofluids have evolved as advanced heat transfer media due to their superior thermal conductivity and energy transport capabilities compared to conventional fluids. They are widely applicable in automotive radiators, electronic cooling systems, and heat exchangers to enhance efficiency and performance. In solar collectors and renewable energy systems, such as biomass and geothermal setups, they improve heat absorption and conversion rates. In nuclear reactors, hybrid nanofluids ensure effective cooling and operational safety. Their precise thermal control makes them valuable in medical devices, particularly hyperthermia treatments for cancer (Gunakala et al., 2025). In microelectronics, they prevent overheating through efficient cooling, while in HVAC systems, they promote energy savings. Applications also extend to the oil and gas sector for drilling and extraction, and to food industries for pasteurization and sterilization, where improved heat transfer reduces processing time. Overall, hybrid nanofluids offer versatile, energy-efficient solutions across diverse industrial and renewable energy applications.

Crane (1970) examined fluid motion surrounding a stretching plate, while Eckert and Drake (1972) analyzed simultaneous heat and mass transfer phenomena. Mabood et al. (2015) studied boundary-layer nanofluid behavior over a deformable surface. Reddy (2014)

**Advanced Computational and Mathematical Approaches in Applied Differential Equations**
**https://doi.org/10.1016/B978-0-44-345743-2.00012-8**

studied magnetohydrodynamic (MHD) and radiative impact on nanofluid flow along an elastic sheet. Hassani et al. (2011) provided an analytical formulation for boundary-layer nanofluid movement across a stretched region. Choi (1995), through his pioneering research on thermal conductivity enhancement using nanoparticles, demonstrated that suspensions containing $Al_2O_3$, CuO, Cu, $SiO_2$, or $TiO_2$ significantly outperform base liquids in energy transport. Özerinç et al. (2010) experimentally explored how temperature variation and particle concentration affect $Fe_3O_4$–Ag/EG bi-nanofluid behavior. Sulochana et al. (2016) examined radiative impacts on magnetohydrodynamic nanofluid motion, while the same authors Sulochana et al. (2018) considered rotational effects combined with Soret diffusion.

Shoaib et al. (2020) analyzed rotating MHD hybrid nanofluid flow incorporating radiation over an extensible surface. Sreedevi et al. (2020) discussed transient hybrid nanofluid convection under radiative influence, and Durga Prasad et al. (2018) evaluated the contribution of absorption on thermal as well as solutal transfer in MHD nanofluid movement. Zainal et al. (2021) inspected hybrid nanofluid transport through a porous mobile wall influenced by magnetism and radiation. Hussain et al. (2021b) numerically examined three-dimensional nanofluid convection with electromagnetic and rotational forces. Arshad et al. (2022) focused on energy transport involving MHD nano-material suspensions at different particle loadings. Babar and Ali (2019) presented an extensive overview addressing synthesis techniques, thermophysical traits, engineering applications, and challenges associated with hybrid nanofluids. Huminic and Huminic (2018) reviewed their performance in advanced heat-transfer technologies, while Sajid and Ali (2018) assessed hybrid nanofluid thermal conductivity comprehensively.

Leong et al. (2017) investigated fabrication routes and thermophysical responses. Nadeem and Haq (2014) evaluated radiative magnetohydrodynamic boundary layers containing nanoparticles. Shah and Ali (2019) critically discussed solar-energy utilization of hybrid nanofluids and practical limitations. Siddique and Jamil (2025) surveyed numerous industrial and biomedical applications, whereas Suneetha et al. (2022) showed that proper combination of nanoparticle types yields substantial enhancement, even at minimal concentrations. Raju et al. (2021) analyzed convective heat exchange in radiative MHD hybrid systems, and Alkuhayli (2023) observed that larger particle content, variable conductivity, and radiation intensity elevate temperature distribution. Allehiany et al. (2023) examined incompressible MHD hybrid nanofluid flow between rotating porous plates, revealing that radial velocity rises with suction yet falls under stretching and rotation. Alqahtani et al. (2023) used numerical computation to study conducting rotational flow of silver–gold hybrid nanofluid between parallel walls. Wahid et al. (2022) investigated MHD hybrid fluid motion past a porous vertical sheet with radiative effects, while Hussain et al. (2021a) analyzed thermal energy behavior of hybrid nanofluids containing single- and multi-walled carbon nanotubes within radiative frameworks. Narayan et al. (2018) explored $MoS_2$–MgO/water nanofluid flow through porous media across an extensible surface with the Cattaneo–Christov heat-flux model and convective boundaries. Foundational principles of heat, mass, and momentum transfer were detailed by Welty et al. (2007). Sundar et al. (2017) discussed Couette and Poiseuille hybrid nanofluid configurations incorporating temperature-dependent viscosity and conductivity, whereas Nazar et al. (2004) described unsteady motion generated by a stretching surface within a rotating medium, showing smooth transition from transient to steady regime.

## 1.2 Objectives of the study

1. To investigate the combined effects of nonlinear thermal radiation, rotation parameter, and first-order chemical reaction on the hydrodynamic, thermal, and solutal behaviors of the hybrid nanofluid.
2. To compare the thermal and mass transfer efficiencies between Ag–$MoS_2$/water and Ag–$MoS_2$/EG hybrid nanofluids, identifying the superior fluid for enhanced heat and mass transport applications.

## 1.3 Formulation in mathematical terms

The present study analyzes the three-dimensional steady boundary-layer behavior of an incompressible, viscous, and electrically conducting hybrid nanofluid flowing over a linearly stretching sheet as shown in Fig. 1.1. The region of interest lies above $Z > 0$, where stretching occurs in the $x$- and $y$-directions following $V_w = by$ and $U_w = ax$. A perpendicular magnetic field $B_o$ is imposed on the surface. The hybrid nanofluid has base fluid as water, silver (Ag), and molybdenum disulfide ($MoS_2$) as nanoparticles. The system experiences uniform rotation about the z-axis at angular speed $\omega^*$. The model incorporates nonlinear radiation and chemical reaction effects while neglecting induced magnetic fields, external electric influences, and viscous dissipation. The Dufour effect is excluded from the formulation, because the heat flux generated by mass diffusion is insignificant in diluted nanofluid mixtures where concentration gradients are minimal. In contrast, the Soret effect is retained, as thermal diffusion substantially influences nanoparticle migration and redistribution in flows primarily driven by temperature gradients (Goud and Reddy, 2022; Mudoi et al., 2025). These assumptions form the basis for developing the governing equations of momentum, heat, and mass transfer.

Governing equations are as follows:

Continuity equation

$$\frac{\partial u}{\partial x} + \frac{\partial v}{\partial y} = -\frac{\partial w}{\partial z} \tag{1.1}$$

Momentum equation

$$u\frac{\partial u}{\partial x} + v\frac{\partial u}{\partial y} = -w\frac{\partial u}{\partial z} + \frac{\mu_{hnf}}{\rho_{hnf}}\frac{\partial^2 u}{\partial z^2} - \frac{\sigma_{hnf} B_0^2 u}{\rho_{hnf}} u + 2\omega^* v \tag{1.2}$$

$$u\frac{\partial v}{\partial x} + v\frac{\partial v}{\partial y} = -w\frac{\partial v}{\partial z} + \frac{\mu_{\mathrm{hnf}}}{\rho_{hnf}}\frac{\partial^2 v}{\partial z^2} - \frac{\sigma_{hnf} B_0^2 v}{\rho_{hnf}} v - 2\omega^* u \tag{1.3}$$

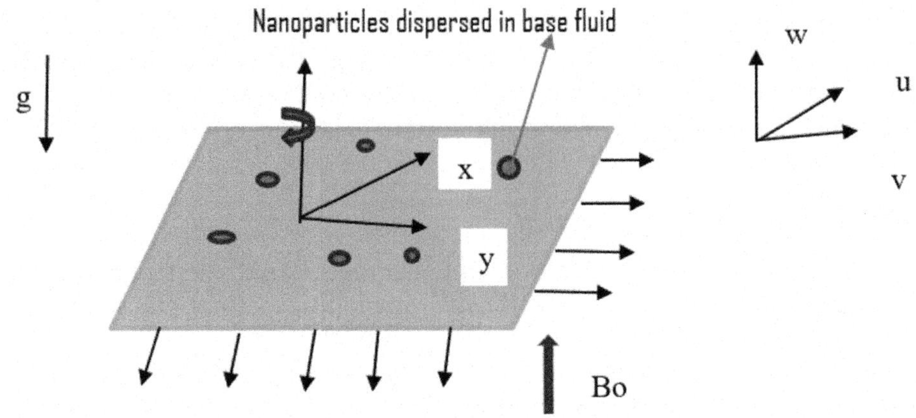


**FIGURE 1.1** Flow geometry.

Energy equation

$$u\frac{\partial T}{\partial x}+v\frac{\partial T}{\partial y}+w\frac{\partial T}{\partial z}=\alpha_{hnf}\frac{\partial^2 T}{\partial z^2}-\frac{1}{(\rho C_p)_{hnf}}\frac{\partial q_r}{\partial z}, \tag{1.4}$$

where

$$\frac{\partial q_r}{\partial z}=\left(-\frac{4\sigma^*}{3k^*}\right)\left[4T_\infty^3\frac{\partial^2 T}{\partial z^2}+\left(\frac{\partial T}{\partial z}\right)^2\left(12T_\infty^2+24T_\infty(T-T_\infty)+12(T-T_\infty)^2\right)\right.$$
$$\left.+\frac{\partial^2 T}{\partial z^2}\left(12T_\infty^2(T-T_\infty)+12T_\infty(T-T_\infty)^2+4(T-T_\infty)^3\right)\right]$$

Concentration equation

$$u\frac{\partial c}{\partial x}+v\frac{\partial c}{\partial y}=-w\frac{\partial c}{\partial z}-K_1(c-c_\infty)+\frac{D_T K_T}{T_m}\frac{\partial^2 T}{\partial z^2} \tag{1.5}$$

The boundary conditions

$$\begin{gathered}u=U_w=ax,\\ v=V_w=by,\\ w=0,\ T=T_w,\quad \text{at } z=0\\ v\to 0,\ u\to 0,\ T\to T_\infty\quad \text{as } z\to\infty\end{gathered} \tag{1.6}$$

Using the similarity transformations:

$$\begin{gathered}u=axm'(\eta),\ v=byn'(\eta),\ \omega=-\sqrt{av_f}\,[m(\eta)+\gamma n(\eta)],\\ \eta=z\sqrt{\frac{a}{v_f}},\ r(\eta)=o(\eta)=\frac{T-T_\infty}{T_\omega-T_\infty},\ s(\eta)=\frac{C-C_\infty}{C_\omega-C_\infty},\end{gathered} \tag{1.7}$$

the equations reduce to:

$$m'''(\eta) - (1-\psi_1)^{2.5}(1-\psi_2)^{2.5}\left\{(1-\psi_2)^{2.5}\left[(1-\psi_1)^{2.5} + \left(\frac{\rho_{s1}}{\rho_f}\right)\psi_1\right] + \psi_2\left(\frac{\rho_{s2}}{\rho_f}\right)\right\}.$$
$$\left[m'^2(\eta) - \{m(\eta)+\gamma n(\eta)\}p''(\eta) - 2\lambda\delta\gamma q'(\eta)\right] - (1-\psi_1)^{2.5}(1-\psi_2)^{2.5}m'(\eta)M^2\frac{\sigma_{hnf}}{\sigma_f} = 0 \quad (1.8)$$

$$n'''(\eta) - (1-\psi_1)^{2.5}(1-\psi_2)^{2.5}\left\{(1-\psi_2)^{2.5}\left[(1-\psi_1)^{2.5} + \left(\frac{\rho_{s1}}{\rho_f}\right)\psi_1\right] + \psi_2\left(\frac{\rho_{s2}}{\rho_f}\right)\right\}.$$
$$\left[n'^2(\eta) - \{m(\eta)+\gamma n(\eta)\}q''(\eta) + \frac{2\lambda\gamma}{\delta}m'(\eta)\right] - (1-\psi_1)^{2.5}(1-\psi_2)^{2.5}n'(\eta)M^2\frac{\sigma_{hnf}}{\sigma_f} = 0 \quad (1.9)$$

$$o''\left[\frac{K_{hnf}}{K_f} + R\left[(r_w-1)^3o^3 + 3(r_w-1)^2o^2 + 3(r_w-1)o + 1\right]\right]$$
$$+ R(o')^2\left[3(r_w-1)^3o^2 + 6(r_w-1)^2o + 3(r_w-1)\right]$$
$$+ \Pr(m+\gamma n)o'\frac{(\rho C_p)_{hnf}}{(\rho C_p)_f} = 0 \quad (1.10)$$

$$s'' + (m+\gamma n)S_c s' - K_r s S_c = 0 \quad (1.11)$$

The boundary conditions (nondimensional):

$$m=0,\ m'=1,\ n=0,\ n'=\gamma,\ o=1,\ \mathrm{s}=1 \quad \text{at } \eta=0.$$
$$m'\to 0,\ n'\to 0,\ o\to 0,\ s\to 0 \quad \text{at } \eta\to\infty. \quad (1.12)$$

$$\lambda = \frac{\omega^*}{a},\ M = \sqrt{\frac{\sigma_f B_0^2}{a\rho_f}},\ \gamma = \frac{b}{a},\ P_r = \frac{(\mu C_\rho)_f}{k_f},\ S_c = \frac{v_f}{D},\ K_r = \frac{K_1}{a} \quad (1.13)$$

Dimensional skin friction and Nusselt number are given by:

$$c_{f_x} = \frac{\mu_{\mathrm{hnf}}\left(\frac{\partial u}{\partial z}\right)_{z=0}}{\rho_f(ax)^2} \quad c_{f_y} = \frac{\mu_{\mathrm{hnf}}\left(\frac{\partial v}{\partial z}\right)_{z=0}}{\rho_f(ax)^2} \quad Nu_x = \frac{-xk_{\mathrm{hnf}}\left(\frac{\partial T}{\partial z}\right)_{z=0}}{k_f(T_w - T_\infty)} \quad Sh_x = \frac{-x\mu_{\mathrm{hnf}}\left(\frac{\partial C}{\partial z}\right)_{z=0}}{Dk_f(C_w - C_\infty)} \quad (1.14)$$

By using similarity transformations this gives:

$$\mathrm{Re}^{1/2}c_{fx} = \frac{1}{(1-\psi_1)^{2.5}(1-\psi_2)^{2.5}}m''(0) \quad \mathrm{Re}^{1/2}c_{fy} = \frac{1}{(1-\psi_1)^{2.5}(1-\psi_2)^{2.5}}n''(0)$$
$$\mathrm{Re}^{-1/2}Nu = \frac{-K_{hnf}}{K_f}o'(0) \quad \mathrm{Re}^{\frac{1}{2}}Sh = -s'(0) \quad (1.15)$$

**Table 1.1** The relation of thermo-physical attributes.

| Attributes | Nanofluid | Hybrid nanofluid |
|---|---|---|
| Dynamic Viscosity ($\mu$) | $\mu_{nf} = \frac{\mu_f}{(1-\psi)^{2.5}}$ | $\mu_{hnf} = \frac{\mu_f}{(1-\psi_1)^{2.5}(1-\psi_2)^{2.5}}$ |
| Density ($\rho$) | $\rho_{nf} = \rho_f(1-\psi) + \psi\rho_s$ | $\rho_{hnf} = (1-\psi_2)[(1-\psi_1)\rho_f + \psi_1\rho_{s1}] + \psi_2\rho_{s2}$ |
| Electrical Conductivity ($\sigma$) | $\frac{\sigma_{nf}}{\sigma_f} = 1 + \frac{3(\sigma_s-\sigma_f)\psi}{(\sigma_s+2\sigma_f)-\psi(\sigma_s-\sigma_f)}$ | $\frac{\sigma_{hnf}}{\sigma_f} = 1 + \left[\frac{3\left(\frac{\sigma_{s1}\psi_1+\sigma_{s2}\psi_2}{\sigma_f}-(\psi_1+\psi_2)\right)}{2+\frac{\sigma_{s1}+\sigma_{s2}}{\sigma_f}-\left(\frac{\sigma_{s1}\psi_1+\sigma_{s2}\psi_2}{\sigma_f}\right)+(\psi_1+\psi_2)}\right]$ |
| Thermal Conductivity ($k$) | $\frac{k_{nf}}{k_f} = \frac{k_s+k_f(n-1)-\psi(n-1)(k_f-k_s)}{k_s+(n-1)k_f+\psi(k_f-k_s)}$ | $\frac{k_{hnf}}{k_f} = \frac{(k_{s1}+k_{s2})+2k_f(1-(\psi_1+\psi_2))+2(\psi_1k_{s1}+\psi_2k_{s2})}{(k_{s1}+k_{s2})+(2+(\psi_1+\psi_2))k_f-(\psi_1k_{s1}+\psi_2k_{s2})}$ |
| Heat Capacity ($\rho C_p$) | $(\rho C_p)_{nf} = \psi(\rho C_p)_s + (1-\psi)(\rho C_p)_f$ | $(\rho C_p)_{hnf} = \psi_2(\rho C_p)_{s2} + (1-\psi_2)[(1-\psi_1)(\rho C_p)_f + \psi_1(\rho C_p)_{s1}]$ |

**Table 1.2** Thermo-physical characteristics.

| Properties | Base fluid ($H_2O$) | Ag | $MoS_2$ | Base fluid (EG) | Cu | $TiO_2$ |
|---|---|---|---|---|---|---|
| $\rho$ | 997.1 | 10500 | 5060 | 1114 | 8933 | 4250 |
| $c_p$ | 4179 | 235 | 397.21 | 2415 | 385 | 686.2 |
| $k$ | 0.613 | 429 | 904.4 | 0.252 | 401 | 8.9538 |
| $\sigma$ | $5.5 \times 10^{-6}$ | $6.3 \times 10^{7}$ | $2.09 \times 10^{-4}$ | $1.07 \times 10^{-6}$ | $5.96 \times 10^{7}$ | $2.6 \times 10^{6}$ |
| Pr | 6.2 | – | – | 204 | – | – |

## 1.4 Numerical methodology

A numerical method is utilized to solve the nonlinear ODE derived from Eqs. (1.1)–(1.5), along with the boundary conditions specified in Eq. (1.6), using MATLAB®. Relations for the thermo-physical attributes and characteristics are given in Table 1.1 and Table 1.2, respectively.

The coupled Eqs. (1.7)–(1.11), together with the conditions (1.12) to (1.15), are converted into an equivalent system of first-order linear equations for computational analysis.

$$\begin{gathered} m = y(1),\ m' = y(2),\ m'' = y(3),\ m''' = y'(3), \\ n = y(4),\ n' = y(5),\ n'' = y(6),\ n''' = y'(6), \\ o = y(7),\ o' = y(8),\ o'' = y'(8),\ s = y(9),\ s' = y(10),\ s'' = y'(10) \end{gathered} \tag{1.16}$$

With boundary conditions

$$\left.\begin{gathered} y(1) = y(4) = 0,\ y(5) = \gamma,\ y(2) = y(7) = y(9) = 1 \quad \text{at } \eta = 0 \\ y(2) = y(5) = y(7) = y(9) = 0 \quad \text{at } \eta = \infty \end{gathered}\right\}. \tag{1.17}$$

## 1.5 Results and discussion

The graphical results shown in Figs. 1.2–1.13 illustrate the influence of key physical parameters for water, Ag–$MoS_2$/EG hybrid nanofluid, and Ag–$MoS_2$/water hybrid nanofluid. From Fig. 1.2 and Fig. 1.3, it is evident that both the primary and secondary velocity gradi-

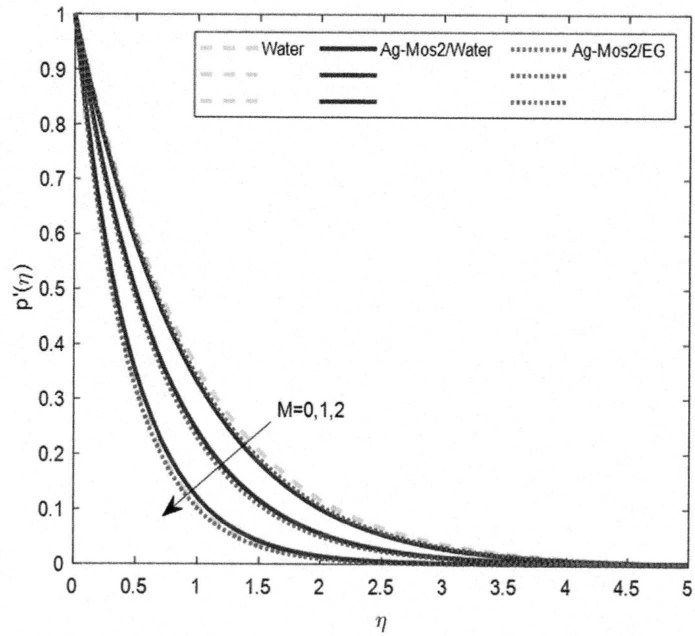


**FIGURE 1.2** $p'(n)$ for M.

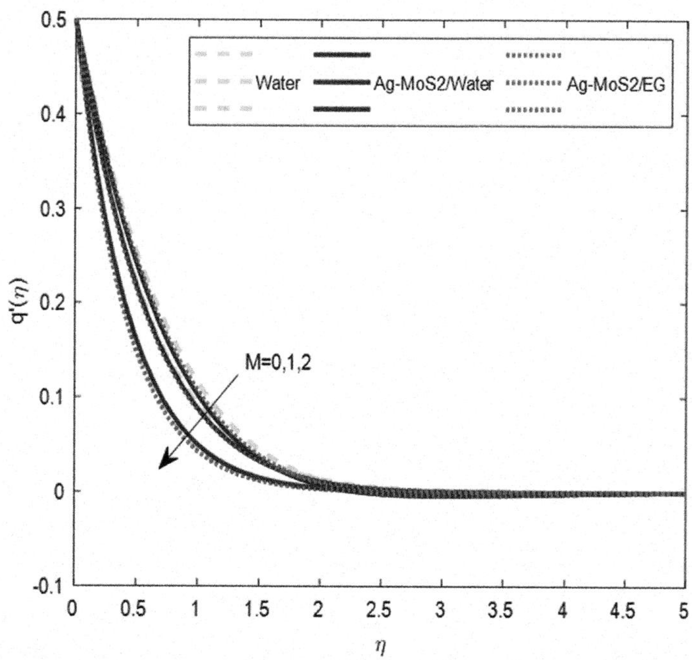


**FIGURE 1.3** $q'(n)$ for M.

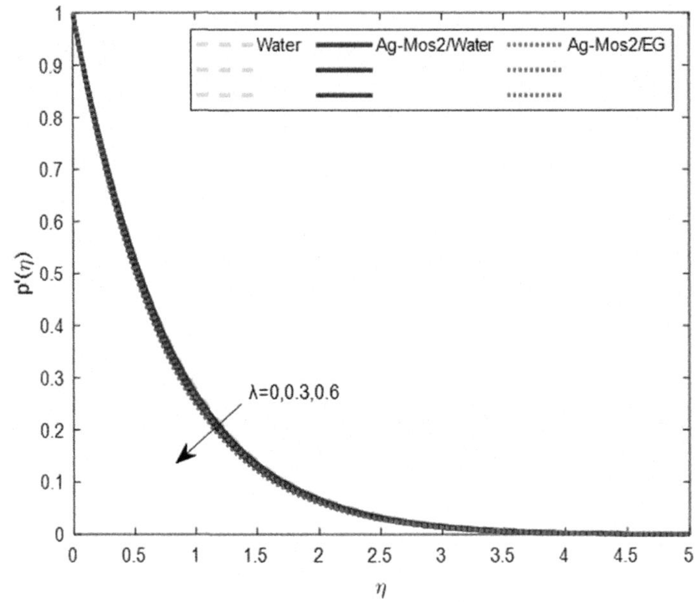


**FIGURE 1.4** $p'(n)$ for λ.

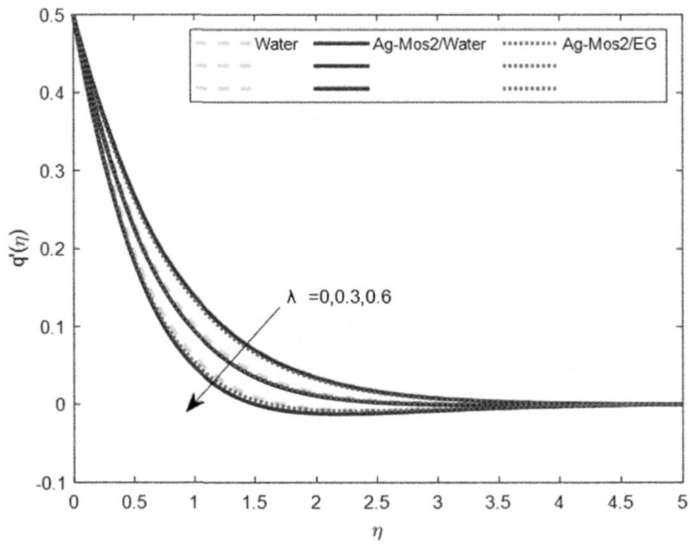


**FIGURE 1.5** $q'(n)$ for λ.

ents $p'(\eta)$ and $q'(\eta)$ decrease with an increase in the magnetic parameter (M). This decline is due to the Lorentz force generated by the applied magnetic field, which exerts a resistive drag on the moving fluid layers, thereby decelerating the flow and reducing the wall shear

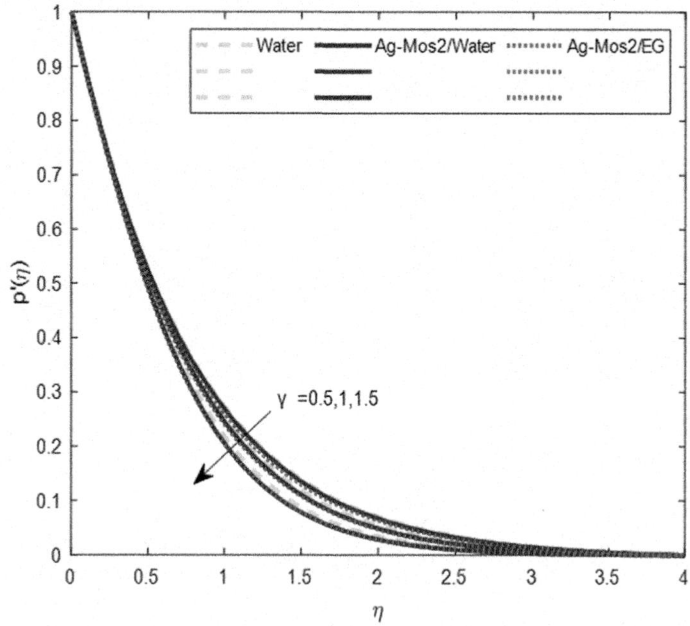


**FIGURE 1.6** $p'(n)$ for $\gamma$.

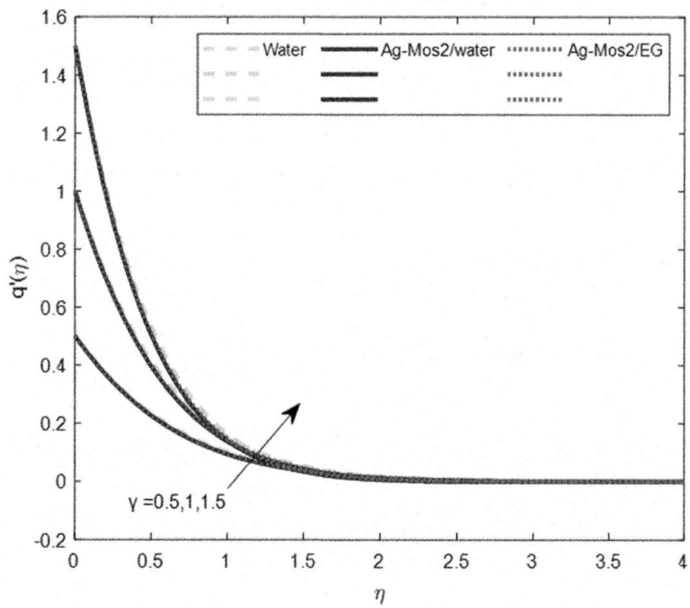


**FIGURE 1.7** $q'(n)$ for $\gamma$.

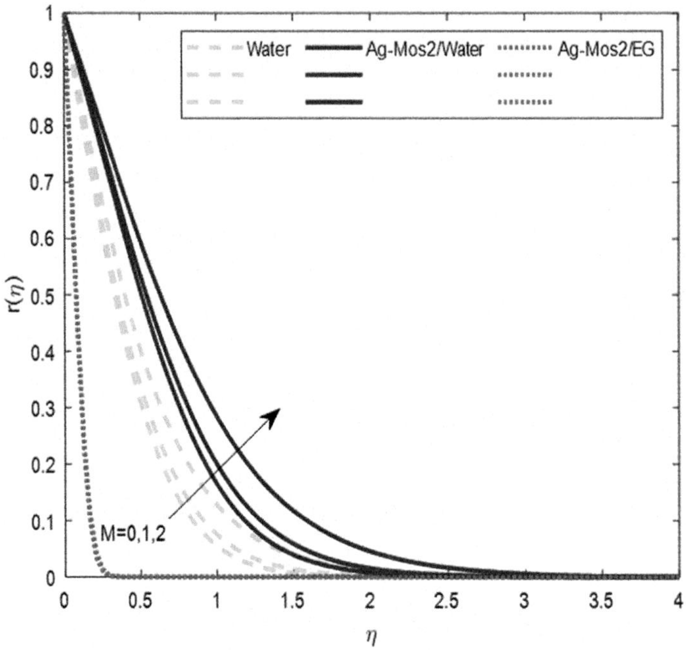


**FIGURE 1.8** M effects on the temperature profile.

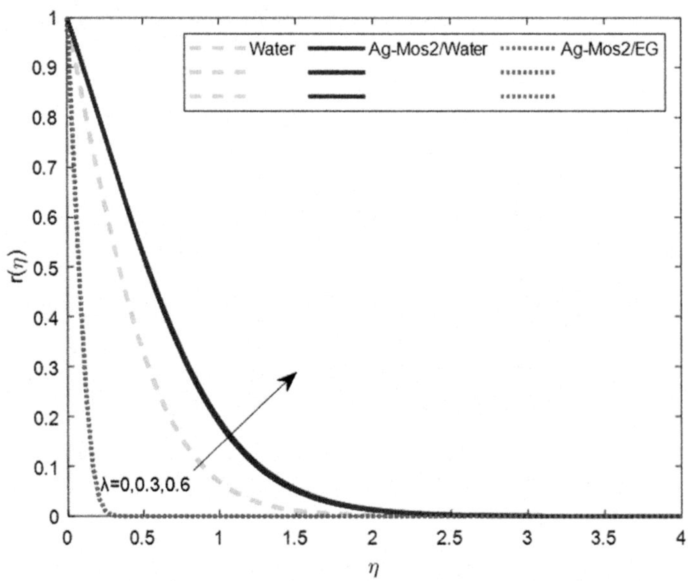


**FIGURE 1.9** λ effects on the temperature profile.

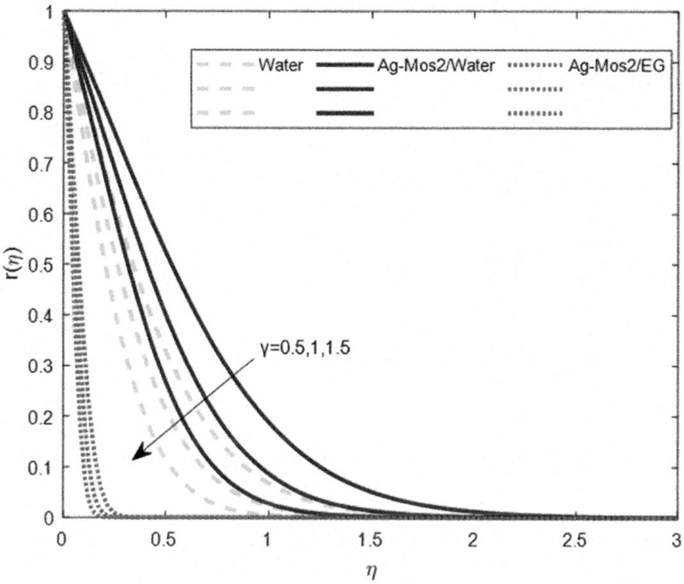


**FIGURE 1.10** Stretching ratio $\gamma$ effects on the temperature profile.

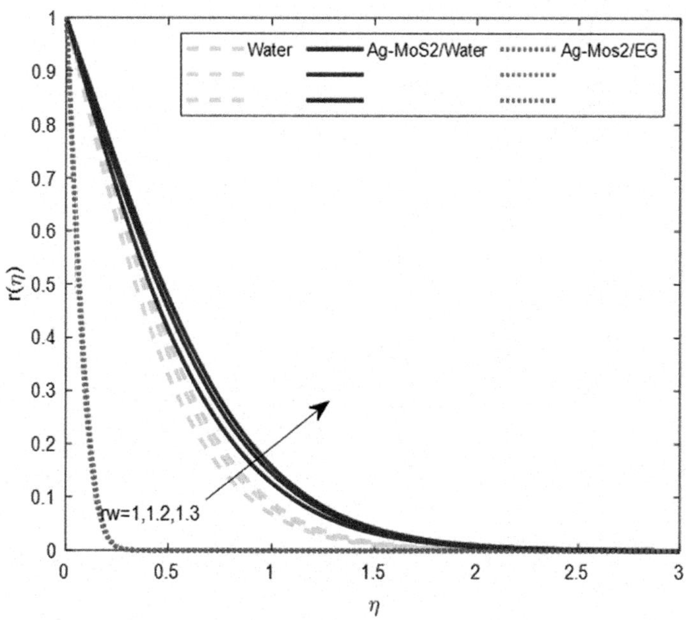


**FIGURE 1.11** rw on temperature profile.

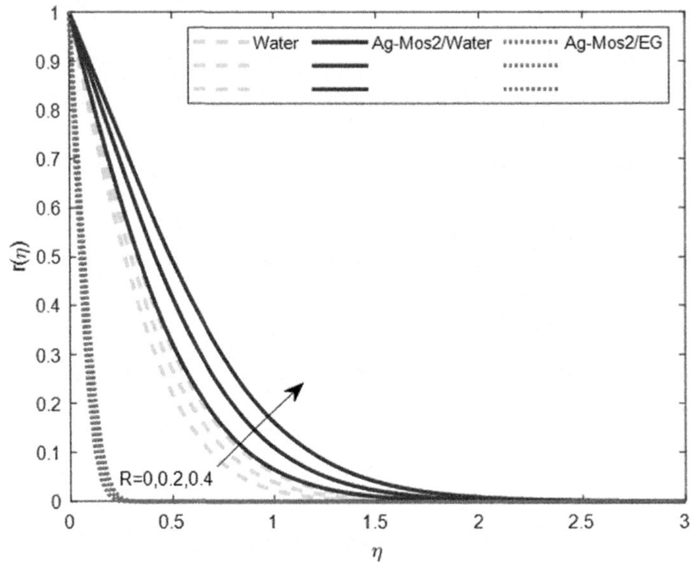


**FIGURE 1.12** R effect on the temperature profile.

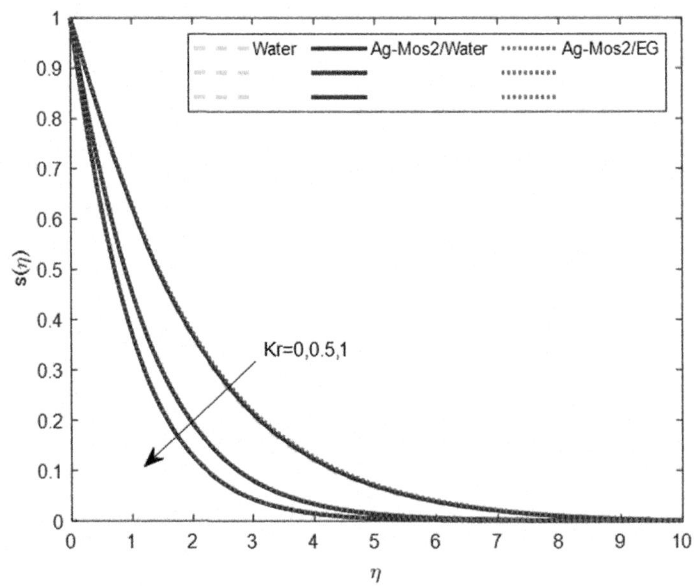


**FIGURE 1.13** Kr effect on concentration profile.

stresses. Similarly, Fig. 1.4 and Fig. 1.5 show that increasing the stretching ratio parameter ($\lambda$) also decreases both $p'(\eta)$ and $q'(\eta)$ indicating that a higher stretching rate leads to a thicker boundary layer and reduced velocity gradients near the wall.

In contrast, Fig. 1.6 and Fig. 1.7 demonstrate the dual effect of the rotation parameter ($\gamma$): while the primary velocity gradient $p'(\eta)$ decreases, the secondary velocity component $q'(\eta)$ increases. This behavior can be attributed to the rotational motion inducing cross-flow components that alter the momentum distribution, reducing the flow in the primary direction but intensifying it transversely.

Fig. 1.8 and Fig. 1.9 depict that the temperature profile increases with both the magnetic parameter (M) and stretching ratio ($\lambda$). The enhancement in temperature is caused by Joule heating from the magnetic field and increased viscous dissipation due to higher stretching rates, both of which contribute to the thermal energy within the boundary layer. However, Fig. 1.10 shows that as the rotation parameter ($\gamma$) increases, the temperature profile decreases, suggesting that rotational motion enhances convective cooling and promotes heat transfer away from the surface.

Fig. 1.11 indicates that an increase in the wall temperature ratio (rw) leads to a rise in the temperature profile, as the higher wall temperature enhances thermal diffusion into the fluid. Fig. 1.12 shows that the thermal radiation parameter (R) intensifies the temperature distribution, demonstrating that radiative heat transfer contributes significantly to the overall energy transport in the flow. Finally, Fig. 1.13 reveals that the chemical reaction parameter (Kr) reduces the concentration profile, implying that stronger chemical reactions accelerate the depletion of species concentration within the boundary layer.

The numerical data in Table 1.3 provides a comparative assessment of Ag–$MoS_2$/water and Ag–$MoS_2$/ethylene glycol (EG) hybrid nanofluids under varying flow and thermal conditions. The analysis reveals that the Nusselt number (Nu), indicative of heat transfer, and the Sherwood number (Sh), representative of mass transfer, exhibit distinct trends for the two base fluids. For the Ag–$MoS_2$/water hybrid nanofluid, an increase in the magnetic and rotational parameters results in a gradual decline in Nu values, signifying that intensified Lorentz and centrifugal forces suppress the thermal boundary layer and marginally reduce heat transfer rates. However, the Sherwood number remains nearly steady, indicating stable mass transport characteristics despite magnetic and rotational influences.

In contrast, the Ag–$MoS_2$/EG hybrid nanofluid demonstrates a similar qualitative behavior but with slightly lower Nu and Sh magnitudes, attributed to the higher viscosity and lower thermal diffusivity of ethylene glycol compared to water. An increase in nanoparticle volume fractions enhances both Nu and Sh, confirming the positive impact of nanoparticle loading on effective thermal conductivity and solutal diffusivity. Furthermore, increasing the rotation parameter ($\lambda$) from 0 to 0.6 leads to a noticeable reduction in Nu, while Sh remains relatively unaffected, highlighting the dominance of rotational inertia on the thermal boundary layer over solutal effects.

Overall, the findings indicate that the Ag–$MoS_2$/water hybrid nanofluid exhibits superior heat and mass transfer performance compared to its ethylene glycol counterpart. The results further confirm that moderate nanoparticle concentration and controlled magnetic and rotational conditions significantly improve thermal–hydrodynamic efficiency, making

**Table 1.3** Nusselt number and Sherwood number for Ag-$Mos_2$/Water and Ag-$Mos_2$/EG nanofluid.

| $\psi_1$ | $\psi_2$ | M | $\lambda$ | $\gamma$ | R | $r_w$ | Pr | Kr | $Nu$ Ag-$Mos_2$/water | $Sh$ Ag-$Mos_2$/water | $Nu$ Ag-$Mos_2$/EG | $Sh$ Ag-$Mos_2$/EG |
|---|---|---|---|---|---|---|---|---|---|---|---|---|
| 0 | 0 | 0.8 | 0.3 | 0.5 | 0.5 | 1 | 6.9 | 0.5 | 1.626394 | 0.717647 | 1.626394 | 0.717647 |
| 0 | 0 | | | | | 1.2 | | | 1.388993 | 0.717647 | 1.388993 | 0.717647 |
| 0 | 0 | | | | | 1.3 | | | 1.274082 | 0.717647 | 1.274082 | 0.717647 |
| 0.02 | 0.01 | | | | | 1.2 | | | 1.077324 | 0.715808 | 7.455539 | 0.714749 |
| 0.02 | 0.02 | | | | | | | | 1.078966 | 0.716395 | 7.457524 | 0.715482 |
| 0.02 | 0.03 | | | | | | | | 1.080621 | 0.716991 | 7.459509 | 0.716221 |
| 0.02 | 0.02 | | | | | 1 | | | 1.614956 | 0.716395 | 10.893646 | 0.715482 |
| 0.02 | 0.02 | | | | | 1.2 | | | 1.392600 | 0.716395 | 9.472109 | 0.715482 |
| 0.02 | 0.02 | | | | | 1.3 | | | 1.282869 | 0.716395 | 8.768299 | 0.715482 |
| 0.02 | 0.02 | | 0 | | | | | | 1.082207 | 0.717761 | 7.459468 | 0.716698 |
| 0.02 | 0.02 | | 0.3 | | | | | | 1.078966 | 0.716395 | 7.457524 | 0.715482 |
| 0.02 | 0.02 | | 0.6 | | | | | | 1.070526 | 0.712859 | 7.452688 | 0.712302 |
| 0.02 | 0.02 | | | 0.2 | | | | | 0.970711 | 0.689962 | 6.785242 | 0.688970 |
| 0.02 | 0.02 | | | 0.4 | | | | | 1.034188 | 0.705274 | 7.177338 | 0.704319 |
| 0.02 | 0.02 | | | 0.6 | | | | | 1.131099 | 0.729664 | 7.785925 | 0.728813 |
| 0.02 | 0.02 | | | | 0 | | | | 2.008675 | 0.716395 | 13.155382 | 0.715482 |
| 0.02 | 0.02 | | | | 0.2 | | | | 1.455851 | 0.716395 | 9.755774 | 0.715482 |
| 0.02 | 0.02 | | | | 0.4 | | | | 1.175421 | 0.716395 | 8.043762 | 0.715482 |
| 0.02 | 0.02 | | | | | | | 0 | 1.078965 | 0.411082 | 7.457524 | 0.409064 |
| 0.02 | 0.02 | | | | | | | 1 | 1.078966 | 0.716395 | 7.457524 | 0.715482 |
| 0.02 | 0.02 | | | | | | | 1.5 | 1.078966 | 0.912496 | 7.457524 | 0.911846 |
| 0.02 | 0.02 | 0 | | | | | | | 1.121824 | 0.732545 | 7.506305 | 0.733813 |
| 0.02 | 0.02 | 1 | | | | | | | 1.057149 | 0.708933 | 7.433264 | 0.707266 |
| 0.02 | 0.02 | 2 | | | | | | | 0.919652 | 0.671809 | 7.279636 | 0.668186 |

the water-based hybrid nanofluid more suitable for high-performance thermal management applications, whereas the EG-based system offers enhanced operational stability in viscous flow environments.

The tabulated data in Table 1.4 presents the variation of local skin friction coefficients for Ag–$MoS_2$/water and Ag–$MoS_2$/ethylene glycol (EG) hybrid nanofluids under different parametric conditions. It is observed that as the nanoparticle volume fractions ($\psi_1$, $\psi_2$) increase, both skin friction values show a gradual rise for both base fluids, indicating that higher particle concentration enhances the momentum boundary layer resistance due to increased effective viscosity and nanoparticle interactions. When the magnetic parameter (M) increases from 0 to 2, the skin friction coefficients initially increase and then decrease, signifying that the Lorentz force initially strengthens the resistance at the wall, but excessive magnetic field suppresses the velocity gradients due to enhanced electromagnetic damping. Similarly, the increase in the stretching ratio parameter ($\lambda$)

**Table 1.4** Skin friction coefficients of Ag-$MoS_2$/water and Ag-$MoS_2$/EG.

| $\psi_1$ | $\psi_2$ | M | $\Lambda$ | $\gamma$ | $c_{fx}$ Ag-$MoS_2$/water | $c_{fy}$ Ag-$MoS_2$/water | $c_{fx}$ Ag-$MoS_2$/EG | $c_{fy}$ Ag-$MoS_2$/EG |
|---|---|---|---|---|---|---|---|---|
| 0.02 | 0.01 | | | | 1.379886 | 0.798897 | 1.400779 | 0.804540 |
| 0.02 | 0.02 | | | | 1.405977 | 0.814846 | 1.425223 | 0.819285 |
| 0.02 | 0.03 | | | | 1.432810 | 0.831253 | 1.450382 | 0.834469 |
| 0.02 | 0.02 | | 0 | | 1.471093 | 0.655914 | 1.488151 | 0.667620 |
| 0.02 | 0.02 | | 0.3 | | 1.405977 | 0.814846 | 1.425223 | 0.819285 |
| 0.02 | 0.02 | | 0.6 | | 1.359991 | 0.977460 | 1.379536 | 0.974613 |
| 0.02 | 0.02 | | | 0.2 | 1.424807 | 0.300137 | 1.443775 | 0.302854 |
| 0.02 | 0.02 | | | 0.4 | 1.413735 | 0.633825 | 1.432881 | 0.638008 |
| 0.02 | 0.02 | | | 0.6 | 1.397008 | 1.005954 | 1.416348 | 1.010340 |
| 0.02 | 0.02 | 0 | | | 1.134943 | 0.717556 | 1.119941 | 0.708072 |
| 0.02 | 0.02 | 1 | | | 1.539446 | 0.869595 | 1.573148 | 0.881030 |
| 0.02 | 0.02 | 2 | | | 2.382069 | 1.251737 | 2.489792 | 1.301194 |

from 0.2 to 0.6 leads to a continuous rise in $C_{fx}$ and $C_{fy}$, implying that stronger surface stretching thickens the momentum boundary layer and boosts wall shear stresses. In contrast, when the rotation parameter ($\gamma$) increases, a notable enhancement in both friction coefficients is observed, suggesting that rotational effects intensify the flow resistance and shear forces along both directions. Furthermore, the Ag–$MoS_2$/EG hybrid nanofluid exhibits slightly higher $C_{fx}$ and $C_{fy}$ values than Ag–$MoS_2$/water, which can be attributed to the higher viscosity and thermal capacity of EG that increase the drag forces at the wall. Overall, the analysis reveals that skin friction is significantly influenced by nanoparticle concentration, magnetic effects, stretching intensity, and rotation, with EG-based hybrid nanofluids showing superior resistance characteristics compared to water-based ones.

Table 1.5 depicts the analysis of the Nu behavior for different hybrid nanofluid combinations under varying nanoparticle volume fractions ($\psi_1, \psi_2$) and reveals distinct heat transfer enhancement trends.

The data presented examines the influence of nanoparticle volume fractions ($\psi_1, \psi_2$) on the local Nusselt number (Nu) for different hybrid nanofluids at a fixed rotational parameter ($\gamma = 1.2$). The analysis reveals that as the nanoparticle volume fractions $\psi_1$ and $\psi_2$ increase, the heat transfer rate, represented by the Nusselt number, also increases for all nanofluid combinations. Among the studied fluids, Ag–$MoS_2/H_2O$ exhibits the highest Nusselt number values, followed closely by $TiO_2$–$MoS_2/H_2O$, while the ethylene glycol (EG)-based hybrid nanofluids show relatively lower values due to EG's lower thermal conductivity compared to water. Additionally, combinations with higher nanoparticle loadings, particularly at $\psi_1 = 0.04$ and $\psi_2 = 0.04$, yield the maximum enhancement in heat transfer performance. This trend demonstrates that hybridization of nanoparticles and increased concentration significantly improve thermal transport characteristics, confirming the superior heat transfer potential of Ag–$MoS_2/H_2O$ hybrid nanofluid under

**Table 1.5** Comparison of different nanofluids.

| $\psi_1$ | $\psi_2$ | $\Gamma$ | $Re^{1/2}Nu$ Ag-$MoS_2/H_20$ | $Re^{1/2}Nu$ Ag-Cu/$H_20$ | $Re^{1/2}Nu$ $TiO_2$-$MoS_2/H_20$ | $Re^{1/2}Nu$ Ag-$MoS_2$/EG | $Re^{1/2}Nu$ Ag-Cu/EG | $Re^{1/2}Nu$ $TiO_2$-$MoS_2$/EG |
|---|---|---|---|---|---|---|---|---|
| 0.02 | 0 | 1.2 | 1.031146 | 9.116566 | 1.033299 | 7.112259 | 9.091896 | 7.104550 |
| 0 | 0.04 | 1.2 | 1.089264 | 9.548803 | 1.089264 | 7.444486 | 9.548803 | 7.444486 |
| 0.02 | 0.02 | 1.2 | 1.078964 | 9.543999 | 1.080798 | 7.457530 | 9.519073 | 7.449159 |
| 0.02 | 0.04 | 1.2 | 1.127319 | 9.979670 | 1.128807 | 7.809344 | 9.954583 | 7.800316 |
| 0 | 0.02 | 1.2 | 1.042792 | 9.127933 | 1.042792 | 7.106756 | 9.127933 | 7.106756 |
| 0.04 | 0 | 1.2 | 1.065921 | 9.524604 | 1.070201 | 7.456736 | 9.473780 | 7.441327 |
| 0.04 | 0.02 | 1.2 | 1.115560 | 9.966828 | 1.119243 | 7.816107 | 9.915583 | 7.799439 |
| 0.04 | 0.04 | 1.2 | 1.165754 | 10.417305 | 1.168790 | 8.182038 | 10.365539 | 8.164120 |

**Table 1.6** Comparison of the present model with the literature for $\psi_1 = 0$, $\psi_2 = 0$, R = 0, rw = 1, M = 0.

| | (Nazar et al., 2004) | | (Arshad et al., 2022) | | Present result | |
|---|---|---|---|---|---|---|
| $\lambda$ | $m''(0)$ | $n''(0)$ | $m''(0)$ | $n''(0)$ | $m''(0)$ | $n''(0)$ |
| 0 | −1.0 | 0.0 | −1.039 | 0.0 | −1.0 | 0 |
| 0.5 | −1.13 | −0.51 | −1.161 | −0.538 | −1.162 | −0.6381 |
| 1.0 | −1.32 | −0.83 | −1.348 | −0.848 | −1.35 | −0.9448 |

**Table 1.7** Range and levels of effective parameters.

| Parameter | Symbol | Level | | |
|---|---|---|---|---|
| | | low | middle | high |
| | | −1 | 0 | 1 |
| $1.2 \le rw \le 1.4$ | A1 | 1.2 | 1.3 | 1.4 |
| $0.2 \le R \le 0.6$ | A2 | 0.2 | 0.4 | 0.6 |
| $0.2 \le \gamma \le 0.4$ | A3 | 0.2 | 0.3 | 0.4 |

the given flow conditions. A comparison of the present model with the literature is presented in Table 1.6.

The combined numerical and graphical analyses comprehensively highlight the influence of various governing parameters on the flow, thermal, and mass transfer characteristics of Ag–$MoS_2$/water and Ag–$MoS_2$/ethylene glycol (EG) hybrid nanofluids over a stretchable rotating surface in the presence of magnetic and thermal effects.

## 1.6 Response surface methodology: RSM

Response surface methodology (RSM) serves as a powerful statistical tool designed to explore the correlation between key variables, namely R, rw, $\gamma$, and outcomes such as $Nu$. The levels and span of these influential parameters are summarized in Table 1.7.

**Table 1.8** Analysis of variance (ANOVA) for Nu ($Ag$-$MoS_2$)/water.

| P1 | P2 | P3 | P4 | P5 | P6 | P7 |
|---|---|---|---|---|---|---|
| Model | 9 | 0.492908 | 0.054768 | 5849.82 | 0.000 | 1.27107 |
| Linear | 3 | 0.479493 | 0.159831 | 17071.87 | 0.000 | |
| rw | 1 | 0.078987 | 0.078987 | 8436.80 | 0.000 | −0.088875 |
| R | 1 | 0.383893 | 0.383893 | 41004.32 | 0.000 | −0.189274 |
| $\gamma$ | 1 | 0.016613 | 0.016613 | 1774.50 | 0.000 | 0.040759 |
| Square | 3 | 0.007309 | 0.002436 | 260.23 | 0.000 | |
| rw*rw | 1 | 0.000004 | 0.000004 | 0.39 | 0.548 | 0.00111 |
| R*R | 1 | 0.003947 | 0.003947 | 421.63 | 0.000 | 0.03441 |
| $\gamma*\gamma$ | 1 | 0.000107 | 0.000107 | 11.39 | 0.007 | 0.00605 |
| 2-Way Interaction | 3 | 0.002211 | 0.000737 | 78.71 | 0.000 | |
| rw*R | 1 | 0.001916 | 0.001916 | 204.60 | 0.000 | −0.01547 |
| rw*$\gamma$ | 1 | 0.000055 | 0.000055 | 5.90 | 0.035 | −0.00263 |
| R*$\gamma$ | 1 | 0.000240 | 0.000240 | 25.64 | 0.000 | −0.00548 |
| Error | 10 | 0.000094 | 0.000009 | | | |
| Lack-of-fit | 5 | 0.000094 | 0.000019 | * | * | |
| Pure error | 5 | 0.000000 | 0.000000 | | | |
| Total | 19 | 0.493001 | | | | |
| $R^2 = 99.98\%$, Adjusted- $R^2 = 99.96\%$ | | | | | | |
| **P1-Source** | **P2-DF** | **P3-Adj SS** | **P4-Adj MS** | **P5-F Value** | **P6-P Value** | **P7-Coefficient** |

## 1.7 Model accuracy

The predictive model is examined through an ANOVA summary and corresponding residual graphics, incorporating various statistical measures. As described in the following section, different validation techniques are utilized to determine its effectiveness and reliability. In Table 1.8 and Table 1.9, the model's significance and data variance are assessed using p-values and F-values. Terms with an F-value $> 1$ and a p-value $< 0.05$ are considered significant and are retained, indicating their relevance to the model.

The normal probability plot of Fig. 1.14 shows that residuals lie close to the straight line, indicating normal distribution. The residuals versus fitted values exhibit random scattering around zero, confirming homoscedasticity. The histogram is nearly symmetric and centered near zero, further supporting normality. Additionally, the residuals versus observation order plot displays a random pattern, indicating no autocorrelation. Overall, these plots validate that the model is statistically reliable, well-fitted, and suitable for predicting the heat transfer performance of the hybrid nanofluid.

Fig. 1.15 illustrates the combined influence of these parameters on the heat transfer rate of the $Ag$–$MoS_2$/water hybrid nanofluid. The color gradient indicates that Nu decreases progressively from green (dark gray in print version) to dark blue (black in print version) as R and rw increase. This trend suggests that higher radiation and rotation parameters lead to a reduction in the rate of heat transfer. Conversely, lower values of R and rw correspond to higher Nu, indicating enhanced convective heat transfer performance.

**Table 1.9** Analysis of variance (ANOVA) for Nu (Ag-$MoS_2$)/EG.

| P1 | P2 | P3 | P4 | P5 | P6 | P7 |
|---|---|---|---|---|---|---|
| Model | 9 | 18.4053 | 2.0450 | 5273.20 | 0.000 | 8.68121 |
| Linear | 3 | 17.8936 | 5.9645 | 15379.77 | 0.000 | |
| rw | 1 | 3.2718 | 3.2718 | 8436.45 | 0.000 | −0.57200 |
| R | 1 | 14.0004 | 14.0004 | 36100.41 | 0.000 | −1.14303 |
| $\gamma$ | 1 | 0.6215 | 0.6215 | 1602.44 | 0.000 | 0.24929 |
| Square | 3 | 0.2747 | 0.0916 | 236.11 | 0.000 | |
| rw*rw | 1 | 0.0001 | 0.0001 | 0.37 | 0.558 | 0.0070 |
| R*R | 1 | 0.1473 | 0.1473 | 379.83 | 0.000 | 0.2102 |
| $\gamma*\gamma$ | 1 | 0.0042 | 0.0042 | 10.91 | 0.008 | 0.0381 |
| 2-way interaction | 3 | 0.0954 | 0.0318 | 81.96 | 0.000 | |
| rw*R | 1 | 0.0855 | 0.0855 | 220.45 | 0.000 | −0.10338 |
| rw*$\gamma$ | 1 | 0.0020 | 0.0020 | 5.05 | 0.048 | −0.01564 |
| R*$\gamma$ | 1 | 0.0079 | 0.0079 | 20.38 | 0.001 | −0.03143 |
| Error | 10 | 0.0039 | 0.0004 | | | |
| Lack-of-fit | 5 | 0.0039 | 0.0008 | * | * | |
| Pure error | 5 | 0.0000 | 0.0000 | | | |
| Total | 19 | 18.4092 | | | | |

$R^2 = 99.98\%$, Adjusted- $R^2 = 99.96\%$

**P1-Source P2-DF P3-Adj SS P4-Adj MS P5-F Value P6-P Value P7-Coefficient**

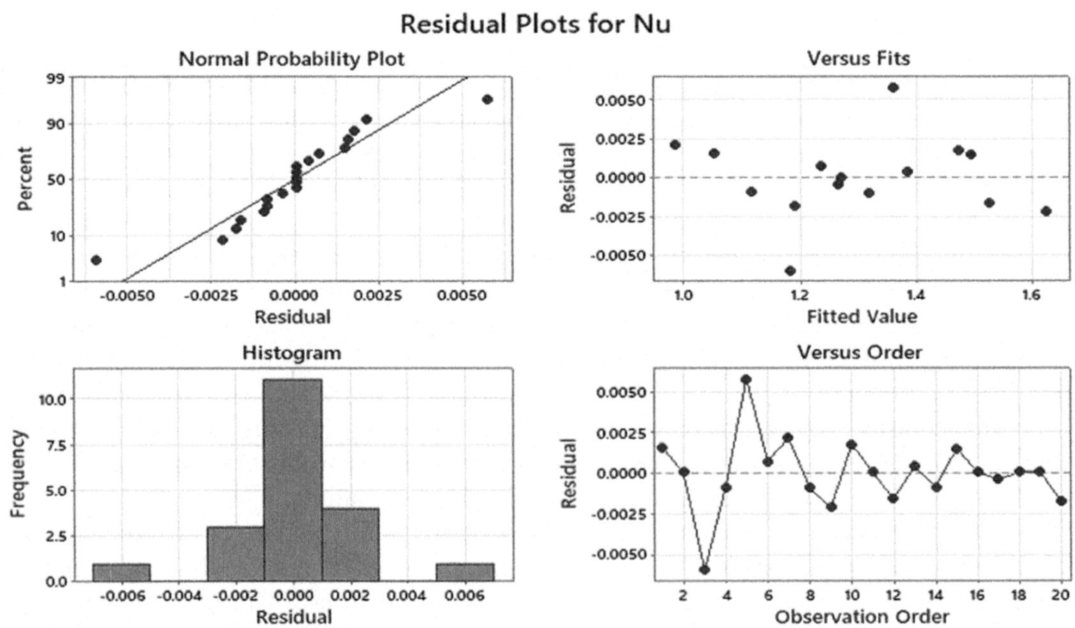


**FIGURE 1.14** The residual analysis for the Nusselt number (Nu) of Ag–$MoS_2$/EG hybrid nanofluid confirms the adequacy of the developed regression model.

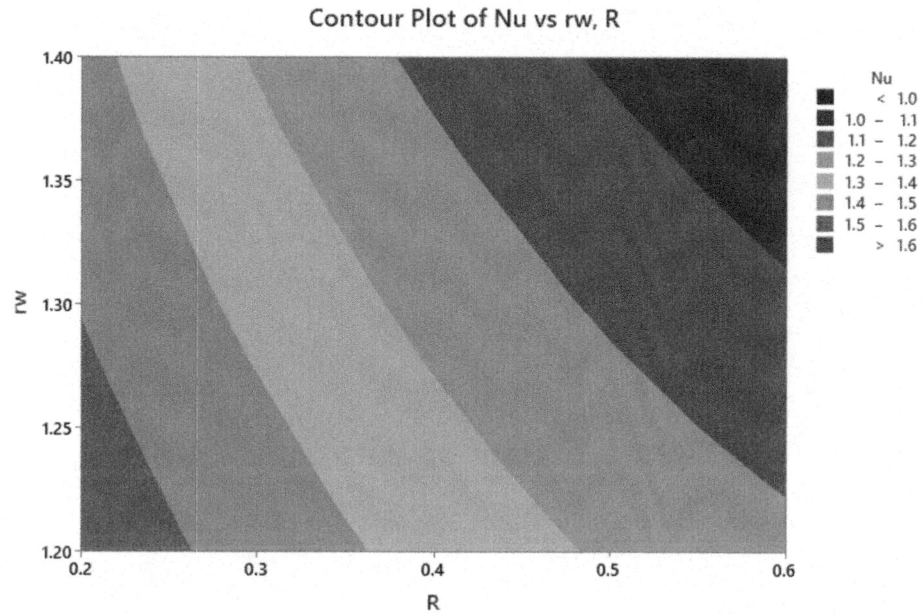


**FIGURE 1.15** Contour plot of the Nusselt number (Nu) as a function of the temperature ratio parameter (rw) and the radiation parameter (R).

Fig. 1.16 illustrates their combined effect on the heat transfer rate of the Ag–$MoS_2$/water hybrid nanofluid. The color variation shows that Nu increases from blue (darker gray in print version) to green (dark gray in print version) as rw decreases and $\gamma$ rise. This indicates that lower rw and higher stretching ratio parameters enhance the Nusselt number, resulting in improved thermal performance.

Fig. 1.17 shows the contour plot of the Nusselt number (Nu) as a function of the radiation parameter (R) and the stretching ratio parameter ($\gamma$). The color transition from dark blue (black in print version) to green (dark gray in print version) indicates that Nu increases with higher values of R and lower values of $\gamma$. This trend reveals that enhancing radiation and reducing stretching ratio parameters leads to improved thermal transport characteristics.

Fig. 1.18 illustrates the combined influence of these parameters on the heat transfer rate of the Ag–$MoS_2$/EG hybrid nanofluid. The color gradient indicates that Nu decreases progressively from green (light gray in print version) to dark green (dark gray in print version) as R and rw increase. This trend suggests that higher radiation and rotation parameters lead to a reduction in the rate of heat transfer. Conversely, lower values of R and rw correspond to higher Nu, indicating enhanced convective heat transfer performance.

Fig. 1.19 illustrates their combined effect on the heat transfer rate of the Ag–$MoS_2$/EG hybrid nanofluid. The color variation shows that Nu increases from green (light gray in print version) to dark green (dark gray in print version) as rw decreases and $\gamma$ rise.

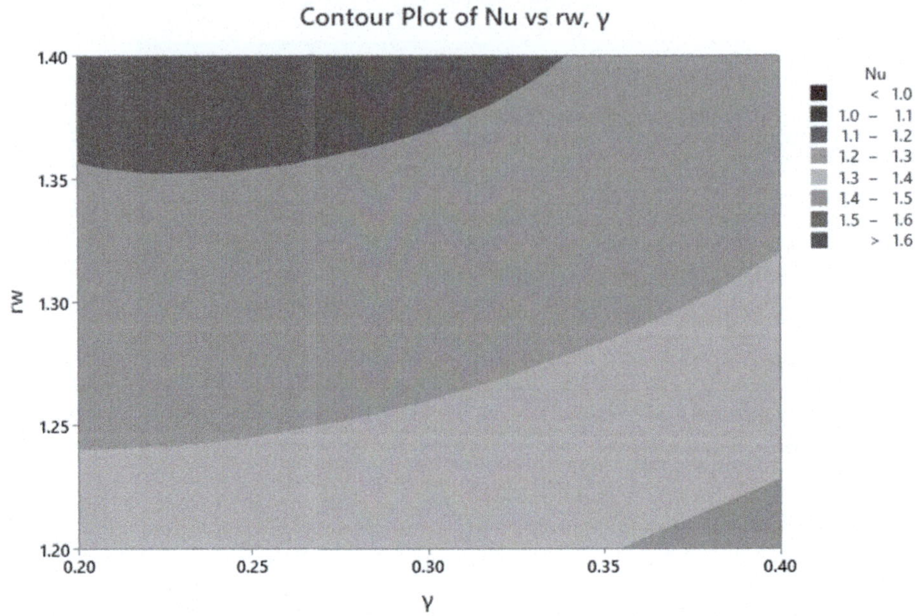


**FIGURE 1.16** Contour plot of the Nusselt number (Nu) as a function of the temperature ratio parameter (rw) and the stretching ratio parameter ($\gamma$).

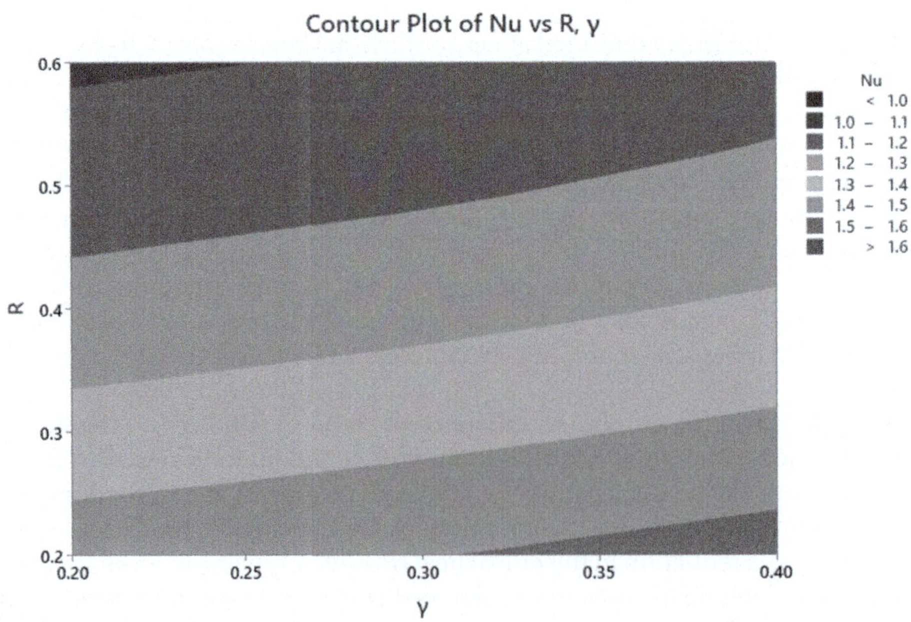


**FIGURE 1.17** Contour plot of the Nusselt number (Nu) as a function of the radiation parameter (R) and the stretching ratio parameter ($\gamma$).

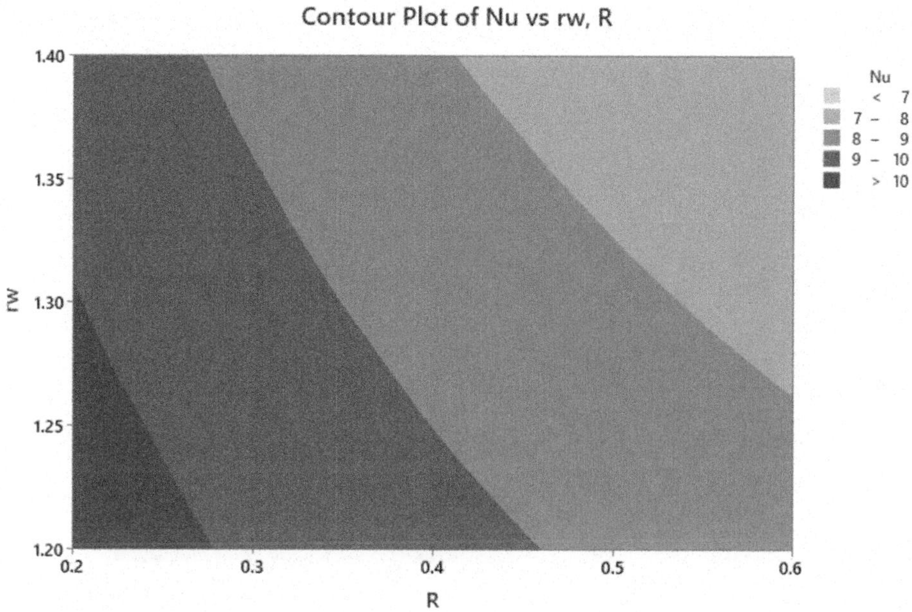


**FIGURE 1.18** Contour plot of the Nusselt number (Nu) as a function of the temperature ratio parameter (rw) and the radiation parameter (R).

This indicates that lower rw and higher stretching ratio parameters enhance the Nusselt number, resulting in improved thermal performance.

Fig. 1.20 shows a contour plot of the Nusselt number (Nu) as a function of the radiation parameter (R) and the stretching ratio parameter ($\gamma$). The color transition from green (light gray in print version) to dark green (dark gray in print version) indicates that Nu increases with higher values of $\gamma$ and lower values of R. This trend reveals that reducing radiation and increasing stretching ratio parameters leads to improved thermal transport characteristics.

## 1.8 Key findings

1. ***Velocity characteristics:*** The primary and secondary velocity gradients $p'(\eta)$ and $q'(\eta)$ decrease with increasing magnetic parameter (M) due to the Lorentz force, which imposes a resistive drag on fluid motion. Similarly, higher stretching ratio ($\lambda$) reduces both velocity gradients, indicating a thickened boundary layer and slower fluid motion near the wall. In contrast, an increase in the rotation parameter ($\gamma$) causes a reduction in the primary velocity component but an enhancement in the secondary velocity component, signifying the strong influence of rotational motion on the transverse momentum transfer.

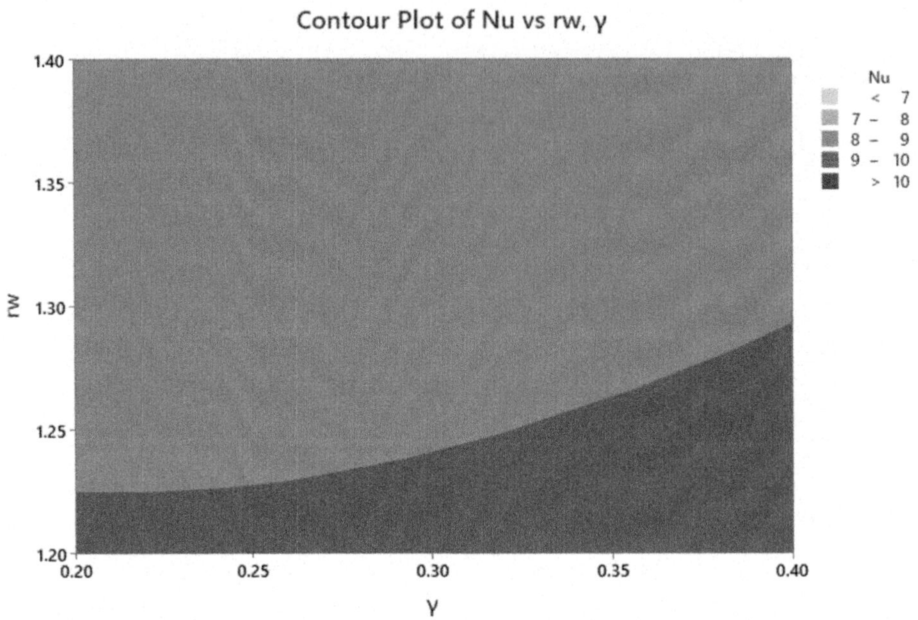


**FIGURE 1.19** Contour plot of the Nusselt number (Nu) as a function of the temperature ratio parameter (rw) and the stretching ratio parameter ($\gamma$).

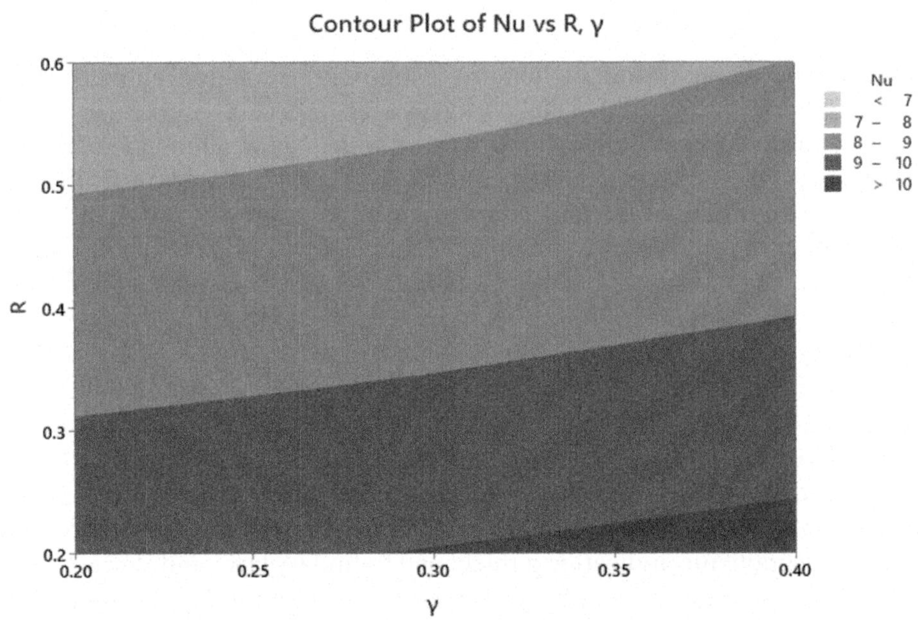


**FIGURE 1.20** Contour plot of the Nusselt number (Nu) as a function of the radiation parameter (R) and the stretching ratio parameter ($\gamma$).

2. ***Thermal behavior:*** The temperature profile rises with increasing magnetic parameter (M) and stretching ratio ($\lambda$) because of the combined effects of Joule heating and viscous dissipation, which enhance thermal energy within the boundary layer. However, increasing the rotation parameter ($\gamma$) reduces the temperature distribution, showing that rotational motion facilitates convective cooling. Similarly, higher wall temperature ratio (rw) and thermal radiation parameter (R) significantly increase the temperature profile, underscoring their roles in promoting heat transport from the wall to the fluid medium.
3. ***Mass transfer characteristics:*** The concentration profile diminishes with an increase in the chemical reaction parameter ($K_r$), revealing that stronger chemical reactions accelerate mass diffusion and species depletion within the boundary layer.
4. ***Skin friction effects:*** The local skin friction coefficients $C_{fx}$ and $C_{fy}$ increase with higher nanoparticle volume fractions ($\psi_1, \psi_2$), reflecting greater wall shear stress due to enhanced viscosity and interparticle interactions. As M increases, the skin friction first rises and then declines, illustrating that moderate magnetic strength increases resistance, whereas excessive magnetization suppresses velocity gradients. Higher $\lambda$ amplifies shear stress, while greater $\gamma$ further intensifies the friction in both directions due to rotational influences. EG-based nanofluids exhibit slightly higher frictional resistance than water-based ones because of their higher viscosity and density.
5. ***Heat and mass transfer rates:*** The Nusselt number (Nu) and Sherwood number (Sh) exhibit distinct variations for the two base fluids. For Ag–$MoS_2$/water, an increase in M and $\gamma$ reduces Nu, implying that Lorentz and centrifugal forces suppress heat transfer, while Sh remains nearly unchanged, indicating stable mass diffusion. In contrast, the Ag–$MoS_2$/EG nanofluid shows similar behavior but with generally lower Nu and Sh values due to its higher viscosity and lower thermal diffusivity. Increasing nanoparticle volume fractions consistently enhances both Nu and Sh, confirming the positive role of nanoparticle loading in improving thermal and solutal transport properties.
6. ***Comparative nanofluid performance:*** Among the studied hybrid nanofluids, Ag–$MoS_2/H_2O$ exhibits the highest Nusselt number and, consequently, the most efficient heat transfer performance, followed by $TiO_2$–$MoS_2/H_2O$. EG-based nanofluids show relatively lower heat transfer rates, which is consistent with the lower thermal conductivity of ethylene glycol.

## 1.9 Conclusions

The investigation demonstrates that the hydrodynamic, thermal, and solutal behaviors of hybrid nanofluids are highly sensitive to magnetic field strength, stretching rate, rotation parameter, thermal radiation, and chemical reaction intensity. Moderate levels of magnetization, stretching, and nanoparticle concentration yield optimal flow stability and thermal efficiency. The Ag–$MoS_2$/water hybrid nanofluid consistently outperforms its EG-based counterpart in heat and mass transfer capabilities due to water's superior thermal conductivity and lower viscosity, making it more suitable for applications requiring enhanced thermal management. Conversely, the Ag–$MoS_2$/EG hybrid nanofluid offers better operational stability in viscous flow environments where high drag and shear resistance are desirable. Overall, the study confirms that tuning the nanoparticle volume fraction and external parameters allows precise control of heat and mass transfer characteristics, optimizing hybrid nanofluids for advanced cooling, energy, and material processing systems.

## 1.10 Nomenclature list

| | |
|---|---|
| $\psi$: | Nanoparticle volume fraction |
| $\omega^*$: | Angular rotation speed (m/s) |
| w, $\infty$: | Represent wall and ambient (free-stream) conditions, respectively |
| $W_{bnf}$: | Water-based nanofluid |
| $\nu$: | Kinematic viscosity ($m^2$/s) |
| $\mu$: | Dynamic viscosity of the base fluid (kg·$m^{-1}$·$s^{-1}$) |
| $\mu_{hnf}$: | Dynamic viscosity of the hybrid nanofluid (kg·$m^{-1}$·$s^{-1}$) |
| $T_m$: | Average (mean) temperature of the fluid (K) |
| $T$: | Temperature of the fluid (K) |
| $\sigma$: | Electrical conductivity (S/m) |
| **Sc:** | Schmidt number |
| $\rho_{hnf}$: | Density of the hybrid nanofluid (kg·$m^{-3}$) |
| $\rho$: | Fluid density (kg·$m^{-3}$) |
| q, p: | Dimensionless velocity components |
| **Pr:** | Prandtl number |
| **Nu:** | Nusselt number |
| **M:** | Magnetic parameter |
| $k$: | Thermal conductivity (W·$m^{-1}$·$K^{-1}$) |
| $K_r$: | Dimensionless chemical reaction rate parameter |
| $K_1$: | Chemical reaction rate constant |
| $\lambda$: | Stretching ratio parameter |
| $\gamma$: | Rotation parameter |
| f: | Base fluid |
| **$EG_{bnf}$:** | Ethylene glycol-based nanofluid |
| **DT:** | Thermal diffusion coefficient |
| **D:** | Mass diffusion coefficient ($m^2$/s) |
| $c_f$: | Coefficient of local skin friction |
| $B_0$: | Magnetic field intensity (kg·$s^{-2}$·$A^{-1}$) |
| b, a: | Stretching rates of the surface along $y$ and $x$ directions, respectively |
| $\alpha_{hnf}$: | Thermal diffusivity of the hybrid nanofluid ($m^2$/s) |
| $\alpha$: | Thermal diffusivity ($m^2$/s) |
| $(W/EG)_{bnf}$: | Water–ethylene glycol mixture nanofluid (50:50 ratio) |
| **hnf:** | Hybrid nanofluid |

**nf:** Nanofluid
$K_T$: Thermal diffusion ratio

## References

Alkuhayli, N.A.M., 2023. Heat transfer analysis of a hybrid nanofluid flow on a rotating disk considering thermal radiation effects. Case Studies in Thermal Engineering (ISSN 2214-157X) 49. https://doi.org/10.1016/j.csite.2023.103131. Elsevier Ltd, United Kingdom, Available from: http://www.journals.elsevier.com/case-studies-in-thermal-engineering/.

Allehiany, F.M., Bilal, M., Alfwzan, W.F., Ali, A., Eldin, S.M., 2023. Numerical solution for the electrically conducting hybrid nanofluid flow between two parallel rotating surfaces subject to thermal radiation. AIP Advances (ISSN 2158-3226) 13 (7). https://doi.org/10.1063/5.0154720. American Institute of Physics Inc., Saudi Arabia, Available from: http://scitation.aip.org/content/aip/journal/adva.

Alqahtani, A.M., Bilal, M., Ali, A., Alsenani, T.R., Eldin, S.M., 2023. Numerical solution of an electrically conducting spinning flow of hybrid nanofluid comprised of silver and gold nanoparticles across two parallel surfaces. Scientific Reports (ISSN 2045-2322) 13 (1). https://doi.org/10.1038/s41598-023-33520-5. Nature Research, Saudi Arabia, Available from: https://www.nature.com/srep/.

Arshad, Mubashar, Hussain, Azad, Hassan, Ali, Wróblewski, Piotr, Elfasakhany, Ashraf, Abdelghany Elkotb, Mohamed, Abdelmohimen, Mostafa A.H., Galal, Ahmed M., 2022. Thermal energy investigation of magneto-hydrodynamic nano-material liquid flow over a stretching sheet: comparison of single and composite particles. Alexandria Engineering Journal (ISSN 1110-0168) 61 (12), 10453–10462. https://doi.org/10.1016/j.aej.2022.03.069.

Babar, H., Ali, H.M., 2019. Towards hybrid nanofluids: preparation, thermophysical properties, applications, and challenges. Journal of Molecular Liquids (ISSN 0167-7322) 281, 598–633. https://doi.org/10.1016/j.molliq.2019.02.102.

Choi, S.U.S., 1995. Enhancing thermal conductivity of fluids with nanoparticles 231. 1995/12 American Society of Mechanical Engineers, Fluids Engineering Division (Publication) FED 99 105, ASME, United States.

Crane, L.J., 1970. Flow past a stretching plate. Zeitschrift für Angewandte Mathematik und Physik 21, 645–647.

Durga Prasad, P., Kiran Kumar, R.V.M.S.S., Varma, S.V.K., 2018. Heat and mass transfer analysis for the MHD flow of nanofluid with radiation absorption. Ain Shams Engineering Journal (ISSN 2090-4479) 9 (4), 801–813. https://doi.org/10.1016/j.asej.2016.04.016.

Eckert, Ernst Rudolf Georg, Drake, Robert M., 1972. Analysis of Heat and Mass Transfer. McGraw-Hill.

Goud, B.S., Reddy, Y.D., 2022. Chemical reaction and Soret effect on an unsteady MHD heat and mass transfer fluid flow along an infinite vertical plate with radiation and heat absorption. Journal of the Indian Chemical Society (ISSN 0019-4522) 99 (11). https://doi.org/10.1016/j.jics.2022.100762. Elsevier B.V., India, Available from: https://www.sciencedirect.com/journal/journal-of-the-indian-chemical-society.

Gunakala, S.R., Job, V.M., Murthy, P.V.S.N., Umakanthan, S., Ramcharitar, V., 2025. A numerical comparison of magnetic nanoparticle hyperthermia in breast, muscle, and prostate tumors. Critical Reviews in Biomedical Engineering (ISSN 1943-619X) 53 (6), 61–83. https://doi.org/10.1615/CritRevBiomedEng.2025055746. Begell House Inc., Trinidad and Tobago, Available from: https://www.dl.begellhouse.com/download/article/532b25944084a140/CRB5306(4)-55746.pdf.

Hassani, M., Mohammad Tabar, M., Nemati, H., Domairry, G., Noori, F., 2011. An analytical solution for boundary layer flow of a nanofluid past a stretching sheet. International Journal of Thermal Sciences (ISSN 1290-0729) 50 (11), 2256–2263. https://doi.org/10.1016/j.ijthermalsci.2011.05.015.

Huminic, Gabriela, Huminic, Angel, 2018. Hybrid nanofluids for heat transfer applications – a state-of-the-art review. International Journal of Heat and Mass Transfer (ISSN 0017-9310) 125, 82–103. https://doi.org/10.1016/j.ijheatmasstransfer.2018.04.059.

Hussain, Azad, Hassan, Ali, Al Mdallal, Qasem, Ahmad, Hijaz, Rehman, Aysha, Altanji, Mohamed, Arshad, Mubashar, 2021a. Heat transport investigation of magneto-hydrodynamics (SWCNT-MWCNT) hybrid nanofluid under the thermal radiation regime. Case Studies in Thermal Engineering (ISSN 2214-157X) 27, 101244. https://doi.org/10.1016/j.csite.2021.101244.

Hussain, Azad, Arshad, Mubashar, Rehman, Aysha, Hassan, Ali, Elagan, S.K., Ahmad, Hijaz, 2021b, Ishan, Amira Three-dimensional water-based magneto-hydrodynamic rotating nanofluid flow over a linear extending sheet and heat transport analysis: a numerical approach. Energies 14 (16), 5133. https://doi.org/10.3390/en14165133.

Leong, K.Y., Ku Ahmad, K.Z., Ong, H.C., Ghazali, M.J., Baharum, A., 2017. Synthesis and thermal conductivity characteristic of hybrid nanofluids – a review. Renewable and Sustainable Energy Reviews (ISSN 1879-0690) 75, 868–878. https://doi.org/10.1016/j.rser.2016.11.068. Elsevier Ltd, Malaysia, Available from: https://www.journals.elsevier.com/renewable-and-sustainable-energy-reviews.

Mabood, F., Khan, W.A., Ismail, A.I.M., 2015. MHD boundary layer flow and heat transfer of nanofluids over a nonlinear stretching sheet: a numerical study. Journal of Magnetism and Magnetic Materials (ISSN 0304-8853) 374, 569–576. https://doi.org/10.1016/j.jmmm.2014.09.013.

Mudoi, S., Sarma, D., Sarma, A.K., 2025. Newtonian fluid flow with heat and mass transfer across an inclined sheet and magnetic field with Soret effects. Journal of the Korean Physical Society (ISSN 1976-8524) 86 (4), 280–291. https://doi.org/10.1007/s40042-024-01278-2. Korean Physical Society, India, Available from: https://www.springer.com/journal/40042.

Nadeem, S., Haq, Rizwan Ul, 2014. Effect of thermal radiation for magnetohydrodynamic boundary layer flow of a nanofluid past a stretching sheet with convective boundary conditions. Journal of Computational and Theoretical Nanoscience (ISSN 1546-1955) 11 (1), 32–40. https://doi.org/10.1166/jctn.2014.3313.

Narayan, P.V.S., Tarakaramu, N., Makinde, O.D., Venkateswarlu, B., Sarojamma, G., 2018. MHD stagnation point flow of viscoelastic nanofluid past a convectively heated stretching surface. Defect and Diffusion Forum (ISSN 1662-9507) 387, 106–120. https://doi.org/10.4028/www.scientific.net/DDF.387.106. Trans Tech Publications Ltd, India, Available from: http://www.scientific.net/.

Nazar, R., Amin, N., Pop, I., 2004. Unsteady boundary layer flow due to a stretching surface in a rotating fluid. Mechanics Research Communications (ISSN 0093-6413) 31 (1), 121–128. https://doi.org/10.1016/j.mechrescom.2003.09.004.

Özerinç, S., Kakaç, S., YazIcIoğlu, A.G., 2010. Enhanced thermal conductivity of nanofluids: a state-of-the-art review. Microfluidics and Nanofluidics (ISSN 1613-4990) 8 (2), 145–170. https://doi.org/10.1007/s10404-009-0524-4.

Raju, C.S.K., Upadhya, S. Mamatha, Seth, Dinesh, 2021. Thermal convective conditions on MHD radiated flow with suspended hybrid nanoparticles. Microsystem Technologies (ISSN 0946-7076) 27 (5), 1933–1942. https://doi.org/10.1007/s00542-020-04971-x.

Reddy, M.G., 2014. Influence of magnetohydrodynamic and thermal radiation boundary layer flow of a nanofluid past a stretching sheet. Journal of Scientific Research (ISSN 2070-0245) 6 (2), 257–272. https://doi.org/10.3329/jsr.v6i2.17233.

Sajid, M.U., Ali, H.M., 2018. Thermal conductivity of hybrid nanofluids: a critical review. International Journal of Heat and Mass Transfer (ISSN 0017-9310) 126, 211–234. https://doi.org/10.1016/j.ijheatmasstransfer.2018.05.021. Elsevier Ltd, Pakistan, Available from: http://www.journals.elsevier.com/international-journal-of-heat-and-mass-transfer/.

Shah, T.R., Ali, H.M., 2019. Applications of hybrid nanofluids in solar energy, practical limitations and challenges: a critical review. Solar Energy (ISSN 0038-092X) 183, 173–203. https://doi.org/10.1016/j.solener.2019.03.012. Elsevier Ltd, Pakistan, Available from: www.elsevier.com/inca/publications/store/3/2/9/index.htt.

Shoaib, Muhammad, Raja, Muhammad Asif Zahoor, Sabir, Muhammad Touseef, Islam, Saeed, Shah, Zahir, Kumam, Poom, Alrabaiah, Hussam, 2020. Numerical investigation for rotating flow of MHD hybrid nanofluid with thermal radiation over a stretching sheet. Scientific Reports 10 (1), 2045–2322. https://doi.org/10.1038/s41598-020-75254-8.

Siddique, Samavia, Jamil, Hifza, 2025. The role of novel nanomaterials in biomedical applications. Journal of Life Sciences and Informatics 1 (1), 60–82.

Sreedevi, P., Sudarsana Reddy, P., Chamkha, Ali, 2020. Heat and mass transfer analysis of unsteady hybrid nanofluid flow over a stretching sheet with thermal radiation. SN Applied Sciences 2 (7), 2523–3963. https://doi.org/10.1007/s42452-020-3011-x.

Sulochana, C., Samrat, S.P., Sandeep, N., 2018. Magnetohydrodynamic radiative nanofluid flow over a rotating surface with Soret effect. Multidiscipline Modeling in Materials and Structures (ISSN 1573-6113) 14 (1), 168–188. https://doi.org/10.1108/MMMS-05-2017-0042. Emerald Group Holdings Ltd.,

India, Available from: http://www.emeraldgrouppublishing.com/products/journals/journals.htm?id=MMMS.

Sulochana, C., Samrat, S.P., Sandeep, N., 2016. Thermal radiation effect on MHD nanofluid flow over a stretching sheet. International Journal of Engineering Research in Africa (ISSN 1663-4144) 23, 89–102. https://doi.org/10.4028/www.scientific.net/JERA.23.89. Trans Tech Publications Ltd, India, Available from: http://www.ttp.net/1663-3571.html.

Sundar, L.S., Sharma, K.V., Singh, M.K., Sousa, A.C.M., 2017. Hybrid nanofluids preparation, thermal properties, heat transfer and friction factor – a review. Renewable & Sustainable Energy Reviews (ISSN 1879-0690) 68, 185–198. https://doi.org/10.1016/j.rser.2016.09.108. Elsevier Ltd, Portugal, Available from: https://www.journals.elsevier.com/renewable-and-sustainable-energy-reviews.

Suneetha, S., Subbarayudu, K., Bala Anki Reddy, P., 2022. Hybrid nanofluids development and benefits: a comprehensive review. Journal of Thermal Engineering (ISSN 2148-7847) 8 (3), 445–455. https://doi.org/10.18186/thermal.1117455.

Wahid, N.S., Md Arifin, N., Khashi'ie, N.S., Pop, I., Bachok, N., Hafidzuddin, M.E.H., 2022. MHD mixed convection flow of a hybrid nanofluid past a permeable vertical flat plate with thermal radiation effect. Alexandria Engineering Journal (ISSN 1110-0168) 61 (4), 3323–3333. https://doi.org/10.1016/j.aej.2021.08.059. Elsevier B.V., Malaysia, Available from: http://www.elsevier.com/wps/find/journaldescription.cws_home/724292/description#description.

Welty, James, Wicks, Charles E., Rorrer, Gregory L., Wilson, Robert E., 2007. Fundamentals of Momentum, Heat, and Mass Transfer. 5. Wiley, Wiley.

Zainal, N.A., Nazar, R., Naganthran, K., Pop, I., 2021. MHD flow and heat transfer of hybrid nanofluid over a permeable moving surface in the presence of thermal radiation. International Journal of Numerical Methods for Heat & Fluid Flow (ISSN 0961-5539) 31 (3), 858–879. https://doi.org/10.1108/HFF-03-2020-0126. Emerald Publishing, Malaysia, Available from: http://www.emeraldinsight.com/info/journals/hff/hff.jsp.

# 2

# Solitary wave solution of generalized Boussinesq equation in fuzzy environment

Rambabu Vana[a], Perumandla Karunakar[b], B. Bala Sai Sankar[b], S. Srinivas[b], and Kuppalapalle Vajravelu[c]

[a]Department of Mathematics, School of Computer Science and Artificial Intelligence, SR University, Warangal, India [b]Department of Mathematics, School of Advanced Sciences, VIT-AP University, Amaravati, India [c]Department of Mathematics, Department of Mechanical, Materials and Aerospace Engineering, University of Central Florida, Orlando, FL, United States

## 2.1 Introduction

The mathematical modeling of many physical systems leads to nonlinear differential equations in different areas of science and engineering. Finding approximate and exact solutions to these equations is important for understanding complex physical phenomena. Many researchers have developed different methods to solve these problems, such as the differential quadrature technique based on shifted Legendre and generalized Laguerre polynomials (Vana and Karunaka, 2025), B-splines based collocation technique (Palav and Pradhan, 2025), multiple-scale probationer method (Ketcheson et al., 2025), multi mesh-free collocation approach (Aourir and Laeli Dastjerdi, 2025), rapidly convergent approximation technique (Das, 2025), Hosoya polynomials collection method (Nirmala and Kumbinarasaiah, 2024), Picard successive approach (Darghoth, 2024), dual reciprocity boundary element approach (Alipour, 2024), optimal auxilary function method (Ismael et al., 2024), modified extended direct algebraic method (Bilal et al., 2024), aboodh transformation decomposition method (Khan et al., 2024), residual power series method (Iqbal and Martínez, 2024), modified variational iteration method (Asaduzzaman et al., 2024), rational sin gordon expansion method (Mamun et al., 2024), modified extended tanh function method (Hussein et al., 2024), and several other techniques (Chakraverty et al., 2019; Chakraverty and Perera, 2018).

He (1999, 2000, 2003, 2005) introduced the homotopy perturbation method (HPM) by combining the perturbation approach with standard homotopy to solve different physical problems. Readers can refer to his work to understand HPM better. Later, HPM was combined with the Laplace transform (LT), creating the HPTM, which became a more effective way to solve nonlinear problems (Karunakar and Chakraverty, 2017, 2018a,b, 2019a,b; Karunakar et al., 2020; Karunakar and Chakraverty, 2021, 2022; Karunakar et al.,

Advanced Computational and Mathematical Approaches in Applied Differential Equations
https://doi.org/10.1016/B978-0-44-345743-2.00013-X

2023; Sankar and Karunakar, 2025). In addition, the FHPTM was developed as an extension for solving fuzzy differential equations (Vana and Karunakar, 2024a,b; Karunakar et al., 2024; Vana et al., 2025; Vana and Karunakar, 2024c,d). This method improves the convergence region and is widely used because of its accuracy and efficiency in solving uncertain nonlinear problems. However, it is important to note that sometimes, the LTM alone may not be enough to solve nonlinear problems, as handling nonlinear terms can be difficult.

Mathematical physicists have been focused more on nonlinear wave equations that models long wave propagation in shallow water, particularly the Boussinesq equation (BE) (Boussinesq, 1871), which explains soliton interactions in shallow water waves. Extensive research has been conducted to find exact and approximation solutions for different forms of BE-type, including the standard BE (Bruzon, 2009), the two-dimensional BE (Christov and Choudhury, 2011), and higher-order BE (Daripa, 2006). The asymptotic behavior of the BE has been analyzed using the Korteweg–de Vries (KdV) (Kordyukova, 2008). Furthermore, studies have explored the nonintegrable versions of the gBE, particularly in the context of two-soliton interactions (Wazwaz, 2010).

In recent years, Razzaq et al. (2024) found optical soliton solutions for the $4^{th}$-order nonlinear gBE utilizing the modified $\left(\frac{\mathcal{G}'}{\mathcal{G}^2}\right)-$ and $\mathcal{F}-$ expansion techniques. Zhu and Qu (2024) obtained smooth periodic wave solutions to the nonlinear gBE using the bifurcation technique. Kai et al. (2024) studied analytic solutions and the dynamic properties of different types of gBE using the complete discrimination system for the polynomial approach. Ding et al. (2024) studied a finite element method to solve the nonstationary gBE. Chen and Li (2024) presented a $(2+1)$-dimensional gBE using the Hirota bilinear approach. The $4^{th}$-order gBE is solved using the extended homogeneous balance method in Chen and Wang (2021). Tian (2020) investigated a $4^{th}$-order nonlinear gBE utilizing the Lie symmetry technique.

The BE is very important in understanding nonlinear lattice waves, sound waves in plasma, and vibrations in a nonlinear string. As the gBE to finds more applications in various fields of mathematical physics, it becomes quite interesting to explore the mathematical structure of these governing equations, including solitary wave solutions in both crisp and uncertain conditions.

In this study, we will concentrate on a $4^{th}$-order nonlinear gBE (Razzaq et al., 2024; Zhu and Qu, 2024; Chen and Wang, 2021; Tian, 2020),

$$\frac{\partial^2 \mathcal{B}}{\partial \tau^2} + a\frac{\partial^2 \mathcal{B}}{\partial \zeta^2} + b\frac{\partial}{\partial \zeta}\left(\mathcal{B}^{\delta}\frac{\partial \mathcal{B}}{\partial \zeta}\right) + c\frac{\partial^4 \mathcal{B}}{\partial \zeta^4} = 0, \tag{2.1}$$

the exact solution of Eq. (2.1) as (Razzaq et al., 2024; Zhu and Qu, 2024; Chen and Wang, 2021; Tian, 2020),

$$\mathcal{B}(\zeta,\tau) = \left(\frac{2\left(\delta^2+3\delta+2\right)ck^2}{b\delta^2}\, sech^2\left(k\zeta - \frac{\sqrt{-a\delta^2k^2-4ck^4}}{\delta}\tau\right)\right)^{\frac{1}{\delta}}. \tag{2.2}$$

Here, $\zeta$ represents space, $\tau$ represents time, and the parameters $k$, $a$, $b$, and $c$ are nonzero constants, where $a < 0$, $\frac{c}{b} > 0$, and $c < 0$, that depend on the depth or velocity of the fluid, denoted by $\mathcal{B}(\zeta, \tau)$.

The gBE has been analyzed using crisp parameter values, but in practical applications, these parameters and factors are affected by variables like fluid velocity and depth, which introduce uncertainty. To solve this uncertain parameter, triangular fuzzy numbers can be replaced with crisp parameter, providing a realistic representation of the uncertainties in nonlinear wave dynamics. This study is important as it presents FHPTM to incorporate uncertainty into the model, resulting in a more accurate representation of wave behavior.

## 2.2 Preliminaries

### 2.2.1 Interval arithmetic operations

An interval is defined as a subset of the real number set $\mathbb{R}$ such that $[\mathcal{Z}_l, \mathcal{Z}_u] = \{\zeta / \mathcal{Z}_l \leq \zeta \leq \mathcal{Z}_u,\ \mathcal{Z}_l, \mathcal{Z}_u \in \mathcal{R}\}$. Let $[z_1, z_2]$, and $[z_3, z_4]$ be two such intervals. The arithmetic operations between these intervals may be defined as following interval analysis (Vana and Karunakar, 2024a; Vana et al., 2025; Vana and Karunakar, 2024c; Moore, 1966):

$$[z_1, z_2] + [z_3, z_4] = [z_1 + z_3, z_2 + z_4],$$

$$[z_1, z_2] - [z_3, z_4] = [z_1 - z_4, z_3 - z_2],$$

$$[z_1, z_2] \times [z_3, z_4] = \begin{cases} [z_1 z_3, z_2 z_4], & \text{for } z_1 \geq 0 \text{ } and \text{ } z_3 \geq 0, \\ [z_2 z_4, z_1 z_3], & \text{for } z_2 < 0 \text{ and } z_4 < 0, \\ [min\{z_1 z_4, z_2 z_3\}, max\{z_1 z_4, z_2 z_3\}], & \text{Otherwise.} \end{cases}$$

$$\frac{1}{[z_1, z_2]} = \begin{cases} \left[\frac{1}{z_2}, \frac{1}{z_1}\right], & \text{if } 0 \notin [z_1, z_2], \\ \left[\frac{1}{z_2}, \infty\right) \cup \left(-\infty, \frac{1}{z_1}\right], & \text{if } z_1 < 0 \text{ and } z_2 > 0, \\ \left[\frac{1}{z_2}, \infty\right), & \text{if } z_1 = 0 \text{ and } z_2 > 0, \\ \left(-\infty, \frac{1}{z_1}\right], & \text{if } z_1 < 0 \text{ and } z_2 = 0. \end{cases}$$

$$\frac{[z_1, z_2]}{[z_3, z_4]} = \begin{cases} \left[\frac{z_1}{z_3}, \frac{z_2}{z_4}\right], & \text{if } z_3 > 0 \text{ and } z_4 > 0, \\ \left[\frac{z_2}{z_3}, \frac{z_1}{z_4}\right], & \text{if } z_3 < 0 \text{ and } z_4 < 0, \\ \left(-\infty, \frac{z_2}{z_3}\right] \cup \left[\frac{z_1}{z_4}, \infty\right), & \text{Otherwise.} \end{cases}$$

### 2.2.2 Fuzzy set

A fuzzy set is defined as a set of ordered pairs, where each pair consists of an element $\zeta \in \mathbb{R}$ and its corresponding membership value $\mu_{\widetilde{Z}}(\zeta) \in [0, 1]$, representing the degree to which $\zeta$ belongs to the set. It is defined as (Karunakar et al., 2020; Karunakar and Chakraverty, 2021),

$$\widetilde{Z} = \left\{\mu_{\widetilde{Z}}(\zeta) : \mathbb{R} \to [0, 1],\ \forall \zeta \in \mathbb{R}\right\}.$$

Here, $\mu_{\widetilde{Z}}(\zeta)$ denotes the membership function of the fuzzy set $\widetilde{Z}$, which is assumed to be piece-wise continuous.

### 2.2.3 Fuzzy number

A fuzzy number (Karunakar et al., 2020; Karunakar and Chakraverty, 2021) is a specific type of fuzzy set on the real line $\mathbb{R}$ that is normalized and convex, with its membership function $\mu_{\widetilde{Z}}$ satisfying the following properties:

- **Normality**: There exists at least one $\zeta \in \mathbb{R}$ such that $\mu_{\widetilde{Z}}(\zeta) = 1$.
- **Convexity**: For all $\zeta_1, \zeta_2 \in \mathbb{R}$ and $\lambda \in [0, 1]$, $\mu_{\widetilde{Z}}(\lambda\zeta_1 + (1-\lambda)\zeta_2) \geq \min\left\{\mu_{\widetilde{Z}}(\zeta_1), \mu_{\widetilde{Z}}(\zeta_2)\right\}$.
- **Piecewise continuity**: The membership function $\mu_{\widetilde{Z}}(\zeta)$ is piecewise continuous on $\mathbb{R}$.

### 2.2.4 Triangular fuzzy numbers (TFN)

The TFN is $\widetilde{Z}_T = [Z_l, Z_c, Z_u]$, whose the membership function $\mu_{\widetilde{Z}_T}$ is defined as follows (Vana and Karunakar, 2024a,b; Vana et al., 2025):

$$\mu_{\widetilde{Z}_T} = \begin{cases} \frac{\zeta - Z_l}{Z_c - Z_l}, & \text{if } Z_l \leq \zeta \leq Z_c \\ \frac{Z_u - \zeta}{Z_u - Z_c}, & \text{if } Z_c \leq \zeta \leq Z_u \\ 0, & \text{Otherwise.} \end{cases}$$

### 2.2.5 Parametric form

The TFN $\widetilde{Z}_T = [Z_l, Z_c, Z_u]$ may be defined in parametric form as (Karunakar and Chakraverty, 2018b, 2019a,b; Karunakar et al., 2020; Karunakar and Chakraverty, 2021, 2022; Karunakar et al., 2023), $\widetilde{Z}_T = \left[\underline{Z}, \overline{Z}\right] = \underline{Z} + 2\beta\Delta\widetilde{Z}$, where $\Delta\widetilde{Z} = \frac{\overline{Z} - \underline{Z}}{2} = \frac{(Z_u - Z_l)(1-r)}{2}$, $0 \leq \beta, r \leq 1$, and $\left[\underline{Z}, \overline{Z}\right] = [Z_l + (Z_c - Z_l)r,\ Z_u + (Z_c - Z_u)r]$.

## 2.3 Fuzzy homotopy perturbation transform method (FHPTM)

The main idea of FHPTM (Vana and Karunakar, 2024a,b; Karunakar et al., 2024; Vana et al., 2025; Vana and Karunakar, 2024c,d) is explained here, along with a discussion on its accuracy, uncertainty in the governing equation or initial condition, and how well it works.

This method is used for solving linear and nonlinear uncertain differential equations of the form

$$\frac{\partial^\alpha}{\partial\tau^\alpha}\left(\widetilde{\mathcal{B}}(\zeta,\tau;\beta,r)\right)+\rho\,\mathbb{L}\left(\widetilde{\mathcal{B}}(\zeta,\tau;\beta,r)\right)+\mathbb{N}\left(\widetilde{\mathcal{B}}(\zeta,\tau;\beta,r)\right)=\widetilde{\mu}(\zeta,\tau;\beta,r), \tag{2.3}$$

subject to initial conditions,

$$\widetilde{\mathcal{B}}^l(\zeta,0,\beta,r)=\frac{\partial^l}{\partial\tau^l}\left(\widetilde{\mathcal{B}}(\zeta,0,\beta,r)\right)=z_l(\zeta),\ l=0,1,\ldots,\alpha-1; \tag{2.4}$$

here, $\alpha$ shows the order of the derivative, and $\rho$ is uncertain parameter. $\mathbb{L}$ and $\mathbb{N}$ are the linear and nonlinear differential operators, and $\widetilde{\mu}(\zeta,\tau;\beta,r)$ is the given source term.

**Definition.** Suppose $\widetilde{\mathcal{B}}$ and its derivatives $\frac{\partial^1\widetilde{\mathcal{B}}}{\partial\tau^1},\frac{\partial^2\widetilde{\mathcal{B}}}{\partial\tau^2},\ldots,\frac{\partial^\alpha\widetilde{\mathcal{B}}}{\partial\tau^\alpha}$ are continuous for $\tau\geq 0$ and exponential order as $\tau\to\infty$. Also, $\frac{\partial^\alpha\widetilde{\mathcal{B}}}{\partial\tau^\alpha}$ is piece-wise continuous for $\tau\geq 0$. The LT $\mathcal{L}$ of a $\alpha^{\text{th}}$-order derivative is given as

$$\mathcal{L}\left(\frac{\partial^\alpha}{\partial\tau^\alpha}\left(\widetilde{\mathcal{B}}(\zeta,\tau;\beta,r)\right)\right)=s^\alpha\mathcal{L}\left[\widetilde{\mathcal{B}}(\zeta,\tau;\beta,r)\right]-\sum_{n=0}^{\alpha-1}s^{\alpha-n-1}\widetilde{\mathcal{B}}^n(\zeta,0;\beta,r). \tag{2.5}$$

The coefficient $\rho$ in Eq. (2.3) or the initial conditions may not always be crisp, because it depends on different measured values from experiments and can be uncertain due to factors such as observational errors. This means $\rho$ may have some uncertainty or ambiguity. To solve this problem, $\rho$ can be instead of intervals or fuzzy numbers to be represented as certain limits using the parametric form. Now, $\rho$ can be expressed as a TFN ($\widetilde{\rho}_T$), defined as $\widetilde{\rho}_T=\left[\underline{\rho},\overline{\rho}\right]=\underline{\rho}+2\beta\Delta\widetilde{\rho}$ in Section 2.2.5, using the $r$-cut technique, which leads to Eq. (2.3) as follows:

$$\frac{\partial^\alpha}{\partial\tau^\alpha}\left(\widetilde{\mathcal{B}}(\zeta,\tau;\beta,r)\right)+\left(\underline{\rho}+2\beta\Delta\widetilde{\rho}\right)\mathbb{L}\left(\widetilde{\mathcal{B}}(\zeta,\tau;\beta,r)\right)+\mathbb{N}\left(\widetilde{\mathcal{B}}(\zeta,\tau;\beta,r)\right)=\widetilde{\mu}(\zeta,\tau;\beta,r). \tag{2.6}$$

Applying the LT to both sides of Eq. (2.6) and using Eq. (2.5), we obtain

$$s^\alpha\mathcal{L}(\mathcal{B})-\sum_{n=0}^{\alpha-1}s^{\alpha-n-1}\widetilde{\mathcal{B}}^n(\zeta,0;\beta,r)+\mathcal{L}\left[\left(\underline{\rho}+2\beta\Delta\widetilde{\rho}\right)\mathbb{L}\widetilde{\mathcal{B}}+\mathbb{N}\widetilde{\mathcal{B}}\right]=\mathcal{L}(\widetilde{\mu}). \tag{2.7}$$

Utilizing the inverse LT ($\mathcal{L}^{-1}$) and simplifying the Eq. (2.7), we get

$$\widetilde{\mathcal{B}}=\mathcal{L}^{-1}\left[\frac{1}{s^\alpha}\left(\sum_{n=0}^{\alpha-1}s^{\alpha-n-1}\widetilde{\mathcal{B}}^n(\zeta,0;\beta,r)+\mathcal{L}(\widetilde{\mu})\right)\right]-\mathcal{L}^{-1}\left[\frac{1}{s^\alpha}\left(\mathcal{L}\left[\left(\underline{\rho}+2\beta\Delta\widetilde{\rho}\right)\mathbb{L}\widetilde{\mathcal{B}}+\mathbb{N}\widetilde{\mathcal{B}}\right]\right)\right]. \tag{2.8}$$

Assuming the solution of Eq. (2.8) as a power series in $p$ (where $0\leq p\leq 1$) as follows:

$$\widetilde{\mathcal{B}}(\zeta,\tau;\beta,r)=\sum_{n=0}^{\infty}p^n\widetilde{\mathcal{B}}_n. \tag{2.9}$$

Next, we decompose the nonlinear terms as

$$\mathbb{N}\widetilde{\mathcal{B}} = \sum_{i=0}^{\infty} p^n \, \mathbb{H}_i(\widetilde{\mathcal{B}}), \tag{2.10}$$

where $\mathbb{H}_i(\widetilde{\mathcal{B}})$ are He's polynomials as follows (Karunakar and Chakraverty, 2017, 2018a,b, 2019a,b; Karunakar et al., 2020; Karunakar and Chakraverty, 2021, 2022; Karunakar et al., 2023; Sankar and Karunakar, 2025):

$$\mathbb{H}_i\left(\widetilde{\mathcal{B}}_0, \widetilde{\mathcal{B}}_1, \ldots, \widetilde{\mathcal{B}}_n\right) = \frac{1}{n!}\frac{\partial^n}{\partial p^n}\left[\mathbb{N}\left(\sum_{n=0}^{\infty} p^n \widetilde{\mathcal{B}}_n\right)\right]_{p=0}. \tag{2.11}$$

Now, by substituting Eqs. (2.9), (2.10), and (2.11) into (2.8), we get the FHPTM formula (Vana and Karunakar, 2024a,b; Karunakar et al., 2024; Vana et al., 2025; Vana and Karunakar, 2024c,d) as follows:

$$\begin{aligned}\sum_{n=0}^{\infty} p^n \widetilde{\mathcal{B}}_n =& \mathcal{L}^{-1}\left[\frac{1}{s^\alpha}\left(\sum_{n=0}^{\alpha-1} s^{\alpha-n-1}\widetilde{\mathcal{B}}^n(\zeta, 0; \beta, r) + \mathcal{L}(\widetilde{\mu})\right)\right] \\ & - p\left(\mathcal{L}^{-1}\left[\frac{1}{s^\alpha}\left(\mathcal{L}\left[\left(\underline{\rho} + 2\beta\Delta\widetilde{\rho}\right)\mathbb{L}\widetilde{\mathcal{B}} + \mathbb{N}\widetilde{\mathcal{B}}\right]\right)\right]\right).\end{aligned} \tag{2.12}$$

The FHPTM sequence is formed by matching the identical power terms of $p$ in Eq. (2.12) as shown below:

$$\begin{aligned}
p^0 &: \widetilde{\mathcal{B}}_0 = \mathcal{L}^{-1}\left[\frac{1}{s^\alpha}\left(\sum_{n=0}^{\alpha-1} s^{\alpha-n-1}\widetilde{\mathcal{B}}^n(\zeta, 0; \beta, r) + \mathcal{L}(\widetilde{\mu})\right)\right], \\
p^1 &: \widetilde{\mathcal{B}}_1 = -\mathcal{L}^{-1}\left[\frac{1}{s^\alpha}\left(\mathcal{L}\left[\left(\underline{\rho} + 2\beta\Delta\widetilde{\rho}\right)\mathbb{L}\widetilde{\mathcal{B}}_0 + \mathbb{H}_0(\widetilde{\mathcal{B}})\right]\right)\right], \\
&\vdots \\
p^n &: \widetilde{\mathcal{B}}_n = -\mathcal{L}^{-1}\left[\frac{1}{s^\alpha}\left(\mathcal{L}\left[\left(\underline{\rho} + 2\beta\Delta\widetilde{\rho}\right)\mathbb{L}\widetilde{\mathcal{B}}_{n-1} + \mathbb{H}_{n-1}(\widetilde{\mathcal{B}})\right]\right)\right].
\end{aligned}$$

The solution of Eq. (2.3) is

$$\widetilde{\mathcal{B}}(\zeta, \tau; \beta, r) = \lim_{p\to 1}\sum_{n=0}^{\infty} p^n \widetilde{\mathcal{B}}_n = \sum_{n=0}^{\infty}\widetilde{\mathcal{B}}_n. \tag{2.13}$$

**Theorem 2.1.** *Let $\widetilde{\mathcal{B}}(\zeta, \tau; \beta, r)$ be defined in a Banach space $\mathbb{B}$ (Chakraverty et al., 2019; Chakraverty and Perera, 2018; Karunakar and Chakraverty, 2019a; Vana and Karunakar, 2024a). Then the series solution given in Eq. (2.13) is convergent if $\|\widetilde{\mathcal{B}}_{n+1}\| \leq \gamma\|\widetilde{\mathcal{B}}_n\|, \forall\, \widetilde{\mathcal{B}}_0 \in \mathbb{B}$, where $0 < \gamma < 1$. Further, if the truncated series, the resulting finite term approximation* *yields a convergent and acceptable approximation to the exact solution within the* $\mathbb{B}$.

$$\widetilde{\mathcal{B}}_j = \sum_{n=0}^{j} \widetilde{\mathcal{B}}_n = \widetilde{\mathcal{B}}_0 + \widetilde{\mathcal{B}}_1 + \cdots + \widetilde{\mathcal{B}}_j,$$

*is used to close solution of Eq. (2.3). We obtain the maximum absolute truncation error as*

$$\|\widetilde{\mathcal{B}} - \sum_{n=0}^{j} \widetilde{\mathcal{B}}_n\| \leq \frac{\gamma^{j+1}\|\widetilde{\mathcal{B}}_0\|}{1-\gamma} \leq \frac{\|\widetilde{\mathcal{B}}_0\|}{1-\gamma}.$$

*Proof.* Let $\widetilde{\mathcal{B}}_0 \in \mathbb{B}$ and $\{S_q\}_{q=0}^{\infty}$ define the sequence of partial sum of Eq. (2.13). We need to show that $\{S_q\}_{q=0}^{\infty}$ is Cauchy sequence in $\mathbb{B}$. For this context, consider the following:

$$\begin{aligned}\|S_{q+1} - S_q\| &= \|\widetilde{\mathcal{B}}_{q+1}\| \\ &\leq \gamma\|\widetilde{\mathcal{B}}_q\| \\ &\leq \gamma^2\|\widetilde{\mathcal{B}}_{q-1}\| \\ &\vdots \\ &\leq \gamma^{q+1}\|\widetilde{\mathcal{B}}_0\|.\end{aligned}$$

For all $q, m \in \mathcal{N}$ with $q \geq m$,

$$\begin{aligned}\|S_q - S_m\| &= \|(S_q - S_{q-1}) + (S_{q-1} - S_{q-2}) + \cdots + (S_{m+1} - S_m)\| \\ &\leq \|S_q - S_{q-1}\| + \|S_{q-1} - S_{q-2}\| + \cdots + \|S_{m+1} - S_m\| \\ &\leq \gamma^q\|\widetilde{\mathcal{B}}_0\| + \gamma^{q-1}\|\widetilde{\mathcal{B}}_0\| + \cdots + \gamma^{m+1}\|\widetilde{\mathcal{B}}_0\| \\ &\leq \gamma^{m+1}\|\widetilde{\mathcal{B}}_0\| \sum_{i=0}^{q-m-1} \gamma^i \\ &= \gamma^{m+1}\|\widetilde{\mathcal{B}}_0\| \frac{1-q^{q-m}}{1-\gamma}\end{aligned}$$

$$\|S_q - S_m\| = \gamma^{m+1}\|\widetilde{\mathcal{B}}_0\| \frac{1-q^{q-m}}{1-\gamma}. \tag{2.14}$$

Since $0 < \gamma < 1$, we get $\lim_{q\to\infty} \|S_q - S_m\| = 0$.

Hence, $\{S_q\}_{q=0}^{\infty}$ is Cauchy's sequence in $\mathbb{B}$, so the series solution Eq. (2.13) converges. Furthermore, from Eq. (2.14), we get

$$\|\widetilde{\mathcal{B}} - \widetilde{\mathcal{B}}_j\| =\leq \frac{\gamma^{j+1}\|\widetilde{\mathcal{B}}_0\|}{1-\gamma} \leq \frac{\|\widetilde{\mathcal{B}}_0\|}{1-\gamma}. \tag{2.15}$$

Here, $\widetilde{\mathcal{B}}_j = \sum_{n=0}^{j} \widetilde{\mathcal{B}}_n$. □

## 2.4 Application of HPTM in crisp environment

In this section, we use the proposed method to solve Eq. (2.1) for the cases $\delta = 1$ and $\delta = 2$.
**Case 1:** Consider equation for $\delta = 1$, given as follows:

$$\frac{\partial^2 \mathcal{B}}{\partial \tau^2} + a\frac{\partial^2 \mathcal{B}}{\partial \zeta^2} + b\left(\frac{\partial \mathcal{B}}{\partial \zeta}\right)^2 + b\left(\mathcal{B}\frac{\partial^2 \mathcal{B}}{\partial \zeta^2}\right) + c\frac{\partial^4 \mathcal{B}}{\partial \zeta^4} = 0, \tag{2.16}$$

subject to initial conditions

$$\begin{aligned} \mathcal{B}(\zeta, 0) &= \frac{12ck^2}{b}\, sech^2(k\zeta), \\ \mathcal{B}_\tau(\zeta, 0) &= \frac{12ck^2}{b}\, \frac{\sqrt{-ak^2 - 4ck^4}}{2}\, sech^2(k\zeta)\, tanh(k\zeta). \end{aligned} \tag{2.17}$$

The exact solution of Eq. (2.16) as (Razzaq et al., 2024; Zhu and Qu, 2024) is

$$\mathcal{B}(\zeta, \tau) = \frac{12ck^2}{b}\, sech^2\left(k\zeta - \sqrt{-ak^2 - 4ck^4}\,\tau\right). \tag{2.18}$$

Using the projected algorithm described in Eq. (2.12), we obtain the following:

$$\begin{aligned} \sum_{n=0}^{\infty} p^n \mathcal{B}_n &= \mathcal{L}^{-1}\left[\frac{1}{s}(\mathcal{B}(\zeta, 0, )) + \frac{1}{s^2}(\mathcal{B}_\tau(\zeta, 0, ))\right] \\ &\quad - p\left(\mathcal{L}^{-1}\left[\mathcal{L}\left(\frac{1}{s^2}\left[a\frac{\partial^2}{\partial \zeta^2}\left(\sum_{n=0}^{\infty} p^n \mathcal{B}_n\right) + c\frac{\partial^4}{\partial \zeta^4}\left(\sum_{n=0}^{\infty} p^n \mathcal{B}_n\right)\right]\right)\right]\right) \\ &\quad - p\left(\mathcal{L}^{-1}\left[\mathcal{L}\left(\frac{1}{s^2}\left[b\left[\sum_{n=0}^{\infty} p^n \left[\mathbb{H}_{1n}(\mathcal{B}) + \mathbb{H}_{2n}(\mathcal{B})\right]\right]\right]\right)\right]\right). \end{aligned} \tag{2.19}$$

A few He's polynomials $\mathbb{H}_{1n}(\mathcal{B})$ for $\left(\frac{\partial \mathcal{B}}{\partial \zeta}\right)^2$ are

$$\begin{aligned} \mathbb{H}_{10}(\mathcal{B}) &= \left(\frac{\partial \mathcal{B}_0}{\partial \zeta}\right)^2, \\ \mathbb{H}_{11}(\mathcal{B}) &= 2\left(\frac{\partial \mathcal{B}_0}{\partial \zeta}\right)\left(\frac{\partial \mathcal{B}_1}{\partial \zeta}\right), \\ \mathbb{H}_{12}(\mathcal{B}) &= 2\left(\frac{\partial \mathcal{B}_1}{\partial \zeta}\right)^2 + 2\left(\frac{\partial \mathcal{B}_1}{\partial \zeta}\right)\left(\frac{\partial \mathcal{B}_2}{\partial \zeta}\right) \\ &\;\;\vdots \end{aligned}$$

A few He's polynomials $\mathbb{H}_{2n}(\mathcal{B})$ for $\left(\mathcal{B}\frac{\partial^2\mathcal{B}}{\partial\zeta^2}\right)$ are

$$\begin{aligned}
\mathbb{H}_{20}(\mathcal{B}) &= \mathcal{B}_0\frac{\partial^2\mathcal{B}_0}{\partial\zeta^2},\\
\mathbb{H}_{21}(\mathcal{B}) &= \mathcal{B}_0\frac{\partial^2\mathcal{B}_1}{\partial\zeta^2}+\mathcal{B}_1\frac{\partial^2\mathcal{B}_0}{\partial\zeta^2},\\
\mathbb{H}_{22}(\mathcal{B}) &= \mathcal{B}_0\frac{\partial^2\mathcal{B}_2}{\partial\zeta^2}+\mathcal{B}_1\frac{\partial^2\mathcal{B}_1}{\partial\zeta^2}+\mathcal{B}_2\frac{\partial^2\mathcal{B}_0}{\partial\zeta^2}.\\
&\vdots
\end{aligned}$$

By comparing the identical power terms of $p$ in Eq. (2.19), we derive the sequence of approach as follows:

$$\begin{aligned}
p^0 : \mathcal{B}_0 &= \mathcal{B}(\zeta,0)+\tau\mathcal{B}_\tau(\zeta,0),\\
&= \frac{12ck^2}{b}\,sech^2(k\zeta)+\frac{12ck^2}{b}\,\frac{\sqrt{-ak^2-4ck^4}}{2}\,sech^2(k\zeta)\,tanh(k\zeta)\\
&= \frac{1}{50\,cosh\left(\frac{\zeta}{10}\right)^2}+\frac{4703015399704855\,\tau\,tanh\left(\frac{\zeta}{10}\right)}{1152921504606846976\,cosh\left(\frac{\zeta}{10}\right)^2},\\
p^1 : \mathcal{B}_1 &= -\mathcal{L}^{-1}\left[\mathcal{L}\left(\frac{1}{s^2}\left[a\frac{\partial^2}{\partial\zeta^2}(\mathcal{B}_0)+c\frac{\partial^4}{\partial\zeta^4}(\mathcal{B}_0)+b\left[\mathbb{H}_{10}(\mathcal{B})+\mathbb{H}_{20}(\mathcal{B}\mathcal{B})\right]\right]\right)\right],\\
&= \tau\left(6\sigma_5^2-6\left(\frac{1}{50\sigma_7}+\frac{4703015399704855\tau\tanh\left(\frac{\zeta}{10}\right)}{1152921504606846976\sigma_7}\right)\sigma_5\right)\\
&+\tau\left(-\frac{23}{62500\sigma_7}+\frac{3\sigma_1}{3125\sigma_2}+\frac{3\sigma_3}{12500\sigma_4}-\frac{21633870838642333\tau\tanh\left(\frac{\zeta}{10}\right)}{28823037615171174400\sigma_7}\right)\\
&+\tau\left(+\frac{19752664678760391\tau\sinh\left(\frac{\zeta}{10}\right)\sigma_6}{144115188075855872000\sigma_8}+\frac{10346633879350681\tau\tanh\left(\frac{\zeta}{10}\right)\sigma_6}{144115188075855872000\sigma_7}\right)\\
&+\tau\left(\frac{2821809239822913\tau\sigma_1\tanh\left(\frac{\zeta}{10}\right)}{144115188075855872000\sigma_2}+\frac{2821809239822913\tau\sigma_3\tanh\left(\frac{\zeta}{10}\right)}{576460752303423488000\sigma_4}\right)\\
&+\tau\left(\frac{940603079940971\tau\sinh\left(\frac{\zeta}{10}\right)\sigma_6^2}{14411518807585587200\sigma_8}+\frac{2821809239822913\tau\sinh^3\left(\frac{\zeta}{10}\right)\sigma_6}{72057594037927936000\cosh^5\left(\frac{\zeta}{10}\right)}\right)\\
&+\tau\left(\frac{940603079940971\tau\tanh\left(\frac{\zeta}{10}\right)\sigma_6^2}{14411518807585587200\sigma_7}+\frac{940603079940971\tau\tanh^3\left(\frac{\zeta}{10}\right)\sigma_6}{28823037615171174400\sigma_7}\right)
\end{aligned}$$

$$+\tau\left(\frac{940603079940971\tau\sinh\left(\frac{\zeta}{10}\right)\sigma_9\sigma_6}{720575940379279360\sigma_8}+\frac{8465427719468739\tau\sigma_1\tanh\left(\frac{\zeta}{10}\right)\sigma_6}{288230376151711744\sigma_2}\right)$$

$$\vdots,$$

where $a=-1, b=-6, c=-1, k=0.1$, and

$$\sigma_1=\sinh^2\left(\frac{\zeta}{10}\right),\quad \sigma_2=\cosh^4\left(\frac{\zeta}{10}\right),\quad \sigma_3=\sinh^4\left(\frac{\zeta}{10}\right),\quad \sigma_4=\cosh^6\left(\frac{\zeta}{10}\right),$$

$$\sigma_5=\frac{\sinh\left(\frac{\zeta}{10}\right)}{250\sigma_8}+\frac{4703015399704855\tau\sigma_6}{1152921504606846976\sigma_7}+\frac{940603079940971\tau\sinh\left(\frac{\zeta}{10}\right)\tanh\left(\frac{\zeta}{10}\right)}{1152921504606846976\sigma_8},$$

$$\sigma_6=\frac{\sigma_9}{10}-\frac{1}{10},\quad \sigma_7=\cosh^2\left(\frac{\zeta}{10}\right),\quad \sigma_8=\cosh^3\left(\frac{\zeta}{10}\right),\quad \sigma_9=\tanh^2\left(\frac{\zeta}{10}\right).$$

Accordingly, the remaining terms $\mathcal{B}_n$ for $n=1,2,3,\ldots$ can be determined.

The series solution of Eq. (2.16) is

$$\mathcal{B}(\zeta,\tau)=\lim_{p\to 1}\sum_{n=0}^{\infty}p^n\mathcal{B}_n=\sum_{n=0}^{\infty}\mathcal{B}_n=\mathcal{B}_0+\mathcal{B}_1+\ldots \tag{2.20}$$

**Case 2:** Consider equation for $\delta=2$, given as follows:

$$\frac{\partial^2\mathcal{B}}{\partial\tau^2}+a\frac{\partial^2\mathcal{B}}{\partial\zeta^2}+2b\mathcal{B}\left(\frac{\partial\mathcal{B}}{\partial\zeta}\right)^2+b\left(\mathcal{B}^2\frac{\partial^2\mathcal{B}}{\partial\zeta^2}\right)+c\frac{\partial^4\mathcal{B}}{\partial\zeta^4}=0, \tag{2.21}$$

subject to initial conditions

$$\begin{aligned}\mathcal{B}(\zeta,0)&=k\sqrt{\frac{6c}{b}}\,sech\,(k\zeta),\\ \mathcal{B}_\tau(\zeta,0)&=k\sqrt{\frac{6c}{b}}\,\sqrt{-k^2(a+ck^2)}\,sech\,(k\zeta)\,tanh\,(k\zeta).\end{aligned} \tag{2.22}$$

The exact solution of Eq. (2.21) as (Chen and Wang, 2021; Tian, 2020):

$$\mathcal{B}(\zeta,\tau)=k\sqrt{\frac{6c}{b}}\,\frac{12ck^2}{b}\,sech^2\left(k\zeta-\sqrt{-k^2(a+ck^2)}\tau\right). \tag{2.23}$$

Using the projected algorithm described in Eq. (2.12), we obtain the following:

$$\sum_{n=0}^{\infty} p^n \mathcal{B}_n = \mathcal{L}^{-1}\left[\frac{1}{s}\left(\mathcal{B}(\zeta,0,)\right)+\frac{1}{s^2}\left(\mathcal{B}_\tau(\zeta,0,)\right)\right]$$
$$-p\left(\mathcal{L}^{-1}\left[\mathcal{L}\left(\frac{1}{s^2}\left[a\frac{\partial^2}{\partial\zeta^2}\left(\sum_{n=0}^{\infty} p^n \mathcal{B}_n\right)+c\frac{\partial^4}{\partial\zeta^4}\left(\sum_{n=0}^{\infty} p^n \mathcal{B}_n\right)\right]\right)\right]\right) \tag{2.24}$$
$$-p\left(\mathcal{L}^{-1}\left[\mathcal{L}\left(\frac{1}{s^2}\left[b\left[\sum_{n=0}^{\infty} p^n\left(2\,\mathbb{H}_{3n}(\mathcal{B})+\mathbb{H}_{4n}(\mathcal{B})\right)\right]\right]\right)\right]\right).$$

A few He's polynomials $\mathbb{H}_{3n}(\mathcal{B})$ for $\mathcal{B}\left(\frac{\partial\mathcal{B}}{\partial\zeta}\right)^2$ are,

$$\mathbb{H}_{30}(\mathcal{B})=\mathcal{B}_0\left(\frac{\partial\mathcal{B}_0}{\partial\zeta}\right)^2,$$
$$\mathbb{H}_{31}(\mathcal{B})=\mathcal{B}_0\left(\frac{\partial\mathcal{B}_1}{\partial\zeta}\right)^2+2\left(\frac{\partial\mathcal{B}_0}{\partial\zeta}\right)\left(\frac{\partial\mathcal{B}_1}{\partial\zeta}\right)\mathcal{B}_0,$$
$$\mathbb{H}_{32}(\mathcal{B})=\mathcal{B}_0\left(\frac{\partial\mathcal{B}_2}{\partial\zeta}\right)^2+2\left(\frac{\partial\mathcal{B}_0}{\partial\zeta}\right)\left(\frac{\partial\mathcal{B}_1}{\partial\zeta}\right)\mathcal{B}_1\left(\frac{\partial\mathcal{B}_1}{\partial\zeta}\right)^2\mathcal{B}_0+2\left(\frac{\partial\mathcal{B}_0}{\partial\zeta}\right)\left(\frac{\partial\mathcal{B}_2}{\partial\zeta}\right)\mathcal{B}_0.$$
$$\vdots$$

A few He's polynomials $\mathbb{H}_{4n}(\mathcal{B})$ for $\left(\mathcal{B}^2\frac{\partial^2\mathcal{B}}{\partial\zeta^2}\right)$ are

$$\mathbb{H}_{40}(\mathcal{B})=\mathcal{B}_0^2\frac{\partial^2\mathcal{B}_0}{\partial\zeta^2},$$
$$\mathbb{H}_{41}(\mathcal{B})=2\mathcal{B}_0\mathcal{B}_1\frac{\partial^2\mathcal{B}_0}{\partial\zeta^2}+\mathcal{B}_0^2\frac{\partial^2\mathcal{B}_1}{\partial\zeta^2},$$
$$\mathbb{H}_{42}(\mathcal{B})=2\mathcal{B}_0\mathcal{B}_2\frac{\partial^2\mathcal{B}_0}{\partial\zeta^2}+\mathcal{B}_1^2\frac{\partial^2\mathcal{B}_0}{\partial\zeta^2}+2\mathcal{B}_0\mathcal{B}_1\frac{\partial^2\mathcal{B}_1}{\partial\zeta^2}+\mathcal{B}_0^2\frac{\partial^2\mathcal{B}_2}{\partial\zeta^2}.$$
$$\vdots$$

By comparing the identical power terms of $p$ in Eq. (2.19), we derive the sequence of approach as follows:

$$p^0:\mathcal{B}_0=\mathcal{B}(\zeta,0)+\tau\mathcal{B}_\tau(\zeta,0)$$
$$=k\sqrt{\frac{6c}{b}}\,sech\,(k\zeta)+k\sqrt{\frac{6c}{b}}\,\sqrt{-k^2(a+ck^2)}\,sech\,(k\zeta)\,\,tanh\,(k\zeta)$$
$$=\frac{1}{10cosh\left(\frac{\zeta}{10}\right)}+\frac{\sqrt{101}\tau^2 tanh\left(\frac{\zeta}{10}\right)}{1000cosh\left(\frac{\zeta}{10}\right)}$$

$$
\begin{aligned}
P^1: \mathcal{B}_1 &= -\mathcal{L}^{-1}\left[\mathcal{L}\left(\frac{1}{s^2}\left[a\frac{\partial^2}{\partial\zeta^2}(\mathcal{B}_0)+c\frac{\partial^4}{\partial\zeta^4}(\mathcal{B}_0)+b\left[2\,\mathbb{H}_{30}(\mathcal{B})+\mathbb{H}_{40}(\mathcal{B}\mathcal{B})\right]\right]\right)\right]\\
&= \frac{\tau}{50000000000\cosh^8\left(\frac{\zeta}{10}\right)}\left(606\sqrt{101}\sinh\left(\frac{\zeta}{10}\right)\tau^6\sigma_2 - 3333\sqrt{101}\sinh\left(\frac{\zeta}{10}\right)\tau^6\sigma_4\right)\\
&+ \frac{\tau}{50000000000\cosh^8\left(\frac{\zeta}{10}\right)}\left(181800\tau^4\sigma_1 - 818100\tau^4\sigma_3\right.\\
&\left.+ 666600\tau^4\cosh\left(\frac{\zeta}{10}\right) + 3000000\sigma_3\right)\\
&+ \frac{\tau}{50000000000\cosh^8\left(\frac{\zeta}{10}\right)}\left(505000\sqrt{101}\sinh\left(\frac{\zeta}{10}\right)t^2\cosh^6\left(\frac{\zeta}{10}\right) - 104000000\sigma_1\right)\\
&+ \frac{\tau}{50000000000\cosh^8\left(\frac{\zeta}{10}\right)}\left(150000\sqrt{101}\sinh\left(\frac{\zeta}{10}\right)\tau^2\sigma_4 + 3030\sqrt{101}\sinh\left(\frac{\zeta}{10}\right)\tau^6\right)\\
&+ \frac{\tau}{50000000000\cosh^8\left(\frac{\zeta}{10}\right)}\left(50500000\cosh^7\left(\frac{\zeta}{10}\right) - 3120000\sqrt{101}\sinh\left(\frac{\zeta}{10}\right)\tau^2\sigma_2\right)\\
&\vdots
\end{aligned}
$$

where,

$$
\sigma_1 = \cosh^5\left(\frac{x}{10}\right),\quad \sigma_2 = \cosh^4\left(\frac{x}{10}\right),\quad \sigma_3 = \cosh^3\left(\frac{x}{10}\right),\quad \sigma_4 = \cosh^2\left(\frac{x}{10}\right).
$$

The series solution of Eq. (2.21) is

$$
\mathcal{B}(\zeta,\tau) = \lim_{p\to 1}\sum_{n=0}^{\infty} p^n\mathcal{B}_n = \sum_{n=0}^{\infty}\mathcal{B}_n = \mathcal{B}_0 + \mathcal{B}_1 + \ldots \tag{2.25}
$$

## 2.5 Application of FHPTM to the fuzzy gBE in fuzzy environment

In this section, the proposed method is applied to solve Eqs. (2.16) and (2.21) in fuzzy scenarios. We focus on finding fuzzy solutions for the generalized fuzzy BEs in Eqs. (2.16) and (2.21). The initial conditions (2.17) and (2.22) are considered as triangular fuzzy numbers (TFNs), because their values depend on different factors that may not always be exact or fixed. These factors include fluid velocity and depth. This method helps to handle uncertainty in the model, allowing the factors to change within a fuzzy number instead of being a crisp values.

The governing gBE (2.16) can be converted into its fuzzy form,

$$\frac{\partial^2 \widetilde{\mathcal{B}}}{\partial \tau^2} + a\frac{\partial^2 \widetilde{\mathcal{B}}}{\partial \zeta^2} + b\left(\frac{\partial \widetilde{\mathcal{B}}}{\partial \zeta}\right)^2 + b\left(\widetilde{\mathcal{B}}\frac{\partial^2 \widetilde{\mathcal{B}}}{\partial \zeta^2}\right) + c\frac{\partial^4 \widetilde{\mathcal{B}}}{\partial \zeta^4} = 0. \tag{2.26}$$

The initial condition (2.17) of Eq. (2.16) as TFN:

$$\begin{aligned} \widetilde{\mathcal{B}}(\zeta, 0; \beta, r) &= \widetilde{\rho}\, sech^2(k\zeta), \\ \widetilde{\mathcal{B}}(\zeta, 0; \beta, r) &= \widetilde{\rho}\, \frac{\sqrt{-ak^2 - 4ck^4}}{2}\, sech^2(k\zeta)\, tanh(k\zeta), \end{aligned} \tag{2.27}$$

where $\widetilde{\rho} = \frac{12ck^2}{b} = [0.001600, 0.001667, 0.001734]$ is represented as TFN.

The following representing the fuzzy initial condition (2.27) in parametric form:

$$\begin{aligned} \widetilde{\mathcal{B}}(\zeta, 0; \beta, r) &= \left(\underline{\rho} + 2\beta\Delta\widetilde{\rho}\right) sech^2(k\zeta), \\ \widetilde{\mathcal{B}}(\zeta, 0; \beta, r) &= \left(\underline{\rho} + 2\beta\Delta\widetilde{\rho}\right) \frac{\sqrt{-ak^2 - 4ck^4}}{2}\, sech^2(k\zeta)\, tanh(k\zeta). \end{aligned} \tag{2.28}$$

The governing gBE (2.21) can be converted into its fuzzy form,

$$\frac{\partial^2 \widetilde{\mathcal{B}}}{\partial \tau^2} + a\frac{\partial^2 \widetilde{\mathcal{B}}}{\partial \zeta^2} + 2b\widetilde{\mathcal{B}}\left(\frac{\partial \widetilde{\mathcal{B}}}{\partial \zeta}\right)^2 + b\left(\widetilde{\mathcal{B}}^2\frac{\partial^2 \widetilde{\mathcal{B}}}{\partial \zeta^2}\right) + c\frac{\partial^4 \widetilde{\mathcal{B}}}{\partial \zeta^4} = 0. \tag{2.29}$$

The initial condition (2.22) of Eq. (2.21) as TFN:

$$\begin{aligned} \widetilde{\mathcal{B}}(\zeta, 0; \beta, r) &= \widetilde{\rho}\, k sech(k\zeta), \\ \widetilde{\mathcal{B}}(\zeta, 0; \beta, r) &= \widetilde{\rho}\, k\sqrt{-k^2(a + ck^2)}\, sech(k\zeta)\, tanh(k\zeta), \end{aligned} \tag{2.30}$$

where $\widetilde{\rho} = \sqrt{\frac{6c}{b}} = [0.5, 1, 1.5]$ is represented as TFN.

The following is the fuzzy initial condition (2.31) in parametric form:

$$\begin{aligned} \widetilde{\mathcal{B}}(\zeta, 0; \beta, r) &= \left(\underline{\rho} + 2\beta\Delta\widetilde{\rho}\right) k sech(k\zeta), \\ \widetilde{\mathcal{B}}(\zeta, 0; \beta, r) &= \left(\underline{\rho} + 2\beta\Delta\widetilde{\rho}\right) k\sqrt{-k^2(a + ck^2)}\, sech(k\zeta)\, tanh(k\zeta). \end{aligned} \tag{2.31}$$

As mentioned earlier, FHPTM is used to solve generalized fuzzy BEs (2.26) and (2.29) with fuzzy initial conditions (2.28) and (2.31).

## 2.6 Results and discussions

This section presents HPTM solutions for Eqs. (2.16) and (2.21). These numerical results are compared with the exact and EADM solutions at $a = -1$, $b = -6$, $c = -1$, and $k = 0.1$.

Further, this section discusses the fuzzy solutions of the generalized fuzzy BEs (Eqs. (2.26) and (2.29)). The uncertainty in the initial conditions is considered in Eqs. (2.27) and (2.29) as TFN, given by [0.001600, 0.001667, 0.001734] and [0.5, 1, 1.5], respectively.

**Crisp case:**

**Table 2.1** Comparison of EADM, HPTM, and exact solutions with absolute errors of Eq. (2.16) at $\tau = 1.5$.

| $\zeta$ | EADM Solution | HPTM Solution | Exact Solution | Absolute Error (EADM) | Absolute Error (HPTM) |
|---|---|---|---|---|---|
| −20 | 0.001111244547151 | 0.001111224547151 | 0.001111216532011 | $2.80 \times 10^{-8}$ | $8.02 \times 10^{-9}$ |
| −15 | 0.002879507917068 | 0.002879497917068 | 0.002879496455373 | $1.15 \times 10^{-8}$ | $1.46 \times 10^{-9}$ |
| −10 | 0.006910732038796 | 0.006910729038796 | 0.006910729884607 | $2.15 \times 10^{-9}$ | $8.46 \times 10^{-10}$ |
| −5 | 0.013862256805686 | 0.013862256765686 | 0.013862256765173 | $4.05 \times 10^{-11}$ | $5.13 \times 10^{-13}$ |
| 0 | 0.019695134793946 | 0.019695134752946 | 0.019695134758182 | $3.58 \times 10^{-11}$ | $5.24 \times 10^{-12}$ |
| 5 | 0.017420305858743 | 0.017420305858343 | 0.017420305858162 | $5.81 \times 10^{-13}$ | $1.81 \times 10^{-13}$ |
| 10 | 0.010077510901697 | 0.010077510901831 | 0.010077510901843 | $1.46 \times 10^{-13}$ | $1.20 \times 10^{-14}$ |
| 15 | 0.004510923238584 | 0.004510923238184 | 0.004510923238173 | $4.11 \times 10^{-13}$ | $1.10 \times 10^{-14}$ |
| 20 | 0.001792860025351 | 0.001792860025028 | 0.001792860025031 | $3.20 \times 10^{-13}$ | $3.00 \times 10^{-15}$ |

**Table 2.2** Comparison of EADM and HPTM, exact solutions with absolute errors of (2.21) at $\tau = 1.5$.

| $\zeta$ | EADM Solution | HPTM Solution | Exact Solution | Absolute Error (EADM) | Absolute Error (HPTM) |
|---|---|---|---|---|---|
| −20 | 0.025321131348940 | 0.025321131144440 | 0.025321131142892 | $2.06 \times 10^{-10}$ | $1.55 \times 10^{-12}$ |
| −15 | 0.040610458764146 | 0.040610453763126 | 0.040610453765967 | $4.998 \times 10^{-9}$ | $2.841 \times 10^{-12}$ |
| −10 | 0.06233997957723 | 0.062339974535593 | 0.062339974537437 | $5.04 \times 10^{-9}$ | $1.844 \times 10^{-12}$ |
| −5 | 0.086561677810281 | 0.086561677916389 | 0.086561677916894 | $1.066 \times 10^{-10}$ | $5.05 \times 10^{-13}$ |
| 0 | 0.0998738826902133 | 0.099873882688281 | 0.099873882689271 | $9.42 \times 10^{-13}$ | $9.90 \times 10^{-13}$ |
| 5 | 0.090673820689094 | 0.090673820639387 | 0.090673820639653 | $4.944 \times 10^{-11}$ | $2.66 \times 10^{-13}$ |
| 10 | 0.067296961117617 | 0.067296961018047 | 0.067296961018083 | $9.953 \times 10^{-11}$ | $3.60 \times 10^{-14}$ |
| 15 | 0.044477255697015 | 0.044477255687799 | 0.044477255687807 | $9.208 \times 10^{-12}$ | $8.00 \times 10^{-15}$ |
| 20 | 0.027896942069159 | 0.027896942067309 | 0.027896942067315 | $1.844 \times 10^{-12}$ | $6.00 \times 10^{-15}$ |

Tables 2.1 and 2.2 display the comparison the of the EADM and HPTM results with exact solution for different values of $\zeta$ at $\tau = 1.5$. Both tables show how close these methods are to the exact solution by calculating absolute errors. In Table 2.1, the HPTM method gives more accurate results than the EADM, because its errors are smaller. Similarly, Table 2.2 also shows that HPTM gives better accuracy than EADM for all $\zeta$ values. Overall, both methods give good approximations, but HPTM is more precise.

Fig. 2.1, compares two solutions of the gBE (2.16) at $\tau = 1$: the HPTM solution and the exact solution. The horizontal axis $\zeta$ represents the spatial coordinate, while the vertical axis $\mathcal{B}(\zeta, \tau)$ indicates the wave velocity. The close agreement between the two solutions demonstrates that HPTM solutions are very close to the exact solution. This highlights the method effectiveness in capturing the behavior of nonlinear dispersive waves. Additionally, the figure reveals that the semianalytical and exact solutions show only a minimal difference. Fig. 2.2 shows a 3D view of the solutions to Eq. (2.16), comparing the HPTM solution (plot (a)) with the exact solution (plot (b)). Both surfaces display a similar

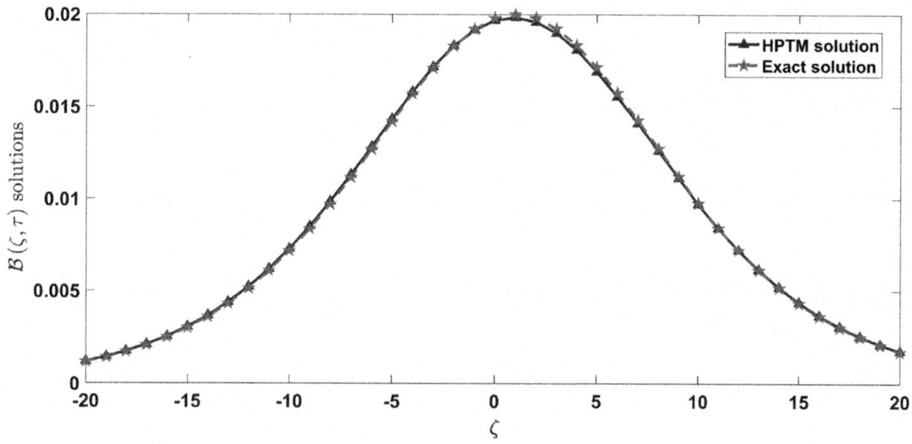


**FIGURE 2.1** Comparison of HPTM; exact solutions for Eq. (2.16) $\tau = 1$.

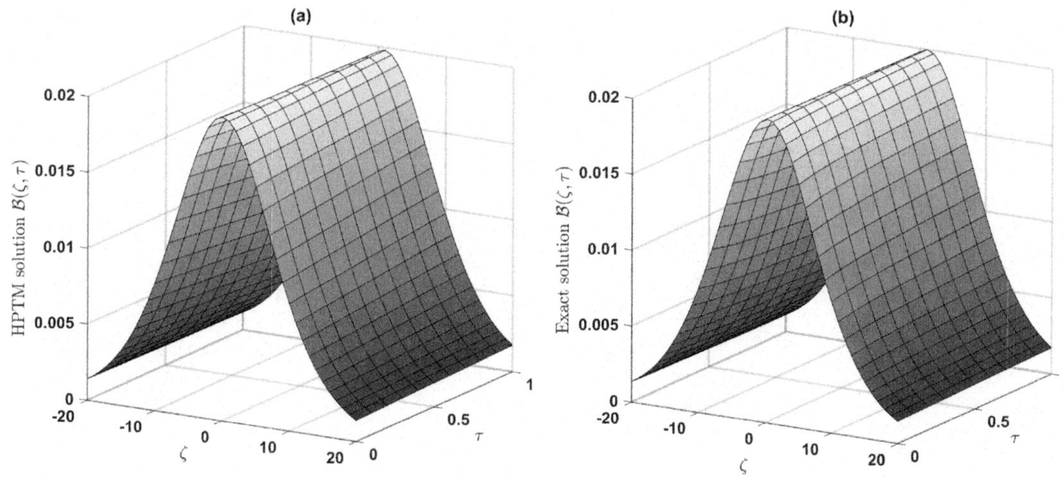


**FIGURE 2.2** 3D view of HPTM; exact solutions of (2.16).

wave pattern, representing the wave profile's evolution over time. The HPTM solution closely matches the exact solution, as seen in their nearly identical shape, amplitude, and curvature. The smooth gradient and structured mesh indicate that HPTM accurately captures the nonlinear wave behavior, maintaining precision in both space and time.

Eq. (2.21) is a nonlinear partial differential equation that describes wave velocity in fluid flow, or energy transfer. It includes effects like time variation, space spreading, and nonlinear interactions. The Fig. 2.3 compares the HPTM solution with the exact solution, showing that both are very close to each other. This means HPTM gives accurate results for solving such complex equations. The shape of the solution looks like a smooth wave, with some spreading near the edges due to higher-order effects. Small differences between the

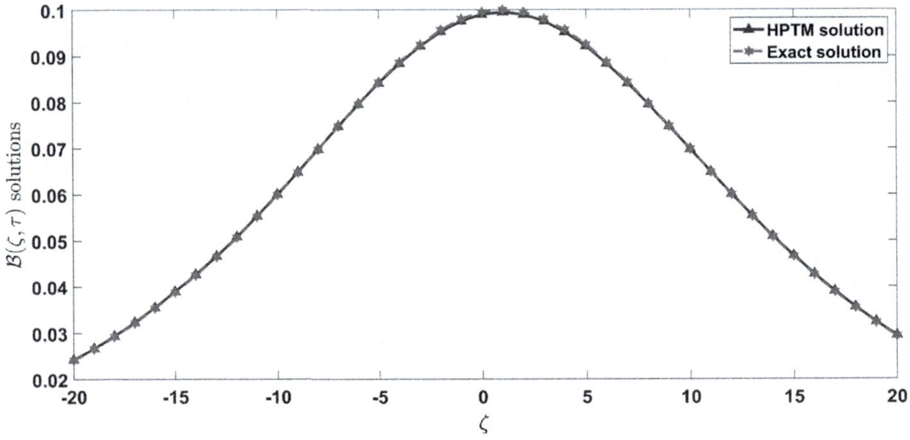


FIGURE 2.3 Comparison of HPTM; exact solutions for Eq. (2.21) at $\tau = 1$.

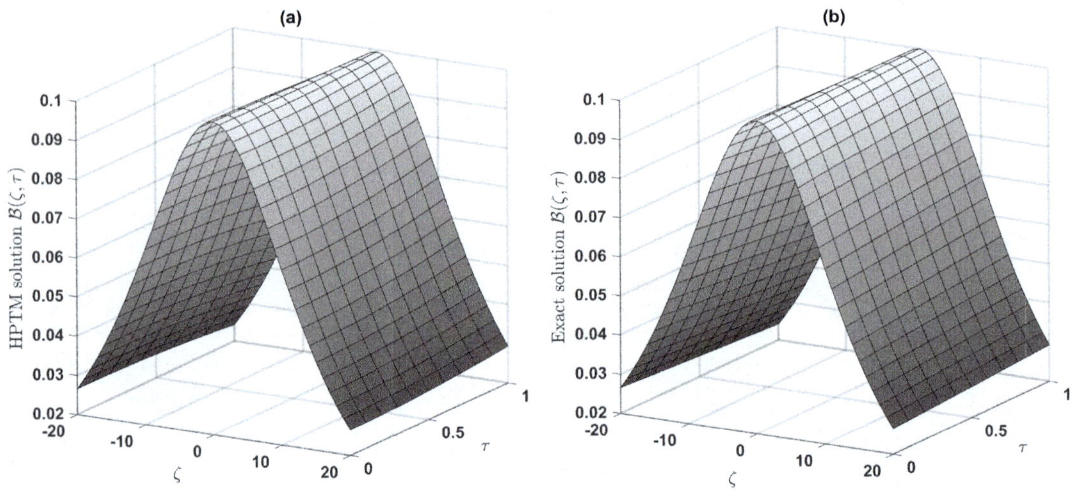


FIGURE 2.4 3D view of HPTM; exact solutions of (2.21).

two solutions come from nonlinear terms. This study proves that HPTM is a useful method for solving nonlinear wave problems, which can be used in areas like fluid mechanics, optics, and heat transfer.

The water wave Eq. (2.21) that describes the movement of nonlinear dispersive waves in fluid dynamics. It includes various physical effects: the second-order time derivative $\frac{\partial^2 \mathcal{B}}{\partial \tau^2}$ represents wave acceleration, the second-order spatial derivative $\frac{\partial^2 \mathcal{B}}{\partial \varsigma^2}$ accounts for linear wave dispersion, and the nonlinear terms model interactions between wave amplitude and velocity. The fourth-order derivative $\frac{\partial^4 \mathcal{B}}{\partial \varsigma^4}$ term helps stabilize the wave profile by considering higher-order dispersion effects. Fig. 2.4 provides a 3D comparison of the

HPTM solution (plot (a)) and the exact solution (plot (b)) for this equation. Both solutions exhibit a similar wave pattern, showing that the HPTM closely approximates the exact solution. The smooth surfaces suggest that HPTM accurately captures wave dynamics, preserving shape, amplitude, and curvature over time. The close agreement between the two plots confirms that HPTM is an effective method for solving higher-order nonlinear wave equations in modeling real-world water wave phenomena.

**Uncertain case:**

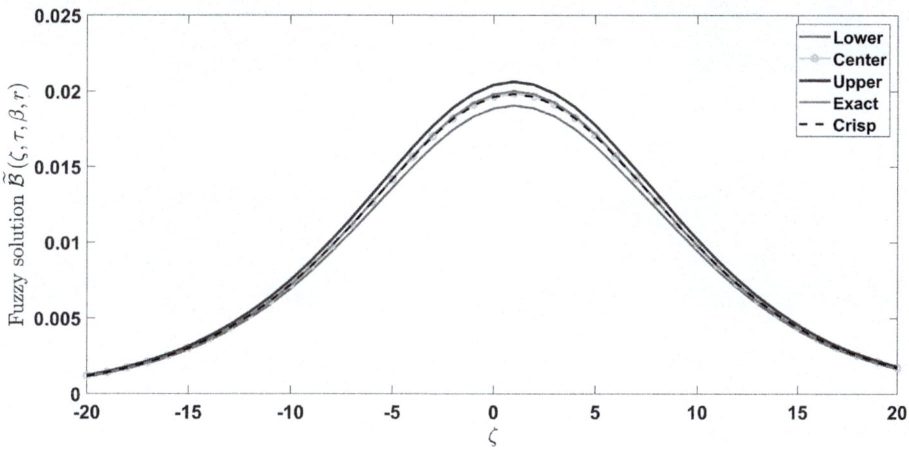


**FIGURE 2.5** Fuzzy solution of (2.26) at $\tau = 1$.

Fig. 2.5, is a comparison of the various solutions of Eq. (2.26) at tau $=$ 1, giving emphasis on fuzzy solutions as well as exact solutions. The horizontal axis ($\zeta$) is the spatial coordinate, and the vertical axis is the fuzzy wave solution $\tilde{\mathcal{B}}(\zeta, \tau; \beta, r)$. The fuzzy solution is specified by the lower (red, dark gray in print version) and upper (blue, darker gray in print version) curves that indicate the potential changes in the cases of uncertainty. The green light gray in print version solution signifies the average profile of the waves in the range. The actual solution (magenta, mid gray in print version) serves as a reference towards the verification of the current results, and crisp solution (black dashed line) serves as a deterministic approximation of the current results without taking into account fuzziness. As it can be observed, the crisp and center solution are very close to the exact solution, which indicates the validity of the current approximation solution. The fact that these solutions are close also indicates that the fuzzy method is very suitable in predicting the dynamics of the waves as well as taking into consideration uncertainties. Fig. 2.6 shows the triangular fuzzy solutions of Eq. (2.26) of various values of $\tau$ (0.5, 1.0, 1.5, and 2.0). The horizontal axes indicate the spatial coordinate, which is denoted as the zeta and the values of the $r$-cut and the vertical axis indicates the fuzzy wave solution which is denoted as $\tilde{\mathcal{B}}(\zeta, \tau, \beta, r)$. Red (mid gray in print version) surface (LBS) is the lower boundary solution and the blue (dark gray in print version) surface (UBS) is the upper boundary solution.

The shape of the wave profile remains the same with a slight change as the value of the tau increases, signifying that there is the evolution of the fuzzy solution as time goes by. The $r$-cut method helps to visualize the various membership values of the fuzzy solution that shows the effect of uncertainty on the wave propagation. Fig. 2.7 presents various

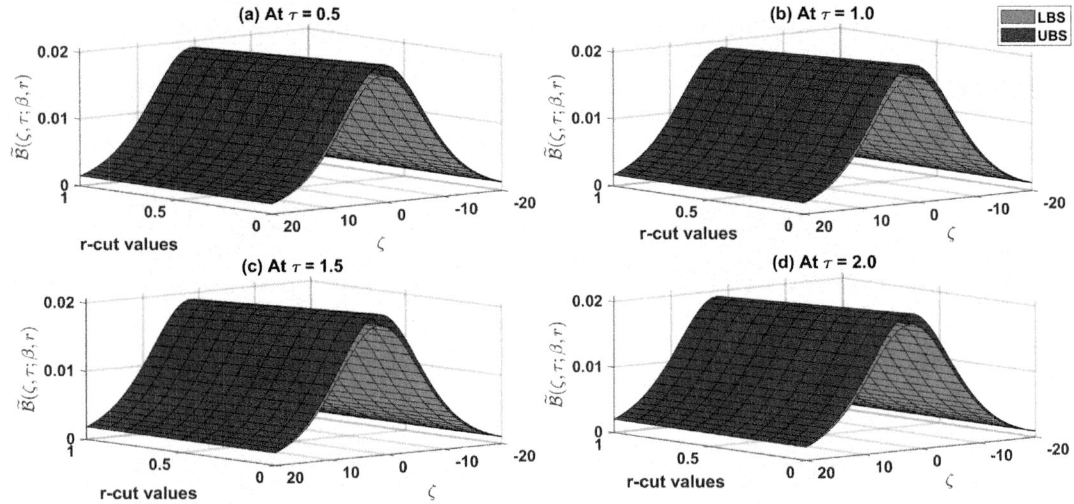


**FIGURE 2.6** 3D view of TFN solutions of (2.26).

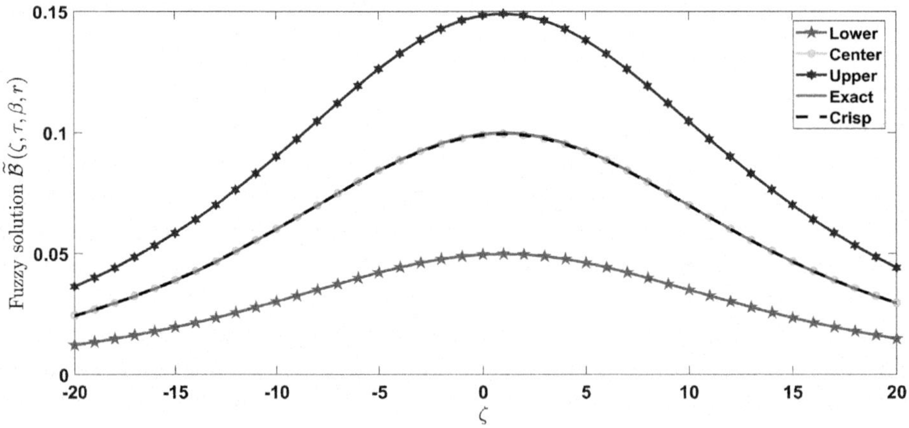


**FIGURE 2.7** Fuzzy solution of (2.29) at $\tau = 1$.

solutions of Eq. (2.29) at $\tau = 1$ indicating both fuzzy and exact solutions. The fuzzy wave has its boundaries, which are the lower and upper solutions, meaning the TFN. The mean behavior of the fuzzy wave is depicted by the center solution. The precise solution is used to reference to the actual wave profile, and the sharp solution is used to give a deterministic

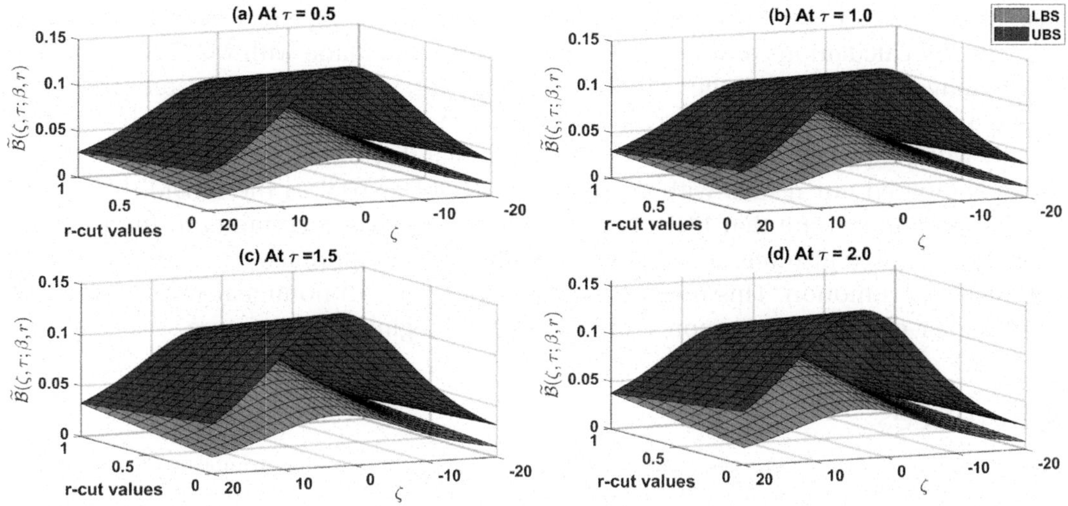


**FIGURE 2.8** 3D view of TFN solutions of (2.29).

approximation of the fuzziness-free way. The fact that the center, crisp, and exact solutions are almost identical indicates that the fuzzy method is good at capturing the behavior of the waves, and that it is able to consider the uncertainty.

Fig. 2.8 shows the trend of the fuzzy triangular solutions of Eq. (2.29) using various values of the parameter of the fuzzy solutions $\tau$ (0.5, 1.0, 1.5, and 2.0). The presence of several nonlinear terms in the equation determines the shape of the wave. The terms that include the mean of $\mathcal{B}(\frac{\partial \mathcal{B}}{\partial \zeta})^2$ and $\mathcal{B}^2 \frac{\partial^2 \mathcal{B}}{\partial \zeta^2}$ are the nonlinear terms which influence the steepness and amplitude of the wave. Also, higher-order derivative term, i.e., the term to provide the dispersion and smoothness of the waves, is the term of providing the derivatives of the $\frac{\partial^4 \mathcal{B}}{\partial \zeta^4}$. With increased values of the tau, the structure of the wave remains the same but changes with the upper and lower boundaries representing the fuzzy region. These nonlinear terms cause the wave shape to be more complicated, showing the effect of nonlinearity in the dynamics of the fuzzy solution.

## 2.7 Conclusion

In this paper, we have employed a longer variant of the homotopy perturbation method (HPTM), the FHPTM, to examine the movement of long waves in shallow water based on the fourth-order nonlinear fuzzy gBE. We have described the FHPTM and its accuracy, the uncertainty of the equation or initial conditions and its effectiveness. With a fuzzy implementation of the gBE, the uncertainties of the velocity of the waves and the depth of the water were taken into account, and this influences the shape of the solitary waves. This is done by comparing the fuzzy solution and the semianalytical solution of the normal case based on a dual parametric approach demonstrated the effect of fuzziness on the solution. All calculations and numeric simulations were made with MATLAB® software. The findings indicated that the FHPTM is an effective, strong, and consistent technique. The technique aids in controlling and manipulation of series solution, which converges rapidly within a small range. The accuracy and reliability of FHPTM was also confirmed as we compared our results with those of the HPTM as well as the EADM. Findings of this paper demonstrate that FHPTM can be used to solve uncertainty in the nonlinear wave equations; it also offers an effective mathematical approach to the analysis of the complex waves motion. This method can be generalized to nonlinear wave models and other higher-order effects in more complex physical regimes in the future.

## References

Alipour, P., 2024. The dual reciprocity boundary element method for one-dimensional nonlinear parabolic partial differential equations. Journal of Mathematical Sciences 280 (2), 131–145.

Aourir, E., Laeli Dastjerdi, H., 2025. Numerical investigation of the mesh-free collocation approach for solving third kind vies with nonlinear vanishing delays. International Journal of Computer Mathematics 102 (1), 112–129.

Asaduzzaman, M., Özger, F., Kilicman, A., 2024. Analytical approximate solutions to the nonlinear Fornberg–Whitham type equations via modified variational iteration method. Partial Differential Equations in Applied Mathematics 9, 100631.

Bilal, M., Iqbal, J., Shah, K., Abdalla, B., Abdeljawad, T., Ullah, I., 2024. Analytical solutions of the space–time fractional Kundu–Eckhaus equation by using modified extended direct algebraic method. Partial Differential Equations in Applied Mathematics 11, 100832.

Boussinesq, J., 1871. Théorie de l'intumescence liquide appelée onde solitaire ou de translation se propageant dans un canal rectangulaire. Comptes Rendus de L'Académie Des Sciences, Paris 72 (755–759), 1871.

Bruzon, M., 2009. Exact solutions of a generalized Boussinesq equation. Theoretical and Mathematical Physics 160, 894–904.

Chakraverty, S., Perera, S., 2018. Recent Advances in Applications of Computational and Fuzzy Mathematics. Springer.

Chakraverty, S., Mahato, N., Karunakar, P., Rao, T.D., 2019. Advanced Numerical and Semi-Analytical Methods for Differential Equations. John Wiley & Sons.

Chen, C., Li, M., 2024. Families of exact solutions of a generalized $(2+1)$-dimensional Boussinesq type equation. Wave Motion 127, 103297.

Chen, C., Wang, Z., 2021. New exact solutions, dynamical and chaotic behaviors for the fourth-order nonlinear generalized Boussinesq water wave equation. Advances in Mathematical Physics 2021 (1), 8409615.

Christov, C., Choudhury, J., 2011. Perturbation solution for the 2d Boussinesq equation. Mechanics Research Communications 38 (3), 274–281.

Darghoth, R.M.H., 2024. Approximate solution of an integrodifferential equation generalized by Harry Dym equation using the Picard successive method. Computational and Mathematical Methods 2024 (1), 7393931.

Daripa, P., 2006. Higher-order Boussinesq equations for two-way propagation of shallow water waves. European Journal of Mechanics. B, Fluids 25 (6), 1008–1021.

Das, P.K., 2025. Exact smooth and nonsmooth solutions for integro-partial differential equations by rapidly convergent approximation method. Theoretical and Mathematical Physics 222 (1), 53–68.

Ding, Q., Hou, Y., He, X., 2024. Optimal error estimates of a second-order fully decoupled finite element method for the nonstationary generalized Boussinesq model. Journal of Computational and Applied Mathematics 450, 116001.

He, J.-H., 1999. Homotopy perturbation technique. Computer Methods in Applied Mechanics and Engineering 178 (3–4), 257–262.
He, J.-H., 2000. A coupling method of a homotopy technique and a perturbation technique for non-linear problems. International Journal of Non-Linear Mechanics 35 (1), 37–43.
He, J.-H., 2003. Homotopy perturbation method: a new nonlinear analytical technique. Applied Mathematics and Computation 135 (1), 73–79.
He, J., 2005. Homotopy perturbation method for nonlinear bifurcation problems. International Journal of Nonlinear Sciences and Numerical Simulation 6 (2), 207–208.
Hussein, H.H., Ahmed, H.M., Alexan, W., 2024. Analytical soliton solutions for cubic-quartic perturbations of the Lakshmanan-Porsezian-Daniel equation using the modified extended tanh function method. Ain Shams Engineering Journal 15 (3), 102513.
Iqbal, S., Martínez, F., 2024. An innovative approach to approximating solutions of fractional partial differential equations. Physica Scripta 99 (6), 065259.
Ismael, M., Hat, S., Alabdali, O., Lone, S.A., Ali, A.H., 2024. Modified optimal auxiliary functions method for approximate-analytical solutions in fractional order nonlinear foam drainage equations. Partial Differential Equations in Applied Mathematics 10, 100700.
Kai, Y., Chen, S., Zhang, K., Yin, Z., 2024. A study of the shallow water waves with some Boussinesq-type equations. Waves Random Complex Media 34 (3), 1251–1268.
Karunakar, P., Chakraverty, S., 2017. Comparison of solutions of linear and non-linear shallow water wave equations using homotopy perturbation method. International Journal of Numerical Methods for Heat & Fluid Flow 27 (9), 2015–2029.
Karunakar, P., Chakraverty, S., 2018a. Solution of interval shallow water wave equations using homotopy perturbation method. Engineering Computations 35 (4), 1610–1624.
Karunakar, P., Chakraverty, S., 2018b. Solving shallow water equations with crisp and uncertain initial conditions. International Journal of Numerical Methods for Heat & Fluid Flow 28 (12), 2801–2815.
Karunakar, P., Chakraverty, S., 2019a. Effect of Coriolis constant on geophysical Korteweg-de Vries equation. Journal of Ocean Engineering and Science 4 (2), 113–121.
Karunakar, P., Chakraverty, S., 2019b. Solutions of time-fractional third-and fifth-order Korteweg–de-Vries equations using homotopy perturbation transform method. Engineering Computations.
Karunakar, P., Chakraverty, S., 2021. Homotopy perturbation method for predicting tsunami wave propagation with crisp and uncertain parameters. International Journal of Numerical Methods for Heat & Fluid Flow 31 (1), 92–105.
Karunakar, P., Chakraverty, S., 2022. Solution of interval-modified Kawahara differential equations using homotopy perturbation transform method. In: Wave Dynamics. World Scientific, pp. 193–202.
Karunakar, P., Biswal, U., Chakraverty, S., 2020. Fluid dynamics problems in uncertain environment. In: Mathematical Methods in Interdisciplinary Sciences, pp. 125–144.
Karunakar, P., Reddy, K.S., Chakraverty, S., 2023. Stability analysis and approximate solution of interval mathematical model for the covid-19 pandemic. Mathematical Methods in the Applied Sciences.
Karunakar, P., Chakraverty, S., Rao, T., Ramesh, K., Hussein, A., 2024. Fuzzy fractional shallow water wave equations: analysis, convergence of solutions, and comparative study with depth as triangular fuzzy number. Physica Scripta 99 (12), 125216.
Ketcheson, D.I., Lóczi, L., Russo, G., 2025. A multiscale model for weakly nonlinear shallow water waves over periodic bathymetry. Multiscale Modeling & Simulation. SIAM Interdisciplinary Journal 23 (1), 397–430.
Khan, Q., Suen, A., Khan, H., 2024. Application of an efficient analytical technique based on Aboodh transformation to solve linear and non-linear dynamical systems of integro-differential equations. Partial Differential Equations in Applied Mathematics 11, 100848.
Kordyukova, S.A., 2008. Korteweg-de Vries hierarchy as an asymptotic limit of the Boussinesq system. Theoretical and Mathematical Physics 154, 250–259.
Mamun, A.-A., Lu, C., Ananna, S.N., Uddin, M.M., 2024. Rational Sine-Gordon expansion method to analyze the dynamical behavior of the time-fractional phi-four and $(2+1)$ dimensional cbs equations. Scientific Reports 14 (1), 9473.
Moore, R.E., 1966. Interval Analysis, vol. 4. Prentice-Hall, Englewood Cliffs.
Nirmala, A., Kumbinarasaiah, S., 2024. Numerical solution of nonlinear Hunter-Saxton equation, Benjamin-Bona Mahony equation, and Klein-Gordon equation using Hosoya polynomial method. Results in Control and Optimization 14, 100388.

Palav, M.S., Pradhan, V.H., 2025. Redefined fourth order uniform hyperbolic polynomial b-splines based collocation method for solving advection-diffusion equation. Applied Mathematics and Computation 484, 128992.

Razzaq, W., Zafar, A., Nazir, A., Awwad, F.A., Ismail, E.A., et al., 2024. Optical soliton solutions of nonlinear differential Boussinesq water wave equation via two analytical techniques. Results in Physics 64, 107898.

Sankar, B.B.S., Karunakar, P., 2025. Solution characteristics of time-fractional geophysical kdv equation with Coriolis parameter through hybrid methods. Journal of Vibration Engineering & Technologies 13 (1), 103.

Tian, S.-F., 2020. Lie symmetry analysis, conservation laws and solitary wave solutions to a fourth-order nonlinear generalized Boussinesq water wave equation. Applied Mathematics Letters 100, 106056.

Vana, R., Karunaka, P., 2025. Numerical solutions of the Benjamin–Bona–Mahony equation using the differential quadrature method with shifted Legendre and generalized Laguerre polynomials. Indian Journal of Pure and Applied Mathematics, 1–14.

Vana, R., Karunakar, P., 2024a. A fuzzy semi-analytical approach for modeling uncertainties in solitary wave solution of coupled nonlinear Boussinesq equations. Physica Scripta 99 (10), 105218.

Vana, R., Karunakar, P., 2024b. Fuzzy uncertainty modeling of generalized Hirota–Satsuma coupled Korteweg–de Vries equation. Physics of Fluids 36 (9).

Vana, R., Karunakar, P., 2024c. Uncertainties in regularized long-wave equation and its modified form: a triangular fuzzy-based approach. Physics of Fluids 36 (4).

Vana, R., Karunakar, P., 2024d. Computational approach and convergence analysis for interval-based solution of the Benjamin–Bona–Mahony equation with imprecise parameters. Engineering Computations 41 (4), 1067–1085.

Vana, R., Karunakar, P., Chakraverty, S., 2025. Semi-analytical and numerical computations for the solution of uncertain fractional Benjamin Bona Mahony equation with triangular fuzzy number. Journal of Computational and Applied Mechanics 56 (1), 196–221.

Wazwaz, A.-M., 2010. Non-integrable variants of Boussinesq equation with two solitons. Applied Mathematics and Computation 217 (2), 820–825.

Zhu, N., Qu, W., 2024. Existence and bifurcation of traveling wave solutions to a generalized Boussinesq equation with nonlinear dispersion. Mathematical Methods in the Applied Sciences 47 (6), 4840–4852.

3

# Implicit cubic spline methods for solving intuitionistic fuzzy parabolic partial differential equations

**Deepak Kumar Sah and Sreenivasulu Ballem**
*Department of Mathematics, Central University of Karnataka, Kalaburagi, India*

## 3.1 Introduction

Partial differential equations (PDEs) are crucial for modeling a variety of real-world phenomena across physics, engineering, finance, and biological systems. However, uncertainties frequently arise due to imprecise measurements, ambiguous initial or boundary conditions, and inherent vagueness in system parameters. Traditional PDE models often fail to adequately capture these uncertainties. To address this, fuzzy set theory (Zadeh, 1965) has been extensively applied, resulting in the development of fuzzy partial differential equations (FPDEs). Among various fuzzy models, intuitionistic fuzzy sets (IFSs), introduced by Atanassov (2012) in 1983, offer an advanced mathematical framework by incorporating degrees of membership, nonmembership, and hesitation, making them particularly suited for handling real-world imprecision. Consequently, there has been growing interest in IFPPDEs as powerful tools for modeling dynamic systems under uncertainty. The study of intuitionistic fuzzy differential equations (IFDEs) has significantly advanced, driven by theoretical progress and sophisticated computational techniques. This paper critically reviews the key developments that have shaped the evolution of IFDEs, focusing on both mathematical formulations and numerical methods. The formal introduction of IFDEs in 2000 by Melliani and Chadli (2000) marked a pivotal step, incorporating intuitionistic fuzzy numbers into differential equations and laying the foundation for further research. In 2015, Mondal and Roy (2015) expanded this framework by solving first-order homogeneous ordinary differential equations using triangular intuitionistic fuzzy numbers, thus broadening the analytical scope of IFDEs. Further advancements emerged in 2018 when Nirmala et al. (2018) enhanced the mathematical framework of IFDEs, solidifying their theoretical basis. In 2018, Akın and Bayeğ (2019) applied the intuitionistic Zadeh extension principle to address second-order fuzzy initial value problems, contributing to methodological progress. A significant breakthrough occurred in 2020 when Amma et al. (2020) rigorously established existence and uniqueness results for boundary value problems using Banach's fixed-point theorem, reinforcing the mathematical robustness of IFDEs. Computational methodologies made substantial progress in 2021, with Man et

Advanced Computational and Mathematical Approaches in Applied Differential Equations
https://doi.org/10.1016/B978-0-44-345743-2.00014-1

al. (2021) introducing an explicit finite difference method to solve the intuitionistic fuzzy heat equation, accompanied by a stability analysis using the Von Neumann method. Later that year, the same authors, along with Anupama, extended their work to the intuitionistic fuzzy Poisson partial differential equation, further validating the applicability of numerical techniques in solving complex IFDEs. The latest advancement, reported in 2023 by Man et al. (2023), involved an explicit finite difference approach to solving the intuitionistic fuzzy wave equation. This contribution highlighted the increasing significance of numerical methods in addressing IFDEs, underscoring their effectiveness in solving complex differential systems. The evolution of IFDEs has been characterized by continuous theoretical refinements and computational innovations. The integration of rigorous mathematical principles with advanced numerical techniques ensures that IFDEs remain at the forefront of applied mathematics and engineering research. This paper aims to propose and analyze an implicit cubic spline method for solving IFPPDEs. The study's objectives include:

- Developing a numerical scheme that effectively handles the uncertainty inherent in IFPPDEs.
- Solving parabolic PDEs under fuzzy conditions with high accuracy and efficiency.
- Creating a numerical scheme using implicit cubic splines to accurately approximate the solution of IFPPDEs.
- To develop a cubic spline method that generates a finite difference method, which should be unconditionally stable.

In this paper, we concentrate on solving IFPPDEs that reflect heat-like processes under uncertainty. These equations are an extension of classical heat equations in which the parameters, initial, and boundary conditions are not exactly known and are instead represented as intuitionistic fuzzy numbers. Such models are extremely useful in real-world applications, such as biomedical engineering, environmental heat transfer, material science, and industrial thermal systems, where precise input data is frequently unavailable due to measurement errors, variability, or insufficient information. By including intuitionistic fuzzy logic, the model better depicts the inherent ambiguity and reluctance found in real-life scenarios.

The paper begins by introducing fundamental terminology essential to the study. It then formulates a finite difference scheme based on the implicit cubic spline method to solve parabolic partial differential equations in a crisp environment. This approach is extended to develop a fuzzy finite difference method for solving IFPPDEs. The stability and convergence of the proposed method are analyzed, followed by validation through a numerical example.

## 3.2 Preliminaries

**Definition 3.1.** (Man et al., 2021) An intuitionistic fuzzy set is defined as a set $\tilde{A} \in \mathrm{X}$ and represented as $\tilde{A} = \left(x, \mu_{\tilde{A}}(x), \nu_{\tilde{A}}(x) \mid x \in X\right)$ and characterized by a membership function

$\mu_{\tilde{A}}(x) : X \to [0, 1]$, and nonmembership function $\nu_{\tilde{A}}(x) : X \to [0, 1]$, for each element $x \in X$ such that $0 \le \mu_{\tilde{A}}(x) + \nu_{\tilde{A}}(x) \le 1$.

**Definition 3.2.** (Man et al., 2021) The $(\boldsymbol{\alpha}, \boldsymbol{\beta})$ cut (level) of an intuitionistic fuzzy set $\tilde{A}$ is represented as $\tilde{A}_{\alpha,\beta}$ and defined as follows:

Let $\alpha \in [0, 1]$, $\beta \in [0, 1]$ be the two fixed numbers such that $\alpha + \beta \le 1$, and $\tilde{A}_{\alpha,\beta} = \{x \in X : \mu_{\tilde{A}}(x) \ge \alpha, \nu_{\tilde{A}}(x) \le \beta\}$, where $\tilde{A}_{\alpha,\beta}$ is a crisp set that represents the $\tilde{A}$ at least to the degree of $\alpha$ and which does not belong to $\tilde{A}$ at most to the degree of $\beta$.

**Definition 3.3.** (Man et al., 2023) An intuitionistic fuzzy set $\tilde{A}$ is said to be an intuitionistic fuzzy number if it holds the following properties:

**(1)** $\exists x_0 \in R$ such that membership value $\mu_{\tilde{A}}(x_0) = 1$ and nonmembership value $\nu_{\tilde{A}}(x_0) = 0$.
**(2)** The membership function $\mu_{\tilde{A}}$ known as the convex fuzzy set if $\mu_{\tilde{A}}(\gamma x_1 + (1-\gamma)x_2) \ge min\{\mu_{\tilde{A}}(x_1), \mu_{\tilde{A}}(x_2)\}; x_1, x_2 \in R, \gamma \in [0, 1]$.
**(3)** The nonmembership function $\nu_{\tilde{A}}$ is known as the concave fuzzy set if $\nu_{\tilde{A}}(\gamma x_1 + (1-\gamma)x_2) \le max\{\nu_{\tilde{A}}(x_1), \nu_{\tilde{A}}(x_2)\}; x_1, x_2 \in R, \gamma \in [0, 1]$.

**Definition 3.4.** (Man et al., 2023) A fuzzy number $\tilde{A} =< (\sigma_1, \sigma_2, \sigma_3), (\sigma_1', \sigma_2, \sigma_3') >$ is known as an intuitionistic triangular fuzzy number if both its membership and nonmembership functions satisfy the conditions mentioned below:

$$\mu_{\tilde{A}}(x; \sigma_1, \sigma_2, \sigma_3) = \begin{cases} 0, & \text{for} \quad x < \sigma_1; \\ \dfrac{x-\sigma_1}{\sigma_2-\sigma_1}, & \text{for} \quad \sigma_1 \le x \le \sigma_2; \\ \dfrac{\sigma_3-x}{\sigma_3-\sigma_2}, & \text{for} \quad \sigma_2 \le x \le \sigma_3; \\ 0, & \text{for} \quad x > \sigma_3; \end{cases}$$

$$\nu_{\tilde{A}}\left(x; \sigma_1', \sigma_2, \sigma_3'\right) = \begin{cases} 1, & \text{for} \quad x < \sigma_1'; \\ \dfrac{\sigma_2-x}{\sigma_2-\sigma_1'}, & \text{for} \quad \sigma_1' \le x \le \sigma_2; \\ \dfrac{x-\sigma_2}{\sigma_3'-\sigma_2}, & \text{for} \quad \sigma_2 \le x \le \sigma_3'; \\ 1, & \text{for} \quad x > \sigma_3', \end{cases}$$

where $\sigma_1' \le \sigma_1 \le \sigma_2 \le \sigma_3 \le \sigma_3'$.

The $(\boldsymbol{\alpha}, \boldsymbol{\beta})$ cut of intuitionistic fuzzy number $\tilde{A} =< (\sigma_1, \sigma_2, \sigma_3), (\sigma_1', \sigma_2, \sigma_3') >$ can be represented as $\tilde{A}_{\alpha,\beta} =< [\underline{A}(\alpha), \overline{A}(\alpha)], \left[\underline{A}'(\beta), \overline{A}'(\beta)\right] >$, where $\underline{A}(\alpha) = \sigma_1 + \alpha(\sigma_2 - \sigma_1)$,

$$\overline{A}(\alpha) = \sigma_3 - \alpha(\sigma_3 - \sigma_2),$$

$$\underline{A}'(\beta) = \sigma_2 - \beta\left(\sigma_2 - \sigma_1'\right),$$

$$\overline{A}'(\beta) = \sigma_2 + \beta(\sigma_3' - \sigma_2).$$

**Definition 3.5.** Fuzzy numbers can be converted into intervals using the parametric form. So, for any fuzzy number $\tilde{\rho} = \left[\underline{\rho}(\alpha), \overline{\rho}(\alpha)\right]$, and $\tilde{\sigma} = \left[\underline{\sigma}(\alpha), \overline{\sigma}(\alpha)\right]$, and scalar $\phi$, we have the interval-based fuzzy arithmetic as

**(i)** $\tilde{\rho} = \tilde{\sigma}$ if and only if $\underline{\rho}(\alpha) = \underline{\sigma}(\alpha)$ and $\overline{\rho}(\alpha) = \overline{\sigma}(\alpha)$,

**(ii)** $\tilde{\rho} + \tilde{\sigma} = \left[\underline{\rho}(\alpha) + \underline{\sigma}(\alpha), \overline{\rho}(\alpha) + \overline{\sigma}(\alpha)\right]$,

**(iii)** $\tilde{\rho} - \tilde{\sigma} = \left[\underline{\rho}(\alpha), \overline{\rho}(\alpha)\right] - \left[\underline{\sigma}(\alpha), \overline{\sigma}(\alpha)\right] = \left[\underline{\rho}(\alpha) - \overline{\sigma}(\alpha), \overline{\rho}(\alpha) - \underline{\sigma}(\alpha)\right]$

**(iv)** $\tilde{\rho} \times \tilde{\sigma} = [min\left(\underline{\rho}(\alpha) \times \underline{\sigma}(\alpha), \underline{\rho}(\alpha) \times \overline{\sigma}(\alpha), \overline{\rho}(\alpha) \times \underline{\sigma}(\alpha), \overline{\rho}(\alpha) \times \overline{\sigma}(\alpha)\right)$,
$max\left(\underline{\rho}(\alpha) \times \underline{\sigma}(\alpha), \underline{\rho}(\alpha) \times \overline{\sigma}(\alpha), \overline{\rho}(\alpha) \times \underline{\sigma}(\alpha), \overline{\rho}(\alpha) \times \overline{\sigma}(\alpha)\right)]$,

**(v)** $\phi\tilde{\rho} = \begin{cases} \left[\phi\underline{\rho}(\alpha), \phi\overline{\rho}(\alpha)\right], & \text{for} \quad \phi \geq 0; \\ \left[\phi\overline{\rho}(\alpha), \phi\underline{\rho}(\alpha)\right], & \text{for} \quad \phi < 0. \end{cases}$

**Definition 3.6.** (Man et al., 2021) The degree of nondeterminacy in an intuitionistic fuzzy set can be calculated as $\pi_{\tilde{A}}(x) = 1 - \mu_{\tilde{A}}(x) - \nu_{\tilde{A}}(x)$, where $x \in X$ and $\tilde{A}$ is the intuitionistic fuzzy set.

## 3.3 Finite difference formulation using the explicit cubic spline method for parabolic partial differential equation in the crisp environment

To derive the finite difference method using the cubic spline method to find the solution of the heat conduction equation, the time derivative is implicitly replaced using the forward difference operator, while the space derivatives are approximated with a natural cubic spline (Sastry, 2012).

A cubic spline approximation to the heat conduction equation produces a finite difference formula. Let $W_{i,j}$ be the discrete approximation of $W(x,t)$ at the point $(x_i, t_j)$. $S_j(x_i)$ represent the cubic spline interpolating the value of $W_{i,j}$ at the $j^{th}$ time level.

Replacing the heat conduction equation $\frac{\partial W}{\partial t} = \frac{\partial^2 W}{\partial x^2}$ by

$$\frac{W_{i,j+1} - W_{i,j}}{k} = \frac{1}{2}\left(M_{i,j} + M_{i,j+1}\right), \tag{3.1}$$

$$M_{i,j} = S''(x_i), \tag{3.2}$$

$$S_j(x_i) = M_{i-1,j}\frac{(x_i - x)^3}{6h} + M_{i,j}\frac{(x - x_{i-1})^3}{6h} + \left(W_{i-1,j} - \frac{h^2}{6}M_{i-1,j}\right)\frac{(x_i - x)}{h} + \left(W_{i,j} - \frac{h^2}{6}M_{i,j}\right)\frac{(x - x_{i-1})}{h}, \tag{3.3}$$

where i = 1,2, 3, ..., N.

$$S_j'\left(x_i^+\right) = -\frac{h}{3}M_{i,j} - \frac{h}{6}M_{i+1,j} + \frac{W_{i+1,j} - W_{i,j}}{h}, \tag{3.4}$$

$$S_j'\left(x_i^-\right) = \frac{h}{3}M_{i,j} + \frac{h}{6}M_{i-1,j} + \frac{W_{i,j} - W_{i-1,j}}{h}, \tag{3.5}$$

so the continuity of the first derivative implies

$$\frac{1}{6}M_{i-1,j} + \frac{2}{3}M_{i,j} + \frac{1}{6}M_{i+1,j} = \left(\frac{W_{i-1,j} - 2W_{i,j} + W_{i+1,j}}{h^2}\right). \tag{3.6}$$

Similarly, at $(j+1)^{th}$ time level,

$$\frac{1}{6}M_{i-1,j+1} + \frac{2}{3}M_{i,j+1} + \frac{1}{6}M_{i+1,j+1} = \left(\frac{W_{i-1,j+1} - 2W_{i,j+1} + W_{i+1,j+1}}{h^2}\right). \tag{3.7}$$

Now, using equations (Eq. (3.1), Eq. (3.6), and Eq. (3.7)), we have the finite difference method

$$\begin{aligned}(1-3\Gamma)\,W_{i-1,j+1} + (4+6\Gamma)\,W_{i,j+1} + (1-3\Gamma)\,W_{i+1,j+1}\\ = (1+3\Gamma)\,W_{i-1,j} + (4-6\Gamma)\,W_{i,j} + (1+3\Gamma)\,W_{i+1,j},\end{aligned}$$

such that $\Gamma = \frac{k}{h^2}$.

The finite difference method presented is derived from the implicit cubic spline approach within a crisp environment, enabling the determination of numerical solutions for crisp parabolic PDEs.

## 3.4 Derivation of a finite difference scheme using the natural cubic spline approach for solving intuitionistic fuzzy parabolic partial differential equation

Let $\tilde{W}$ be a fuzzy function of the independent crisp variable $x$ and $t$, which satisfies the IFPPDEs as

$$\left(D_t - b^2 D_x^2\right)\tilde{W} = \tilde{0}; \tag{3.8}$$

the boundary condition is given as $\tilde{W}(0,t) = \tilde{W}(l,t) = \tilde{0}$, also the initial condition is given as $\tilde{W}(x,0) = \tilde{f}(x)$. We define the domain as $I = \{(x,t) : 0 \le x \le l, 0 \le t \le T\}$. An $(\alpha,\beta)$ cut of $\tilde{W}(x,t)$ and its parametric form will be $\tilde{W}(x,t)[\alpha,\beta] = [\underline{W}(x,t;\alpha), \overline{W}(x,t;\alpha)], [\underline{W'}(x,t;\beta), \overline{W}'(x,t;\beta)]$. We assume that the continuous partial derivative of $\underline{W}(x,t;\alpha)$, $\overline{W}(x,t;\alpha)$, and $\underline{W'}(x,t;\beta)$, $\overline{W}'(x,t;\beta)$ with respect to $x$ and $t$ exists. Thus for any $(x,t) \in I$, where $\alpha \in [0,1]$, $\beta \in [0,1]$, $\left(D_t - b^2 D_x^2\right)\underline{W}(x,t,\alpha)$, $\left(D_t - b^2 D_x^2\right)\overline{W}(x,t,\alpha)$, $\left(D_t - b^2 D_x^2\right)\underline{W'}(x,t,\beta)$, $\left(D_t - b^2 D_x^2\right)\overline{W}'(x,t,\beta)$ are continuous for $\alpha \in [0,1]$ and $\beta \in [0,1]$. As we know, to find the finite difference method, the time derivative is implicitly replaced using the forward difference operator, while the space derivatives are approximated with a natural cubic spline. So now using the standard finite difference method and definition of Taylor's theorem (Allahviranloo, 2002),

$$D_x^2 \underline{W}(x,t,\alpha)|_{i,j} \approx \frac{\underline{W}_{i+1,j} - 2\underline{W}_{i,j} + \underline{W}_{i+1,j}}{h^2}, \tag{3.9}$$

$$D_x^2 \overline{W}(x,t,\alpha)|_{i,j} \approx \frac{\overline{W}_{i+1,j} - 2\overline{W}_{i,j} + \overline{W}_{i+1,j}}{h^2}, \tag{3.10}$$

$$D_x^2 \underline{W}'(x,t,\beta)|_{i,j} \approx \frac{\underline{W}'_{i+1,j} - 2\underline{W}'_{i,j} + \underline{W}'_{i+1,j}}{h^2}, \tag{3.11}$$

$$D_x^2 \overline{W}'(x,t,\beta)|_{i,j} \approx \frac{\overline{W}'_{i+1,j} - 2\overline{W}'_{i,j} + \overline{W}'_{i+1,j}}{h^2}. \tag{3.12}$$

The error of the above finite difference scheme (3.9)–(3.12) is $O\left(h^2\right)$. The forward difference scheme for $(D_t)\tilde{W}$, we have

$$D_t \underline{W}(x,t;\alpha)|_{i,j} \approx \frac{\underline{W}_{i,j+1} - \underline{W}_{i,j}}{k}, \tag{3.13}$$

$$D_t \overline{W}(x,t;\alpha)|_{i,j} \approx \frac{\overline{W}_{i,j+1} - \overline{W}_{i,j}}{k}, \tag{3.14}$$

$$D_t \underline{W}'(x,t;\beta)|_{i,j} \approx \frac{\underline{W}'_{i,j+1} - \underline{W}'_{i,j}}{k}, \tag{3.15}$$

$$D_t \overline{W}'(x,t;\beta)|_{i,j} \approx \frac{\overline{W}'_{i,j+1} - \overline{W}'_{i,j}}{k}. \tag{3.16}$$

The error of the above finite difference schemes (Eq. (3.13))–(Eq. (3.16)) is $O(k)$.

Now, the time derivative is implicitly replaced using the forward difference operator:

$$\frac{\tilde{W}_{i,j+1} - \tilde{W}_{i,j}}{k} = \frac{1}{2}\left(\tilde{M}_{i,j} + \tilde{M}_{i,j+1}\right), \tag{3.17}$$

$$\frac{\tilde{W}_{i+1,j+1} - \tilde{W}_{i+1,j}}{k} = \frac{1}{2}\left(\tilde{M}_{i+1,j} + \tilde{M}_{i+1,j+1}\right), \tag{3.18}$$

$$\frac{\tilde{W}_{i-1,j+1} - \tilde{W}_{i-1,j}}{k} = \frac{1}{2}\left(\tilde{M}_{i-1,j} + \tilde{M}_{i-1,j+1}\right). \tag{3.19}$$

The fuzzy linear parabolic partial differential equation is solved using the cubic spline method:

$$\frac{1}{6}\tilde{M}_{i-1,j} + \frac{2}{3}\tilde{M}_{i,j} + \frac{1}{6}\tilde{M}_{i+1,j} = \left(\frac{\tilde{W}_{i-1,j} - 2\tilde{W}_{i,j} + \tilde{W}_{i+1,j}}{h^2}\right), \qquad i = 1,2,.,N-1. \tag{3.20}$$

Similarly, at $(j+1)^{th}$ time level,

$$\frac{1}{6}\tilde{M}_{i-1,j+1}+\frac{2}{3}\tilde{M}_{i,j+1}+\frac{1}{6}\tilde{M}_{i+1,j+1}=\left(\frac{\tilde{W}_{i-1,j+1}-2\tilde{W}_{i,j+1}+\tilde{W}_{i+1,j+1}}{h^2}\right), \qquad i=1,2,.,N-1. \tag{3.21}$$

Now implementing (Eq. (3.17)), (Eq. (3.18)), and (Eq. (3.19)) in (Eq. (3.20)) and (Eq. (3.21)) we have

$$\begin{aligned}&(1-3\Gamma)\,\tilde{W}_{i-1,j+1}+(4+6\Gamma)\,\tilde{W}_{i,j+1}+(1-3\Gamma)\,\tilde{W}_{i+1,j+1}\\&=(1+3\Gamma)\,\tilde{W}_{i-1,j}+(4-6\Gamma)\,\tilde{W}_{i,j}+(1+3\Gamma)\,\tilde{W}_{i+1,j},\end{aligned} \tag{3.22}$$

where $\Gamma=\frac{k}{h^2}$. Now, taking the $(\alpha,\beta)$ level of the above finite difference form

$$\begin{aligned}&(1-3\Gamma)\,\underline{W}_{i-1,j+1}+(4+6\Gamma)\,\underline{W}_{i,j+1}+(1-3\Gamma)\,\underline{W}_{i+1,j+1}\\&=(1+3\Gamma)\,\underline{W}_{i-1,j}+(4-6\Gamma)\,\underline{W}_{i,j}+(1+3\Gamma)\,\underline{W}_{i+1,j},\end{aligned} \tag{3.23}$$

$$\begin{aligned}&(1-3\Gamma)\,\overline{W}_{i-1,j+1}+(4+6\Gamma)\,\overline{W}_{i,j+1}+(1-3\Gamma)\,\overline{W}_{i+1,j+1}\\&=(1+3\Gamma)\,\overline{W}_{i-1,j}+(4-6\Gamma)\,\overline{W}_{i,j}+(1+3\Gamma)\,\overline{W}_{i+1,j},\end{aligned} \tag{3.24}$$

$$\begin{aligned}&(1-3\Gamma)\,\underline{W}'_{i-1,j+1}+(4+6\Gamma)\,\underline{W}'_{i,j+1}+(1-3\Gamma)\,\underline{W}'_{i+1,j+1}\\&=(1+3\Gamma)\,\underline{W}'_{i-1,j}+(4-6\Gamma)\,\underline{W}'_{i,j}+(1+3\Gamma)\,\underline{W}'_{i+1,j},\end{aligned} \tag{3.25}$$

$$\begin{aligned}&(1-3\Gamma)\,\overline{W}'_{i-1,j+1}+(4+6\Gamma)\,\overline{W}'_{i,j+1}+(1-3\Gamma)\,\overline{W}'_{i+1,j+1}\\&=(1+3\Gamma)\,\overline{W}'_{i-1,j}+(4-6\Gamma)\,\overline{W}'_{i,j}+(1+3\Gamma)\,\overline{W}'_{i+1,j}.\end{aligned} \tag{3.26}$$

## 3.5 Convergence analysis

Assume $\tilde{W}\left(x_i,t_j\right)$ and $\tilde{W}_{i,j}$ is the exact solution and approximate solution at the same point $(x_i,t_j)$ is obtained by the proposed finite difference method respectively. At the point

$$(x_i,t_j),\ \text{error is } \tilde{\varepsilon}_{i,j}=\tilde{W}_{i,j}-\tilde{W}\left(x_i,t_j\right).$$

The difference between $\tilde{W}_{i,j}$ and $\tilde{W}\left(x_i,t_j\right)$ is a local truncation error, which is given by

$$\begin{aligned}&(1-3\Gamma)\,\tilde{\varepsilon}_{i-1,j+1}+(4+6\Gamma)\,\tilde{\varepsilon}_{i,j+1}+(1-3\Gamma)\,\tilde{\varepsilon}_{i+1,j+1}\\&=(1+3\Gamma)\,\tilde{\varepsilon}_{i-1,j}+(4-6\Gamma)\,\tilde{\varepsilon}_{i,j}+(1+3\Gamma)\,\tilde{\varepsilon}_{i+1,j}+O\left(k^2+h^2\right)H\,(x,t),\end{aligned} \tag{3.27}$$

where $\tilde{\varepsilon}_{i,j}=\tilde{W}_{i,j}-\tilde{W}\left(x_i,t_j\right)$ and $H\,(x,t)$ is the remainder term, i.e., the derivative of $\tilde{W}\left(x_i,t_j\right)$. When both step sizes, $h$ and $k$, approach zero, the error function also approaches zero. To verify this, we show that the norm of the error function at the discretized points is zero.

The properties of $\tilde{\varepsilon}\left(x_i, t_j\right)$ and the solution function $\tilde{W}\left(x_i, t_j\right)$ are identical and have the same property. Hence

$$\partial_t \tilde{\varepsilon}(x, t)\mid_{i,j} - \partial_{xx} \tilde{\varepsilon}(x, t)\mid_{i,j} = (k^2 + h^2) H_{i,j}, \tag{3.28}$$

$$H_{i,j} \propto \left( \frac{\partial^4 \tilde{W}}{\partial x^4} + \frac{\partial^4 \tilde{W}}{\partial x^2 \partial t^2} \right). \tag{3.29}$$

Let $\hbar_{i,j} = \left(k^2 + h^2\right) H_{i,j}$ and consider

$$M = max \left\{ \left| \frac{\partial^4 \tilde{W}(x,t)}{\partial x^4} \right|_{i,j}, \left| \frac{\partial^4 \tilde{W}(x,t)}{\partial x^2 \partial t^2} \right|_{i,j} \right\}, \tag{3.30}$$

$$\hbar_{i,j} \propto M\left(k^2 + h^2\right). \tag{3.31}$$

Now, we have a system of linear equations that can be generated by using an error equation, which can be expressed in matrix form as $AX = B$, where $A = \begin{pmatrix} M_1 & M_2 \\ M_2 & M_1 \end{pmatrix}$, $X = \begin{pmatrix} \frac{E}{E} \end{pmatrix}$, and $B = (k^2 + h^2)\begin{pmatrix} \frac{\hbar}{\hbar} \end{pmatrix}$, where $M_1$ is both the block matrix and the diagonal sparse matrix. If the matrix A be nonsingular, then the vector $X = \begin{pmatrix} \frac{E}{E} \end{pmatrix}$ can be found by the determinant of $A = det\left(M_1 - M_2\right).det\left(M_1 + M_2\right)$.

$$\begin{pmatrix} \frac{E}{E} \end{pmatrix} = \left(k^2 + h^2\right)\begin{pmatrix} M_1 & M_2 \\ M_2 & M_1 \end{pmatrix}^{-1}\begin{pmatrix} \frac{\hbar}{\hbar} \end{pmatrix}. \tag{3.32}$$

It is sufficient to show that the convergence is under the vector and matrix infinity norm.

$$\left| \frac{E}{E} \right|_{\infty} \leq \left(k^2 + h^2\right)\left| \begin{pmatrix} M_1 & M_2 \\ M_2 & M_1 \end{pmatrix}^{-1} \right|_{\infty} \left| \begin{pmatrix} \frac{\hbar}{\hbar} \end{pmatrix} \right|_{\infty}. \tag{3.33}$$

If

$$\left\| \begin{pmatrix} M_1 & M_2 \\ M_2 & M_1 \end{pmatrix}^{-1} \right\|_{\infty} = \text{N and } m = \left| \begin{pmatrix} \frac{\hbar}{\hbar} \end{pmatrix} \right|_{\infty}, \text{ then } 0 < \left| \frac{E}{E} \right|_{\infty} \leq Nm\left(k^2 + h^2\right). \tag{3.34}$$

This means that as $h \to 0$, $k \to 0$, $\Rightarrow \left| \frac{E}{E} \right|_{\infty} \to 0$. Hence the considered system is convergence as $h \to 0$, $k \to 0$.

## 3.6 Stability analysis

Before analyzing the stability of the proposed method, let us first review some theorems.

**Theorem 3.6.1.** *Let the matrix M have the special structure as* $\begin{pmatrix} K & L \\ L & K \end{pmatrix}$, *then the eigenvalues of the matrix M are the union of eigenvalues* $K+L$ *and eigenvalues of* $K-L$.

**Theorem 3.6.2.** *The difference equation is stable if and only if it meets the necessary and sufficient condition* $\eta\,(M) \le 1$ *(Smith, 1985).*

**Remark.** Here the largest eigenvalue of the matrix is represented by $\eta\,(M)$.

Now let us discuss the stability of the proposed method. For stability, it is sufficient to show that $\eta\,(M) \le 1$. Our matrix representation of the system of equations is as follows:

$$A\begin{pmatrix} \underline{W}_{j+1} \\ \overline{W}_{j+1} \end{pmatrix} = B\begin{pmatrix} \underline{W}_{j} \\ \overline{W}_{j} \end{pmatrix}, \tag{3.35}$$

where $A = \begin{pmatrix} K & L \\ L & K \end{pmatrix}$ and $B = \begin{pmatrix} P & N \\ N & P \end{pmatrix}$. Here A and B are the block matrices.

To show the stability, we need to find the eigenvalue of $K+L$, $K-L$, $P+N$, $P-N$ matrices. Since

$$K+L = K-L = \begin{pmatrix} 4+6\Gamma & 1-3\Gamma & 0 & . & 0 \\ 1-3\Gamma & 4+6\Gamma & 1-3\Gamma & 0. & 0 \\ 0 & 1-3\Gamma & 4+6\Gamma & 1-3\Gamma. & 0 \\ . & . & . & . & . \\ 0 & 0 & . & 1-3\Gamma & 4+6\Gamma \end{pmatrix},$$

$$P+N = P-N = \begin{pmatrix} (4-6\Gamma) & (1+3\Gamma) & 0 & . & 0 \\ (1+3\Gamma) & (4-6\Gamma) & (1+3\Gamma) & 0. & 0 \\ 0 & (1+3\Gamma) & (4-6\Gamma) & (1+3\Gamma). & 0 \\ . & . & . & . & . \\ 0 & 0 & . & (1+3\Gamma) & (4-6\Gamma) \end{pmatrix},$$

$K+L$, $K-L$, $P+N$, $P-N$ are the $(n-1 \times n-1)$ square tridiagonal matrices.

So for eigenvalues $\lambda_{K+L}$ and $\lambda_{K-L}$, let $a = 4+6\Gamma$, $b = 1-3\Gamma$, $c = 1-3\Gamma$. Since by matrix method of stability, eigenvalues are given by

$$\lambda_{K+L} = a + 2\sqrt{bc}cos\left(\frac{r\pi}{N+1}\right), \qquad r = 1, 2, 3, ., N. \tag{3.36}$$

Hence $\lambda_{K+L} = (4+6\Gamma) + 2(1-3\Gamma)\,cos\left(\frac{r\pi}{N+1}\right)$, or $\lambda_{K+L} = 6 - 4(1-3\Gamma)\,sin^2\left(\frac{r\pi}{2(N+1)}\right)$. Similarly,

$$\lambda_{K-L} = 6 - 4sin^2\left(\frac{r\pi}{2(N+1)}\right). \tag{3.37}$$

For $\lambda_{P+N}$ and $\lambda_{P-N}$; let $a = 4 - 6\Gamma$, $b = 1 + 3\Gamma$, $c = 1 + 3\Gamma$, here $\Gamma = \frac{k}{h^2}$.

$$\lambda_{P+N} = (4 - 6\Gamma) + 2(1 + 3\Gamma)\cos\left(\frac{\pi r}{N+1}\right), \qquad r = 1, 2, 3, .., N.$$

$$\lambda_{P+N} = 6 - 4(1 + 3\Gamma)\sin^2\left(\frac{r\pi}{2(N+1)}\right). \tag{3.38}$$

Similarly,

$$\lambda_{P-N} = 6 - 4(1 + 3\Gamma)\sin^2\left(\frac{r\pi}{2(N+1)}\right), \tag{3.39}$$

where $\Gamma = \frac{k}{h^2}$. As we know, when the $\eta(M) = \eta\left(A^{-1}B\right) < 1$, then the finite difference method exhibits stability. Since $\eta(M) = max_r \left|\frac{6-4(1+3\Gamma)Sin^2\left(\frac{r\pi}{2(N+1)}\right)}{6-4(1-3\Gamma)Sin^2\left(\frac{r\pi}{2(N+1)}\right)}\right| < 1, \forall\Gamma \geq 0$.

Hence the proposed finite difference method is unconditionally stable.

## 3.7 Numerical illustrations and discussions

To test the efficiency of the proposed numerical scheme, we applied the scheme to an example of an IFPPDE.

$$\left(D_t - D_x^2\right)\tilde{W} = \tilde{0}, \qquad x \in [0, 1], \qquad t > 0. \tag{3.40}$$

Subject to boundary conditions are $\tilde{W}(0, t) = \tilde{W}(1, t) = 0$, and the initial condition is $\tilde{W}(x, 0) = \tilde{g}(x) = \tilde{\kappa}\cos\left(\pi x - \frac{\pi}{2}\right)$. Here $\tilde{\kappa}$ is an intuitionistic triangular fuzzy number. The parametric form of the $\tilde{\kappa}$ in term of $(\boldsymbol{\alpha}, \boldsymbol{\beta})$-level form is given as Man et al. (2021):

$$\tilde{\kappa}[\alpha, \beta] = <\left[\underline{\kappa}(\alpha), \overline{\kappa}(\alpha)\right], \left[\underline{\kappa}(\beta), \overline{\kappa}(\beta)\right]> = <[\alpha - 1, 1 - \alpha], [-1.25\beta, 1.25\beta]>$$

The exact solution for

$$\frac{\partial \underline{W}}{\partial t}(x, t, \alpha) = \frac{\partial \underline{W}}{\partial x^2}(x, t, \alpha), \tag{3.41}$$

$$\frac{\partial \overline{W}}{\partial t}(x, t, \alpha) = \frac{\partial \overline{W}}{\partial x^2}(x, t, \alpha), \tag{3.42}$$

$$\frac{\partial \underline{W}'}{\partial t}(x, t, \beta) = \frac{\partial \underline{W}'}{\partial x^2}(x, t, \beta), \tag{3.43}$$

$$\frac{\partial \overline{W}'}{\partial t}(x, t, \beta) = \frac{\partial \overline{W}'}{\partial x^2}(x, t, \beta), \tag{3.44}$$

where $x \in [0, 1]$ and $t > 0$ are, respectively.

$$\underline{W}(x, t; \alpha) = \underline{\kappa}(\alpha)e^{-\pi^2 t}\cos\left(\pi x - \frac{\pi}{2}\right), \tag{3.45}$$

$$\overline{W}(x,t;\alpha) = \overline{\kappa}(\alpha)\, e^{-\pi^2 t} cos\left(\pi x - \frac{\pi}{2}\right), \tag{3.46}$$

$$\underline{W}'(x,t;\beta) = \underline{\kappa}(\beta)\, e^{-\pi^2 t} cos\left(\pi x - \frac{\pi}{2}\right), \tag{3.47}$$

$$\overline{W}'(x,t;\beta) = \overline{\kappa}(\beta)\, e^{-\pi^2 t} cos\left(\pi x - \frac{\pi}{2}\right). \tag{3.48}$$

The above four equations can also be written like

$$\underline{W}(x,t;\alpha) = (\alpha - 1)\, e^{-\pi^2 t} cos\left(\pi x - \frac{\pi}{2}\right), \tag{3.49}$$

$$\overline{W}(x,t;\alpha) = (1 - \alpha)\, e^{-\pi^2 t} cos\left(\pi x - \frac{\pi}{2}\right), \tag{3.50}$$

$$\underline{W}'(x,t;\beta) = (-1.25\beta)\, e^{-\pi^2 t} cos\left(\pi x - \frac{\pi}{2}\right), \tag{3.51}$$

$$\overline{W}'(x,t;\beta) = (1.25\beta)\, e^{-\pi^2 t} cos\left(\pi x - \frac{\pi}{2}\right). \tag{3.52}$$

Now we will implement the finite difference, which we derived using the cubic spline method to approximate the IFPPDEs. Let us consider the step size along $x$ axis is $h = 0.1$ and along $t$ axis is $k = 0.000001$ for the following system of finite difference schemes:

$$\begin{aligned}(1-3\Gamma)\,\underline{W}_{i-1,j+1} + (4+6\Gamma)\,\underline{W}_{i,j+1} + (1-3\Gamma)\,\underline{W}_{i+1,j+1}\\ = (1+3\Gamma)\,\underline{W}_{i-1,j} + (4-6\Gamma)\,\underline{W}_{i,j} + (1+3\Gamma)\,\underline{W}_{i+1,j},\end{aligned} \tag{3.53}$$

$$\begin{aligned}(1-3\Gamma)\,\overline{W}_{i-1,j+1} + (4+6\Gamma)\,\overline{W}_{i,j+1} + (1-3\Gamma)\,\overline{W}_{i+1,j+1}\\ = (1+3\Gamma)\,\overline{W}_{i-1,j} + (4-6\Gamma)\,\overline{W}_{i,j} + (1+3\Gamma)\,\overline{W}_{i+1,j},\end{aligned} \tag{3.54}$$

$$\begin{aligned}(1-3\Gamma)\,\underline{W}'_{i-1,j+1} + (4+6\Gamma)\,\underline{W}'_{i,j+1} + (1-3\Gamma)\,\underline{W}'_{i+1,j+1}\\ = (1+3\Gamma)\,\underline{W}'_{i-1,j} + (4-6\Gamma)\,\underline{W}'_{i,j} + (1+3\Gamma)\,\underline{W}'_{i+1,j},\end{aligned} \tag{3.55}$$

$$\begin{aligned}(1-3\Gamma)\,\overline{W}'_{i-1,j+1} + (4+6\Gamma)\,\overline{W}'_{i,j+1} + (1-3\Gamma)\,\overline{W}'_{i+1,j+1}\\ = (1+3\Gamma)\,\overline{W}'_{i-1,j} + (4-6\Gamma)\,\overline{W}'_{i,j} + (1+3\Gamma)\,\overline{W}'_{i+1,j},\end{aligned} \tag{3.56}$$

where $\Gamma = \frac{k}{h^2}$. After solving the above system of finite difference schemes, we obtained the following results Table 3.1, Table 3.2, Table 3.3, Table 3.4, Table 3.5, Table 3.6, Table 3.7, Table 3.8.

Fig. 3.1 compares the intuitionistic fuzzy cubic spline's solution $W_{1,1}$ and exact solution W(1,1) for different values of $\alpha$ and $\beta$ levels at h = 0.1, k = 0.000001 showing the nature of solutions.

Fig. 3.2 compares the intuitionistic fuzzy cubic spline's solution $W_{2,1}$ and exact solution W(2,1) for different values of $\alpha$ and $\beta$ levels at h = 0.1, k = 0.000001 showing the nature of solutions.

Tables 3.1–3.4 give a detailed explanation of the solution of the IFPPDE for the approximate solution $W_{1,1}$ and exact solution $W(1,1)$ at each upper and lower $(\boldsymbol{\alpha}, \boldsymbol{\beta})$ levels

**Table 3.1** Comparison of the intuitionistic fuzzy heat equation's approximate solutions $W_{1,1}$ and the exact solutions W(1,1) at different lower $\alpha$ levels at h = 0.1 and k = 0.000001.

| Sr. no | $\alpha$ levels | $W_{1,1}$ lower | Exact $W(1,1)$ lower | Absolute error |
|---|---|---|---|---|
| 1 | 0 | −0.309013919333556 | −0.309013944514510 | 0.0000000251809540 |
| 2 | 0.1 | −0.27811252740020 | −0.278112550063059 | 0.0000000226628590 |
| 3 | 0.2 | −0.247211135466844 | −0.247211155611608 | 0.0000000201447640 |
| 4 | 0.3 | −0.216309743533489 | −0.216309761160157 | 0.0000000176266680 |
| 5 | 0.4 | −0.185408351600133 | −0.185408366708706 | 0.0000000151085730 |
| 6 | 0.5 | −0.154506959666778 | −0.154506972257255 | 0.0000000125904770 |
| 7 | 0.6 | −0.123605567733422 | −0.123605577805804 | 0.0000000100723820 |
| 8 | 0.7 | −0.092704175800067 | −0.092704183354353 | 0.0000000075542860 |
| 9 | 0.8 | −0.061802783866711 | −0.061802788902902 | 0.0000000050361910 |
| 10 | 0.9 | −0.030901391933356 | −0.030901394451451 | 0.0000000025180950 |
| 11 | 1 | 0 | 0 | 0.0000000000000000 |

**Table 3.2** Comparison of the intuitionistic fuzzy heat equation's approximate solutions $W_{1,1}$ and the exact solutions W(1,1) at different upper $\alpha$ levels at h = 0.1 and k = 0.000001.

| Sr. no | $\alpha$ levels | $W_{1,1}$ upper | Exact $W(1,1)$ upper | Absolute error |
|---|---|---|---|---|
| 1 | 0 | 0.309013919333556 | 0.309013944514510 | 0.0000000251809540 |
| 2 | 0.1 | 0.27811252740020 | 0.278112550063059 | 0.0000000226628590 |
| 3 | 0.2 | 0.247211135466844 | 0.247211155611608 | 0.0000000201447640 |
| 4 | 0.3 | 0.216309743533489 | 0.216309761160157 | 0.0000000176266680 |
| 5 | 0.4 | 0.185408351600133 | 0.185408366708706 | 0.0000000151085730 |
| 6 | 0.5 | 0.154506959666778 | 0.154506972257255 | 0.0000000125904770 |
| 7 | 0.6 | 0.123605567733422 | 0.123605577805804 | 0.0000000100723820 |
| 8 | 0.7 | 0.092704175800067 | 0.092704183354353 | 0.0000000075542860 |
| 9 | 0.8 | 0.061802783866711 | 0.061802788902902 | 0.0000000050361910 |
| 10 | 0.9 | 0.030901391933356 | 0.030901394451451 | 0.0000000025180950 |
| 11 | 1 | 0 | 0 | 0.0000000000000000 |

at $\boldsymbol{h} = 0.1$ and, $\boldsymbol{k} = 0.000001$ (see Table 3.1, Table 3.2, Table 3.3, Table 3.4). These tables also give the absolute error of the approximate solution at each $(\boldsymbol{\alpha}, \boldsymbol{\beta})$ levels. It is also clear from Tables 3.1–3.4 that the solution at each $\alpha$ level gives seven decimal places of accuracy where the solution at each $\beta$ levels give five decimal places of accuracy. Hence the explicit cubic spline for the finite difference method is an effective method that can be used to approximate the solution of the parabolic partial differential equation. Similarly, Tables 3.5–3.8 gives good approximate solutions for $W(2,1)$ and $W_{2,1}$ as the same accuracy as we got for $W(1,1)$ and $W_{1,1}$ (see Table 3.5, Table 3.6, Table 3.7, Table 3.8).

Since our problem contains the initial and boundary values in terms of the intuitionistic triangular fuzzy number, Figs. 3.1 and 3.2 represent the nature and properties of the fuzzy solutions (see Fig. 3.1, Fig. 3.2). Both figures represent the solution to the problem

**Table 3.3** Comparison of the intuitionistic fuzzy heat equation's approximate solutions $W_{1,1}$ and the exact solutions W(1,1) at different lower $\beta$ levels at h = 0.1 and k = 0.000001.

| Sr. no | $\beta$ levels | $W_{1,1}$ lower | Exact $W(1,1)$ lower | Absolute error |
|---|---|---|---|---|
| 1 | 0 | 0 | 0 | 0 |
| 2 | 0.1 | −0.038626739916694 | −0.038626361835522 | 0.000000378081172 |
| 3 | 0.2 | −0.077253479833389 | −0.077252723671043 | 0.000000756162346 |
| 4 | 0.3 | −0.115880219750083 | −0.115879085506565 | 0.000001134243518 |
| 5 | 0.4 | −0.154506959666778 | −0.154505447342087 | 0.000001512324691 |
| 6 | 0.5 | −0.193133699583472 | −0.193131809177609 | 0.000001890405863 |
| 7 | 0.6 | −0.231760439500167 | −0.231758171013130 | 0.000002268487037 |
| 8 | 0.7 | −0.270387179416861 | −0.270384532848652 | 0.000002646568209 |
| 9 | 0.8 | −0.309013919333556 | −0.309010894684174 | 0.000003024649382 |
| 10 | 0.9 | −0.347640659250250 | −0.347637256519695 | 0.000003402730555 |
| 11 | 1 | −0.386267399166945 | −0.386263618355217 | 0.000003780811728 |

**Table 3.4** Comparison of the intuitionistic fuzzy heat equation's approximate solutions $W_{1,1}$ and the exact solutions W(1,1) at different upper $\beta$ levels at h = 0.1 and k = 0.000001.

| Sr. no | $\beta$ levels | $W_{1,1}$ upper | Exact $W(1,1)$ upper | Absolute error |
|---|---|---|---|---|
| 1 | 0 | 0 | 0 | 0 |
| 2 | 0.1 | 0.038626739916694 | 0.038626361835522 | 0.000000378081172 |
| 3 | 0.2 | 0.077253479833389 | 0.077252723671043 | 0.000000756162346 |
| 4 | 0.3 | 0.115880219750083 | 0.115879085506565 | 0.000001134243518 |
| 5 | 0.4 | 0.154506959666778 | 0.154505447342087 | 0.000001512324691 |
| 6 | 0.5 | 0.193133699583472 | 0.193131809177609 | 0.000001890405863 |
| 7 | 0.6 | 0.231760439500167 | 0.231758171013130 | 0.000002268487037 |
| 8 | 0.7 | 0.270387179416861 | 0.270384532848652 | 0.000002646568209 |
| 9 | 0.8 | 0.309013919333556 | 0.309010894684174 | 0.000003024649382 |
| 10 | 0.9 | 0.347640659250250 | 0.347637256519695 | 0.000003402730555 |
| 11 | 1 | 0.386267399166945 | 0.386263618355217 | 0.000003780811728 |

considered in terms of the intuitionistic triangular fuzzy number. Also we can observe that solution width or fuzziness of the solution increases as we move from $W_{1,1}$ to $W_{2,1}$ (see Fig. 3.1, Fig. 3.2).

## 3.8 Conclusion

This study offers a comprehensive numerical analysis of the one-dimensional intuitionistic fuzzy parabolic partial differential equations (IFPPDE) using an implicit cubic spline method. A detailed convergence analysis confirms the reliability of the proposed method. Moreover, the implicit cubic spline method, which generates fuzzy finite differences, is

**Table 3.5** Comparison of the intuitionistic fuzzy heat equation's approximate solutions $W_{2,1}$ and the exact solutions W(2,1) at different lower $\alpha$ levels at h = 0.1 and k = 0.000001.

| Sr. no | $\alpha$ levels | $W_{2,1}$ lower | Exact $W$ (2, 1) lower | Absolute error |
|---|---|---|---|---|
| 1 | 0 | −0.587779403216166 | −0.587779451113188 | 0.00000004789702202 |
| 2 | 0.1 | −0.529001462894550 | −0.529001506001869 | 0.00000004310731905 |
| 3 | 0.2 | −0.470223522572933 | −0.470223560890550 | 0.00000003831761702 |
| 4 | 0.3 | −0.411445582251316 | −0.411445615779232 | 0.00000003352791600 |
| 5 | 0.4 | −0.352667641929700 | −0.352667670667913 | 0.00000002873821298 |
| 6 | 0.5 | −0.293889701608083 | −0.293889725556594 | 0.00000002394851101 |
| 7 | 0.6 | −0.235111761286466 | −0.235111780445275 | 0.00000001915880901 |
| 8 | 0.7 | −0.176333820964850 | −0.176333835333956 | 0.00000001436910599 |
| 9 | 0.8 | −0.117555880643233 | −0.117555890222638 | 0.00000000957940501 |
| 10 | 0.9 | −0.058777940321617 | −0.058777945111319 | 0.00000000478970200 |
| 11 | 1 | 0 | 0 | 0.00000000000000000 |

**Table 3.6** Comparison of the intuitionistic fuzzy heat equation's approximate solutions $W_{2,1}$ and the exact solutions W(2,1) at different upper $\alpha$ levels at h = 0.1 and k = 0.000001.

| Sr. no | $\alpha$ levels | $W_{2,1}$ upper | Exact $W$ (2, 1) upper | Absolute error |
|---|---|---|---|---|
| 1 | 0 | 0.587779403216166 | 0.587779451113188 | 0.00000004789702202 |
| 2 | 0.1 | 0.529001462894550 | 0.529001506001869 | 0.00000004310731905 |
| 3 | 0.2 | 0.470223522572933 | 0.470223560890550 | 0.00000003831761702 |
| 4 | 0.3 | 0.411445582251316 | 0.411445615779232 | 0.00000003352791600 |
| 5 | 0.4 | 0.352667641929700 | 0.352667670667913 | 0.00000002873821298 |
| 6 | 0.5 | 0.293889701608083 | 0.293889725556594 | 0.00000002394851101 |
| 7 | 0.6 | 0.235111761286466 | 0.235111780445275 | 0.00000001915880901 |
| 8 | 0.7 | 0.176333820964850 | 0.176333835333956 | 0.00000001436910599 |
| 9 | 0.8 | 0.117555880643233 | 0.117555890222638 | 0.00000000957940501 |
| 10 | 0.9 | 0.058777940321617 | 0.058777945111319 | 0.00000000478970200 |
| 11 | 1 | 0 | 0 | 0.00000000000000000 |

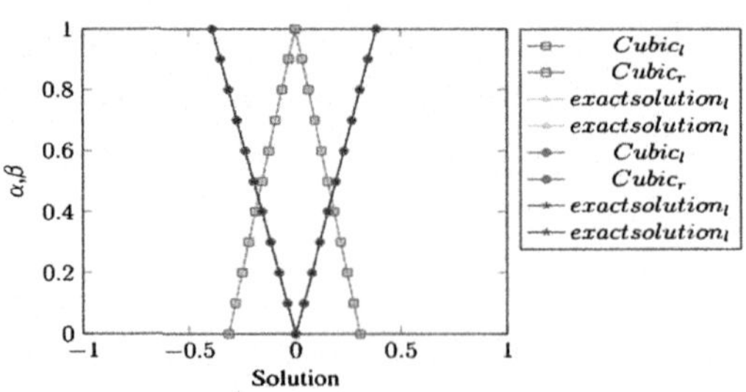


**FIGURE 3.1** Comparison of intuitionistic fuzzy cubic spline's solution $W_{1,1}$ and exact solutions W(1,1) at different $\alpha$ and $\beta$ levels at h = 0.1, k = 0.000001.

**Table 3.7** Comparison of the intuitionistic fuzzy heat equation's approximate solutions $W_{2,1}$ and the exact solutions W(2,1) at different lower $\beta$ levels at h = 0.1 and k = 0.000001.

| Sr. no | $\beta$ levels | $W_{2,1}$ lower | Exact $W(2,1)$ lower | Absolute error |
|---|---|---|---|---|
| 1 | 0 | 0 | 0 | 0.00000000000000000 |
| 2 | 0.1 | −0.073472425402021 | −0.073471706248895 | 0.00000071915312599 |
| 3 | 0.2 | −0.146944850804041 | −0.146943412497789 | 0.00000143830625199 |
| 4 | 0.3 | −0.220417276206062 | −0.220415118746684 | 0.00000215745937798 |
| 5 | 0.4 | −0.293889701608083 | −0.293886824995579 | 0.00000287661250398 |
| 6 | 0.5 | −0.367362127010104 | −0.367358531244474 | 0.00000359576563003 |
| 7 | 0.6 | −0.440834552412125 | −0.440830237493368 | 0.00000431491875696 |
| 8 | 0.7 | −0.514306977814145 | −0.514301943742263 | 0.00000503407188202 |
| 9 | 0.8 | −0.587779403216166 | −0.587773649991158 | 0.00000575322500795 |
| 10 | 0.9 | −0.661251828618187 | −0.661245356240052 | 0.00000647237813500 |
| 11 | 1 | −0.734724254020208 | −0.734717062488947 | 0.00000719153126105 |

**Table 3.8** Comparison of the intuitionistic fuzzy heat equation's approximate solutions $W_{2,1}$ and the exact solutions W(2,1) at different upper $\beta$ levels at h = 0.1 and k = 0.000001.

| Sr. no | $\beta$ levels | $W_{2,1}$ upper | Exact $W(2,1)$ upper | Absolute error |
|---|---|---|---|---|
| 1 | 0 | 0 | 0 | 0.00000000000000000 |
| 2 | 0.1 | 0.073472425402021 | 0.073471706248895 | 0.00000071915312599 |
| 3 | 0.2 | 0.146944850804041 | 0.146943412497789 | 0.00000143830625199 |
| 4 | 0.3 | 0.220417276206062 | 0.220415118746684 | 0.00000215745937798 |
| 5 | 0.4 | 0.293889701608083 | 0.293886824995579 | 0.00000287661250398 |
| 6 | 0.5 | 0.367362127010104 | 0.367358531244474 | 0.00000359576563003 |
| 7 | 0.6 | 0.440834552412125 | 0.440830237493368 | 0.00000431491875696 |
| 8 | 0.7 | 0.514306977814145 | 0.514301943742263 | 0.00000503407188202 |
| 9 | 0.8 | 0.587779403216166 | 0.587773649991158 | 0.00000575322500795 |
| 10 | 0.9 | 0.661251828618187 | 0.661245356240052 | 0.00000647237813500 |
| 11 | 1 | 0.734724254020208 | 0.734717062488947 | 0.00000719153126105 |

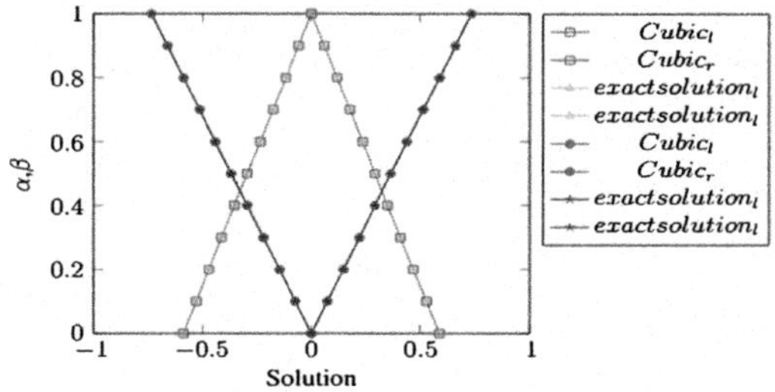


**FIGURE 3.2** Comparison of intuitionistic fuzzy cubic spline's solution $W_{2,1}$ and exact solutions W(2,1) at different $\alpha$ and $\beta$ levels at h = 0.1, k = 0.000001.

proven to be unconditionally stable through the matrix stability method. The effectiveness of the approach is further demonstrated through a numerical example, where solutions at various levels are compared using graphical and tabular representations. These findings underscore the robustness and accuracy of the method in handling intuitionistic fuzzy uncertainty, making it a valuable tool for solving IFPPDEs.

Future research will likely focus on improving stability, accuracy, and computational efficiency, further solidifying the relevance and applicability of IFDEs across various scientific disciplines.

## References

Akın, Ö., Bayeğ, S., 2019. Intuitionistic fuzzy initial value problems-an application. Hacettepe Journal of Mathematics and Statistics (ISSN 2651-477X) 48 (6), 1682–1694. https://doi.org/10.15672/HJMS.2018.598. Hacettepe University, Turkey, Available from: https://dergipark.org.tr/en/pub/hujms/issue/50516/656668.

Allahviranloo, Tofigh, 2002. Difference methods for fuzzy partial differential equations. Computational Methods in Applied Mathematics (ISSN 1609-9389) 2 (3), 233–242. https://doi.org/10.2478/cmam-2002-0014.

Amma, B.B., Melliani, S., Chadli, L.S., 2020. Intuitionistic fuzzy partial functional differential equations with integral boundary conditions. In: 2020 IEEE 6th International Conference on Optimisation and Applications (ICOA), pp. 1–5.

Atanassov, K.T., 2012. On intuitionistic fuzzy sets theory. Studies in Fuzziness and Soft Computing (ISSN 1434-9922) 283, 1–334. https://doi.org/10.1007/978-3-642-29127-2.

Man, S., Saw, B.C., Hazra, S.B., 2021. An explicit method for solving intuitionistic fuzzy heat equation. Advances in Fuzzy Mathematics 16, 27–39.

Man, S., Saw, B.C., Bairagi, A., Hazra, S.B., 2023. Finite difference method for intuitionistic fuzzy partial differential equations. Computer Sciences and Mathematics Forum 7 (1), 1–6.

Melliani, S., Chadli, L.S., 2000. Intuitionistic fuzzy differential equation. Notes on Intuitionistic Fuzzy Sets 6 (2), 37–41.

Mondal, S.P., Roy, T.K., 2015. System of differential equation with initial value as triangular intuitionistic fuzzy number and its application. International Journal of Applied and Computational Mathematics (ISSN 2199-5796) 1 (3), 449–474. https://doi.org/10.1007/s40819-015-0026-x. Springer, India, Available from: https://link.springer.com/journal/40819.

Nirmala, V., Parimala, V., Rajarajeswari, P., 2018. Application of Runge-Kutta method for finding multiple numerical solutions to intuitionistic fuzzy differential equations. Journal of Physics. Conference Series (ISSN 1742-6588) 1139 (1), 012012. https://doi.org/10.1088/1742-6596/1139/1/012012.

Sastry, Shankar S., 2012. Introductory Methods of Numerical Analysis. PHI Learning Pvt. Ltd.

Smith, G.D., 1985. Numerical Solution of Partial Differential Equations: Finite Difference Methods. Oxford University Press.

Zadeh, L.A., 1965. Fuzzy sets. Information and Control (ISSN 0019-9958) 8 (3), 338–353. https://doi.org/10.1016/s0019-9958(65)90241-x.

# 4

# On the analysis of the three-species food chain model involving fractional derivatives with nonsingular kernels

Rajarama Mohan Jena[a], Subrat Kumar Jena[b], and S. Chakraverty[c]
[a]ITER, Siksha 'O' Anusandhan University, Department of Mathematics, Bhubaneswar, Odisha, India
[b]University of Cagliari, Department of Civil, Environmental and Architecture, Cagliari, Italy [c]National Institute of Technology Rourkela, Department of Mathematics, Rourkela, India

Fractional differential equations (FDEs) have gained enough prominence over the past few years due to their wide applications in various fields, such as vibration (R.M. Jena and Chakraverty, 2019c), option pricing (R.M. Jena and Chakraverty, 2019a), damped beam (Jena et al., 2019c), Bagley-Torvik model (R.M. Jena and Chakraverty, 2019b), marriage model (Jena et al., 2019b), Shallow water wave (Jena et al., 2019a), vibration model (Rajarama Mohan Jena and Chakraverty, 2019e), etc. Fractional partial differential equations (FPDEs) are essential for understanding many disciplines, such as biochemistry, biology, diffusion processes, fluid flows, etc. The most popular topic in ecology is the complex relationship between predator and prey.

In 1925–1926, a model describing the population dynamics between the two prey and predator species in differential equations (DEs) was proposed by Italian mathematicians Lotka and Volterra (Volterra, 1926; Lotka, 1925). Two trophic levels (Dubey and Upadhyay, 2004) of standard three-species food chain models (FCMs) are inadequate to achieve realistic dynamics. In recent years, FCMs with tritrophic levels have gained much attention from many mathematicians and ecologists. The mouse-snake-owl and worm-robin-falcon are examples of three-species ecosystems. The chaotic behavior of three species in the food chain models has been studied by Klebanoff, Hastings, and Powell (Hastings and Powell, 1991; Klebanoff and Hastings, 1994). Some researchers have analyzed the food chain system for three species, which involves two functional reactions (Baek et al., 2011) and Lotka-Volterra three FCM (Chauvet et al., 2002). Merdan (Merdan and Khaniyev, 2008; Merdan, 2010) has employed the variational iteration method and the homotopy perturbation method (HPM) to achieve the approximate solution for the three-species food chain model. The differential transformation scheme is used for the three species

Advanced Computational and Mathematical Approaches in Applied Differential Equations
https://doi.org/10.1016/B978-0-44-345743-2.00015-3

FCMs with the functional response from Holling type II and the functional response from Beddington-DeAngelis (Merdan et al., 2011). Rao and Narayan have studied the stability of a three-species food chain model (TFCM) with harvesting (Rao and Narayan, 2017). Wu et al. (2010) examined the chaos regulation and synchronization of a TFCM with Holling functional response. Maiti et al. (2006) investigated the complex nature of the FCM with a mixed type of functional response, such as stability, persistence, and bifurcation. In terms of stability analysis, Sanjaya et al. (2012) performed an investigation on the mathematical model of the interaction of the TFCM with Holling type-III functional response (Sanjaya et al., 2013) and with a mixed functional response. In 2016, Zibaei and Namjoo (2016) introduced a nonstandard finite difference scheme to solve fractional-order Lotka-Volterra FCM. Guo and Fang (Guo and Fang, 2016) addressed the stability and numerical solutions of three competitors' fractional Lotka-Volterra equations by implementing HPM.

In recent years, fractional calculus has extensively used inherited properties, dissipative effects, and memory in real-world problems. The best definitions of fractional calculus are Riemann–Liouville (RL) and Caputo fractional operators, defined by the convolution and the power decay function as kernel (Baleanu et al., 2012). Several mathematicians have extended the principles of fractional differential and integral operators to the fields related to science and engineering using power-law distribution. Nevertheless, this power-law distribution has no statistical significance when the fractional order is less than 1. The fractional differential operators based on the power-law kernel meet certain classical conditions, including index law, classical mechanical law, and singular kernels (Caputo and Fabrizio, 2015; Atangana and Baleanu, 2016; Atangana and Gómez-Aguilar, 2018; Jena et al., 2021; Saad et al., 2018). It suggests that those operators based on power law are physically weak and cannot deal with more complex phenomena. Another problem is the singularity, which means objects cannot be described mathematically. Hence it is a challenge to explain many natural phenomena with singularity.

The recent development of fractional calculus has driven the development of two key operators, namely Caputo–Fabrizio and Atangana–Baleanu. These modern derivatives do not work under a power distribution but have nonsingular kernels. A generalized Mittag-Leffler function is used for Atangana–Baleanu derivative, and the Caputo–Fabrizio operator relies on the exponential law. This research aims to find solutions for the titled problem using the newly introduced fractional derivative and demonstrate the existence, uniqueness, and stability by applying the fixed-point theorem defined in CFC and ABC sense. The article also discusses contact analysis of prey, mid-predator, and top predator based on time and fractional parameter values through graphs.

## 4.1 Preliminaries

This segment addresses the basic definitions and notation of fractional calculus.

**Definition 4.1.1.** The fractional derivative of a function $u(x,t)$ defined in the Caputo sense is written as (Podlubny, 1999)

$$ {}_{0}^{C}D_{t}^{\alpha}u(x,t) = \frac{1}{\Gamma(m-\alpha)}\int_{0}^{t}(t-\tau)^{m-\alpha-1}u^{m}(x,\tau)\,d\tau, \qquad m-1<\alpha<m, \qquad m\in N. \tag{4.1} $$

If $0 < \alpha \leq 1$, then the Laplace transform (LT) for the Liouville-Caputo fractional derivatives is defined as follows:

$$ L\left\{{}_{0}^{C}D_{t}^{\alpha}u(x,t)\right\}(s) = s^{\alpha}L\{u(x,t)\}(s) - s^{\alpha-1}\{u(x,0)\}. \tag{4.2} $$

**Definition 4.1.2.** The fractional derivatives in Caputo–Fabrizio-Caputo sense (CFC) is defined as (Caputo and Fabrizio, 2015)

$$ {}^{CFC}_{\ \ \ 0}D_{t}^{\alpha}u(x,t) = \frac{C(\alpha)}{1-\alpha}\int_{0}^{t}exp\left(-\alpha\frac{(t-\tau)}{1-\alpha}\right)u'(x,\tau)\,d\tau, \qquad 0<\alpha\leq 1, \tag{4.3} $$

where $C(\alpha)$ is a normalization function and $C(0) = C(1) = 1$.

The following integral is derived from Nieto and Losada (Losada and Nieto, 2015).

**Definition 4.1.3.** Let $0 < \alpha < 1$. The Caputo–Fabrizio integral operator is defined as

$$ {}^{CF}_{\ \ 0}I_{t}^{\alpha}(u(t)) = \frac{2(1-\alpha)}{(2-\alpha)C(\alpha)}u(t) + \frac{2\alpha}{(2-\alpha)C(\alpha)}\int_{0}^{t}u(\tau)\,d\tau, t\geq 0. \tag{4.4} $$

**Note 4.1.1.** According to Nieto and Losada (Losada and Nieto, 2015), the following results must hold:

$$ \frac{2(1-\alpha)}{(2-\alpha)C(\alpha)}u(t) + \frac{2\alpha}{(2-\alpha)C(\alpha)} = 1, \tag{4.5} $$

which implies that $C(\alpha) = \frac{2}{2-\alpha}$, $0 < \alpha < 1$.

**Definition 4.1.4.** The Atangana–Baleanu fractional derivative in Caputo sense (ABC) is defined as (Atangana and Baleanu, 2016)

$$ {}^{ABC}_{\ \ \ \ 0}D_{t}^{\alpha}u(x,t) = \frac{A(\alpha)}{(m-\alpha)}\int_{0}^{t}E_{\alpha}\left(-\alpha\frac{(t-\tau)^{\alpha}}{m-\alpha}\right)u^{m}(x,\tau)\,d\tau, \qquad m-1<\alpha<m, \tag{4.6} $$

where $A(\alpha) = 1 - \alpha + \frac{\alpha}{\Gamma(\alpha)}$ and $E_{\alpha}(t) = \sum_{i=0}^{\infty}\frac{t^{i}}{\Gamma(1+i\alpha)}$ is the Mittag-Leffler function (MLF).

**Definition 4.1.5.** The Atangana–Baleanu fractional integral of order $\alpha$ is defined as (Atangana and Baleanu, 2016)

$$ {}^{AB}_{\ \ 0}I_{t}^{\alpha}(u(t)) = \frac{(1-\alpha)}{A(\alpha)}u(t) + \frac{\alpha}{\Gamma(\alpha)A(\alpha)}\int_{0}^{t}u(\tau)(t-\tau)^{\alpha-1}\,d\tau. \tag{4.7} $$

**Definition 4.1.6.** Suppose $0 < \alpha \leq 1$, then the LT of CFC and ABC fractional derivatives are defined respectively as (Caputo and Fabrizio, 2015; Atangana and Baleanu, 2016)

$$ L\left\{{}^{CFC}_{\ \ \ 0}D_{t}^{\alpha}u(x,t)\right\}(s) = C(\alpha)\left(\frac{sL\{u(x,t)\}(s) - u(x,0)}{s+\alpha(1-s)}\right), \tag{4.8} $$

and

$$L\left\{{}^{ABC}_{\ 0}D_t^{\alpha}u(x,t)\right\}(s) = A(\alpha)\left(\frac{s^{\alpha}L\{u(x,t)\}(s) - s^{\alpha-1}u(x,0)}{s^{\alpha}(1-\alpha)+\alpha}\right). \tag{4.9}$$

**Definition 4.1.7** (Elzaki transform). Elzaki transform (ET) of a function $u(x,t)$ is defined as (Elzaki et al., 2012)

$$E[u(x,t)](s) = E(s) = s\int_0^{\infty} u(x,t)e^{\frac{-t}{s}}dt, \qquad t>0. \tag{4.10}$$

**Theorem 4.1.1** (Laplace and Elzaki transforms duality). *Let $F(s)$ be the Laplace transform of $u(x,t)$, then ET $E(s)$ of $u(x,t)$ is given as (R.M. Jena and Chakraverty, 2019d)*

$$F(s) = sE\left(\frac{1}{s}\right), \tag{4.11}$$

*or*

$$E(s) = sF\left(\frac{1}{s}\right). \tag{4.12}$$

In the following theorems, we discuss the ET of ABC and CFC fractional derivatives:

**Theorem 4.1.2.** *The ET of CFC fractional derivatives is given as*

$$E\left\{{}^{CFC}_{\ \ 0}D_t^{\alpha}u(x,t)\right\}(s) = C(\alpha)\left(\frac{E(s) - s^2u(x,0)}{1+\alpha(s-1)}\right). \tag{4.13}$$

*Proof.* Using Eqs. (4.8) and (4.11), we obtain

$$E\left\{{}^{CFC}_{\ \ 0}D_t^{\alpha}u(x,t)\right\}(s) = sC(\alpha)\left(\frac{\frac{1}{s}F\left(\frac{1}{s}\right) - u(x,0)}{\frac{1}{s}+\alpha\left(1-\frac{1}{s}\right)}\right) = C(\alpha)\left(\frac{\frac{1}{s}E(s) - su(x,0)}{\frac{1}{s}+\alpha\left(1-\frac{1}{s}\right)}\right).$$

From the above equation, we have

$$E\left\{{}^{CFC}_{\ \ 0}D_t^{\alpha}u(x,t)\right\}(s) = C(\alpha)\left(\frac{E(s) - s^2u(x,0)}{1+\alpha(s-1)}\right).$$

□

**Theorem 4.1.3.** *The ET of ABC fractional derivatives may be written as*

$$E\left\{{}^{ABC}_{\ 0}D_t^{\alpha}u(x,t)\right\}(s) = A(\alpha)\left(\frac{E(s) - s^2u(x,0)}{1+\alpha(s^{\alpha}-1)}\right). \tag{4.14}$$

*Proof.* Using Eqs. (4.9) and (4.11), we get

$$E\left\{{}^{ABC}_{0}D^{\alpha}_{t}u(x,t)\right\}(s) = sA(\alpha)\left(\frac{\left(\frac{1}{s}\right)^{\alpha}F\left(\frac{1}{s}\right)-\left(\frac{1}{s}\right)^{\alpha-1}u(x,0)}{\left(\frac{1}{s}\right)^{\alpha}(1-\alpha)+\alpha}\right)$$

$$= A(\alpha)\left(\frac{\left(\frac{1}{s}\right)^{\alpha}E(s)-\left(\frac{1}{s}\right)^{\alpha-1}su(x,0)}{\left(\frac{1}{s}\right)^{\alpha}(1-\alpha)+\alpha}\right).$$

The above equation is rewritten as

$$E\left\{{}^{ABC}_{0}D^{\alpha}_{t}u(x,t)\right\}(s) = A(\alpha)\left(\frac{E(s)-s^2u(x,0)}{1+\alpha(s^{\alpha}-1)}\right).$$

□

## 4.2 Mathematical formulation for fractional-order TFCM

The model equations for a fractional-order TFCM are given by Merdan and Khaniyev (2008); Merdan (2010)

$$\begin{cases} D^{\alpha}_{t}\xi(t) = \xi(t)(a_1 - b_1\xi(t) - c_1\psi(t)), \\ D^{\alpha}_{t}\psi(t) = \psi(t)(-a_2 + b_2\xi(t) - d_1\zeta(t)), \\ D^{\alpha}_{t}\zeta(t) = \zeta(t)(-a_3 + c_2\psi(t)), \end{cases} \tag{4.1}$$

with initial conditions

$$\xi(0) = \xi_0,\ \psi(0) = \psi_0, \quad \text{and} \quad \zeta(0) = \zeta_0, \tag{4.2}$$

where $0 < \alpha \le 1$ is the order of the fractional derivative. Here $\xi(t)$, $\psi(t)$, $\zeta(t)$ signify the population densities of prey, mid-predator, top-predator, respectively, and parameters $a_1$, $b_1$, $c_1$, $a_2$, $b_2$, $d_1$, $a_3$, and $c_2$ are positive constants. As population densities of these species are positive, so the solutions of Eq. (4.1) will be restricted to nonnegative, that is, $\{(\xi,\psi,\zeta)\,|\,\xi \ge 0,\ \psi \ge 0, \zeta \ge 0\}$. Hence considering this, the above system is to be solved with proper initial conditions as $\xi(0) \ge 0$, $\psi(0) \ge 0$, and $\zeta(0) \ge 0$. The overhead structure $a_1$ defines the rate of growth of the prey population $\xi$; $b_1$ represents the rate of prey decrease due to inner competition; $c_1$ is the per capita mid-predator predation rate $\psi$; $a_2$ denotes the mid-predator $\psi$ natural mortality rate in the absence of prey $\xi$; $b_2$ is the conversion rate of a consumed prey into a mid-predator $\psi$; $d_1$ is per capita predation rate of the top-predator $\zeta$; $a_3$ represents the natural death rate of top-predator, and $c_2$ signifies the rate of conversion of a consumed predator to a top-predator.

In this study, we have considered $\xi + \psi + \zeta$ as constant. Let us take $B$ to be the Banach space of real-valued continuous function on the interval $P$ with the norm $\|(\xi,\psi,\zeta)\| =$

$\|\xi\| + \|\psi\| + \|\zeta\|$, where $\|\xi\| = sup\{|\xi(t)| : t \in P\}$, $\|\psi\| = sup\{|\psi(t)| : t \in P\}$, and $\|\zeta\| = sup\{|\zeta(t)| : t \in P\}$. Mainly $B = C(p) \times C(p)$, where $C(p)$ is the Banach space of continuous real-valued function on $P$ having supremum norm.

### 4.2.1 The TFCM in the exponential law kernel sense

The fractional-order TFCM in CFC sense is written as

$$\begin{cases} {}^{CFC}_{0}D_t^{\alpha}\xi(t) = \xi(t)(a_1 - b_1\xi(t) - c_1\psi(t)), \\ {}^{CFC}_{0}D_t^{\alpha}\psi(t) = \psi(t)(-a_2 + b_2\xi(t) - d_1\zeta(t)), \\ {}^{CFC}_{0}D_t^{\alpha}\zeta(t) = \zeta(t)(-a_3 + c_2\psi(t)), \end{cases} \tag{4.3}$$

subject to the initial condition Eq. (4.2), where ${}^{CFC}_{0}D_t^{\alpha}$ is the fractional derivative of CFC type, and $\alpha$ with $0 < \alpha \leq 1$. In this section, we show the existence of the present system by using the fixed-point theory. By using the Losada and Nieto (Losada and Nieto, 2015) fractional integral operator on Eq. (4.3), we get

$$\begin{aligned} \xi(t) - \xi(0) &= {}^{CF}_{0}I_t^{\alpha}\left\{a_1\xi(t) - b_1(\xi(t))^2 - c_1\xi(t)\psi(t)\right\}, \\ \psi(t) - \psi(0) &= {}^{CF}_{0}I_t^{\alpha}\{-a_2\psi(t) + b_2\xi(t)\psi(t) - d_1\zeta(t)\psi(t)\}, \\ \zeta(t) - \zeta(0) &= {}^{CF}_{0}I_t^{\alpha}\{-a_3\zeta(t) + c_2\psi(t)\zeta(t)\}. \end{aligned} \tag{4.4}$$

Using the result given in Eq. (4.4), we have

$$\begin{aligned} \xi(t) &= \xi_0 + \frac{2(1-\alpha)}{(2-\alpha)C(\alpha)}\left\{a_1\xi(t) - b_1(\xi(t))^2 - c_1\xi(t)\psi(t)\right\} \\ &\quad + \frac{2\alpha}{(2-\alpha)C(\alpha)}\int_0^t \left\{a_1\xi(\tau) - b_1(\xi(\tau))^2 - c_1\xi(\tau)\psi(\tau)\right\}d\tau, \\ \psi(t) &= \psi_0 + \frac{2(1-\alpha)}{(2-\alpha)C(\alpha)}\{-a_2\psi(t) + b_2\xi(t)\psi(t) - d_1\zeta(t)\psi(t)\} + \\ \zeta(t) &= \zeta_0 + \frac{2(1-\alpha)}{(2-\alpha)C(\alpha)}\{-a_3\zeta(t) + c_2\psi(t)\zeta(t)\} \\ &\quad + \frac{2\alpha}{(2-\alpha)C(\alpha)}\int_0^t \{-a_3\zeta(\tau) + c_2\psi(\tau)\zeta(\tau)\}d\tau. \end{aligned} \tag{4.5}$$

We now consider the following kernels:

$$\begin{aligned} \phi_1(t,\xi) &= a_1\xi - b_1\xi^2 - c_1\xi\psi, \\ \phi_2(t,\psi) &= -a_2\psi + b_2\xi\psi - d_1\zeta\psi, \\ \phi_3(t,\zeta) &= -a_3\zeta + c_2\psi\zeta. \end{aligned} \tag{4.6}$$

**Theorem 4.2.1.** *The kernels $\phi_1$, $\phi_2$, and $\phi_3$ satisfy the conditions and contractions of Lipschitz if the inequality $0 \leq a_1 + 2k_1b_1 + k_2c_1 < 1$ holds.*

*Proof.* Let us start with $\phi_1$. Suppose $\xi$ and $\xi_1$ are two functions, then we obtain

$$\|\phi_1(t,\xi)-\phi_1(t,\xi_1)\| = \left\|a_1\xi(t)-b_1\xi^2(t)-c_1\xi(t)\psi(t)-a_1\xi_1(t)+b_1\xi_1^2(t)+c_1\xi_1(t)\psi(t)\right\|. \tag{4.7}$$

Applying Cauchy's inequality on Eq. (4.7), we get

$$\begin{aligned}
\|\phi_1(t,\xi)-\phi_1(t,\xi_1)\| &\le \|a_1\{\xi(t)-\xi_1(t)\}\| + \left\|b_1\left\{\xi^2(t)-\xi_1^2(t)\right\}\right\| + \|c_1\psi(t)\{\xi(t)-\xi_1(t)\}\|,\\
&\le a_1\|\xi(t)-\xi_1(t)\| + b_1\|\xi(t)-\xi_1(t)\|\,\|\xi(t)+\xi_1(t)\| + c_1\psi(t)\|\xi(t)-\xi_1(t)\|,\\
&\le a_1\|\xi(t)-\xi_1(t)\| + b_1\|\xi(t)-\xi_1(t)\|\,\|\xi(t)+\xi_1(t)\| + c_1\|\psi(t)\|\,\|\xi(t)-\xi_1(t)\|,\\
&\le a_1\|\xi(t)-\xi_1(t)\| + b_1\|\xi(t)-\xi_1(t)\|\,(\|\xi(t)\|+\|\xi_1(t)\|) + c_1\|\psi(t)\|\,\|\xi(t)-\xi_1(t)\|.
\end{aligned} \tag{4.8}$$

Assuming $\|\xi(t)\| \le k_1$, $\|\xi_1(t)\| \le k_1$ $\|\psi(t)\| \le k_2$, and $\|\zeta(t)\| \le k_3$ are bounded functions, we have

$$\|\phi_1(t,\xi)-\phi_1(t,\xi_1)\| \le \{a_1+2k_1b_1+k_2c_1\}\|\xi(t)-\xi_1(t)\| = \gamma_1\|\xi(t)-\xi_1(t)\|, \tag{4.9}$$

where $\gamma_1 = a_1 + 2k_1b_1 + k_2c_1$. So, $\phi_1$ satisfies the Lipschitz condition. If we restrict $0 \le a_1 + 2k_1b_1 + k_2c_1 < 1$, then it will also be a contraction. Similarly $\phi_2$ and $\phi_3$ will satisfy the Lipschitz condition, which is given below:

$$\begin{aligned}
\|\phi_2(t,\psi)-\phi_2(t,\psi_1)\| &\le \gamma_2\|\psi(t)-\psi_1(t)\|,\\
\|\phi_3(t,\zeta)-\phi_3(t,\zeta_1)\| &\le \gamma_3\|\zeta(t)-\zeta_1(t)\|.
\end{aligned} \tag{4.10}$$

Substituting Eq. (4.6) into Eq. (4.5), we obtain

$$\begin{aligned}
\xi(t) &= \xi_0 + \frac{2(1-\alpha)}{(2-\alpha)C(\alpha)}\phi_1(t,\xi) + \frac{2\alpha}{(2-\alpha)C(\alpha)}\int_0^t \phi_1(\tau,\xi)\,d\tau,\\
\psi(t) &= \psi_0 + \frac{2(1-\alpha)}{(2-\alpha)C(\alpha)}\phi_2(t,\psi) + \frac{2\alpha}{(2-\alpha)C(\alpha)}\int_0^t \phi_2(\tau,\psi)\,d\tau,\\
\zeta(t) &= \zeta_0 + \frac{2(1-\alpha)}{(2-\alpha)C(\alpha)}\phi_3(t,\zeta) + \frac{2\alpha}{(2-\alpha)C(\alpha)}\int_0^t \phi_3(\tau,\zeta)\,d\tau.
\end{aligned} \tag{4.11}$$

The following recursive expressions are obtained from Eq. (4.11):

$$\begin{aligned}
\xi_n(t) &= \frac{2(1-\alpha)}{(2-\alpha)C(\alpha)}\phi_1(t,\xi_{n-1}) + \frac{2\alpha}{(2-\alpha)C(\alpha)}\int_0^t \phi_1(\tau,\xi_{n-1})\,d\tau,\\
\psi_n(t) &= \frac{2(1-\alpha)}{(2-\alpha)C(\alpha)}\phi_2(t,\psi_{n-1}) + \frac{2\alpha}{(2-\alpha)C(\alpha)},\\
\zeta_n(t) &= \frac{2(1-\alpha)}{(2-\alpha)C(\alpha)}\phi_3(t,\zeta_{n-1}) + \frac{2\alpha}{(2-\alpha)C(\alpha)}\int_0^t \phi_3(\tau,\zeta_{n-1})\,d\tau,
\end{aligned} \tag{4.12}$$

with the initial conditions Eq. (4.2).

The difference between the two successive terms may be expressed as

$$\begin{aligned}
\Phi_{1(n)}(t) &= \xi_n(t) - \xi_{n-1}(t) \\
&= \frac{2(1-\alpha)}{(2-\alpha)C(\alpha)}\{\phi_1(t,\xi_{n-1}) - \phi_1(t,\xi_{n-2})\} \\
&\quad + \frac{2\alpha}{(2-\alpha)C(\alpha)}\int_0^t \{\phi_1(\tau,\xi_{n-1}) - \phi_1(\tau,\xi_{n-2})\}d\tau,
\end{aligned}$$

$$\Phi_{2(n)}(t) = \psi_n(t) - \psi_{n-1}(t) = \frac{2(1-\alpha)}{(2-\alpha)C(\alpha)}\{\phi_2(t,\psi_{n-1}) - \phi_2(t,\psi_{n-2})\} + \tag{4.13}$$

$$\begin{aligned}
\Phi_{3(n)}(t) &= \zeta_n(t) - \zeta_{n-1}(t) \\
&= \frac{2(1-\alpha)}{(2-\alpha)C(\alpha)}\{\phi_3(t,\zeta_{n-1}) - \phi_3(t,\zeta_{n-2})\} \\
&\quad + \frac{2\alpha}{(2-\alpha)C(\alpha)}\int_0^t \{\phi_3(\tau,\zeta_{n-1}) - \phi_3(\tau,\zeta_{n-2})\}d\tau.
\end{aligned}$$

It is worth mentioning that

$$\xi_n(t) = \sum_{i=0}^{\infty}\Phi_{1(i)},\ \psi_n(t) = \sum_{i=0}^{\infty}\Phi_{2(i)},\quad \text{and}\quad \zeta_n(t) = \sum_{i=0}^{\infty}\Phi_{3(i)}. \tag{4.14}$$

Now from Eq. (4.14), we have

$$\left\|\Phi_{1(n)}(t)\right\| = \left\|\xi_n(t) - \xi_{n-1}(t)\right\| = \left\|\begin{array}{l}\frac{2(1-\alpha)}{(2-\alpha)C(\alpha)}\{\phi_1(t,\xi_{n-1}) - \phi_1(t,\xi_{n-2})\} + \\ \frac{2\alpha}{(2-\alpha)C(\alpha)}\int_0^t \{\phi_1(\tau,\xi_{n-1}) - \phi_1(\tau,\xi_{n-2})\}d\tau\end{array}\right\|. \tag{4.15}$$

Applying triangular inequality, Eq. (4.15) gives

$$\begin{aligned}
\left\|\xi_n(t) - \xi_{n-1}(t)\right\| &\le \frac{2(1-\alpha)}{(2-\alpha)C(\alpha)}\left\|\phi_1(t,\xi_{n-1}) - \phi_1(t,\xi_{n-2})\right\| \\
&\quad + \frac{2\alpha}{(2-\alpha)C(\alpha)}\left\|\int_0^t \{\phi_1(\tau,\xi_{n-1}) - \phi_1(\tau,\xi_{n-2})\}d\tau\right\|.
\end{aligned} \tag{4.16}$$

Since the kernel $\phi_1$ satisfies the Lipschitz condition, so

$$\left\|\xi_n(t) - \xi_{n-1}(t)\right\| \le \frac{2(1-\alpha)}{(2-\alpha)C(\alpha)}\gamma_1\left\|\xi_{n-1} - \xi_{n-2}\right\| + \frac{2\alpha}{(2-\alpha)C(\alpha)}\gamma_1\int_0^t \left\|\xi_{n-1} - \xi_{n-2}\right\|d\tau. \tag{4.17}$$

Now, we have

$$\left\|\Phi_{1(n)}(t)\right\| \le \frac{2(1-\alpha)}{(2-\alpha)C(\alpha)}\gamma_1\left\|\Phi_{1(n-1)}(t)\right\| + \frac{2\alpha}{(2-\alpha)C(\alpha)}\gamma_1\int_0^t \left\|\Phi_{1(n-1)}(t)\right\|d\tau. \tag{4.18}$$

Likewise, the following results may be obtained:

$$\|\Phi_{2(n)}(t)\| \le \frac{2(1-\alpha)}{(2-\alpha)C(\alpha)}\gamma_2 \|\Phi_{2(n-1)}(t)\| + \frac{2\alpha}{(2-\alpha)C(\alpha)}\gamma_2 \int_0^t \|\Phi_{2(n-1)}(t)\| d\tau,$$

$$\|\Phi_{3(n)}(t)\| \le \frac{2(1-\alpha)}{(2-\alpha)C(\alpha)}\gamma_3 \|\Phi_{3(n-1)}(t)\| + \frac{2\alpha}{(2-\alpha)C(\alpha)}\gamma_3 \int_0^t \|\Phi_{3(n-1)}(t)\| d\tau. \quad (4.19)$$

□

Considering the above results, we have derived the following theorem:

**Theorem 4.2.2.** *The fractional TFCM Eq. (4.3) has an exact solution if there exists a* $t_0$ *such that*

$$\left(\frac{2(1-\alpha)}{(2-\alpha)C(\alpha)}\gamma_1 + \frac{2\alpha}{(2-\alpha)C(\alpha)}\gamma_1 t_0\right) < 1.$$

*Proof.* We know that the functions $\xi(t)$, $\psi(t)$, and $\zeta(t)$ are bounded. Further, we have proved that the kernels follow the Lipschitz condition. Thus taking Eqs. (4.18) and (4.19) into consideration, the following results are derived by using the recursive method:

$$\|\phi_{1(n)}(t)\| \le \|\xi_n(0)\| \left[\frac{2(1-\alpha)}{(2-\alpha)C(\alpha)}\gamma_1 + \frac{2\alpha}{(2-\alpha)C(\alpha)}\gamma_1 t\right]^n,$$

$$\|\phi_{2(n)}(t)\| \le \|\psi_n(0)\| \left[\frac{2(1-\alpha)}{(2-\alpha)C(\alpha)}\gamma_2 + \frac{2\alpha}{(2-\alpha)C(\alpha)}\gamma_2 t\right]^n, \quad (4.20)$$

$$\|\phi_{3(n)}(t)\| \le \|\zeta_n(0)\| \left[\frac{2(1-\alpha)}{(2-\alpha)C(\alpha)}\gamma_3 + \frac{2\alpha}{(2-\alpha)C(\alpha)}\gamma_3 t\right]^n.$$

Therefore the solutions, as mentioned above, exist and are continuous. So to prove that Eq. (4.20) is the solution of Eq. (4.3), we assume

$$\xi(t) - \xi(0) = \xi_n(t) - A_n(t),$$

$$\psi(t) - \psi(0) = \psi_n(t) - B_n(t), \quad (4.21)$$

$$\zeta(t) - \zeta(0) = \zeta_n(t) - C_n(t).$$

Thus we obtain

$$\begin{aligned}\|A_n(t)\| &= \left\| \frac{2(1-\alpha)}{(2-\alpha)C(\alpha)}\{\phi_1(t,\xi) - \phi_1(t,\xi_{n-1})\} \right.\\ &\quad \left. + \frac{2\alpha}{(2-\alpha)C(\alpha)}\int_0^t \{\phi_1(\tau,\xi) - \phi_1(\tau,\xi_{n-1})\} d\tau \right\|,\\ &\le \frac{2(1-\alpha)}{(2-\alpha)C(\alpha)}\|\phi_1(t,\xi) - \phi_1(t,\xi_{n-1})\|\\ &\quad + \frac{2\alpha}{(2-\alpha)C(\alpha)}\int_0^t \|\phi_1(\tau,\xi) - \phi_1(\tau,\xi_{n-1})\| d\tau,\end{aligned}$$

$$\leq \frac{2(1-\alpha)}{(2-\alpha)C(\alpha)}\gamma_1 \|\xi - \xi_{n-1}\| + \frac{2\alpha}{(2-\alpha)C(\alpha)} t\gamma_1 \|\xi - \xi_{n-1}\|. \tag{4.22}$$

Using the above process recursively, we have

$$\|A_n(t)\| \leq \left(\frac{2(1-\alpha)}{(2-\alpha)C(\alpha)} + \frac{2\alpha}{(2-\alpha)C(\alpha)} t\right)^{n+1} \gamma_1^{n+1} k_1. \tag{4.23}$$

Then at $t = t_0$, we get

$$\|A_n(t)\| \leq \left(\frac{2(1-\alpha)}{(2-\alpha)C(\alpha)} + \frac{2\alpha}{(2-\alpha)C(\alpha)} t_0\right)^{n+1} \gamma_1^{n+1} k_1. \tag{4.24}$$

Now, taking a limit $n \to \infty$ in Eq. (4.24), we have

$$\|A_n(t)\| \to 0$$

Similarly, we may get

$$\|B_n(t)\| \to 0 \quad \text{and} \quad \|C_n(t)\| \to 0.$$

Hence it is proved. □

Next, we prove the uniqueness of the system of solution Eq. (4.3).

Let us consider that there exists another system of solution of Eq. (4.3) viz. $\xi_1(t)$, $\psi_1(t)$, and $\zeta_1(t)$, then

$$\xi(t) - \xi_1(t) = \frac{2(1-\alpha)}{(2-\alpha)C(\alpha)} \{\phi_1(t,\xi) - \phi_1(t,\xi_1)\} + . \tag{4.25}$$

Applying the norm on Eq. (4.25), we get

$$\|\xi(t) - \xi_1(t)\| \leq \frac{2(1-\alpha)}{(2-\alpha)C(\alpha)} \|\phi_1(t,\xi) - \phi_1(t,\xi_1)\| + . \tag{4.26}$$

Employing Lipschitz condition on the kernels, we have

$$\|\xi(t) - \xi_1(t)\| \leq \frac{2(1-\alpha)}{(2-\alpha)C(\alpha)}\gamma_1 \|\xi(t) - \xi_1(t)\| + \frac{2\alpha}{(2-\alpha)C(\alpha)}\gamma_1 t \|\xi(t) - \xi_1(t)\|. \tag{4.27}$$

It gives

$$\|\xi(t) - \xi_1(t)\| \left\{1 - \frac{2(1-\alpha)}{(2-\alpha)C(\alpha)}\gamma_1 + \frac{2\alpha}{(2-\alpha)C(\alpha)}\gamma_1 t\right\} \leq 0. \tag{4.28}$$

**Theorem 4.2.3.** *Eq. (4.3) provides a unique solution if the following inequality holds:*

$$\left\{1 - \frac{2(1-\alpha)}{(2-\alpha)C(\alpha)}\gamma_1 + \frac{2\alpha}{(2-\alpha)C(\alpha)}\gamma_1 t\right\} > 0. \tag{4.29}$$

*Proof.* Assume that the condition Eq. (4.29) holds, then from Eq. (4.28), we obtain

$$\|\xi(t)-\xi_1(t)\|\left\{1-\frac{2(1-\alpha)}{(2-\alpha)C(\alpha)}\gamma_1+\frac{2\alpha}{(2-\alpha)C(\alpha)}\gamma_1 t\right\}\leq 0,\quad \text{implies that}$$
$$\|\xi(t)-\xi_1(t)\|=0 \quad \text{or} \quad \xi(t)=\xi_1(t). \tag{4.30}$$

Using the procedure likewise, we obtain

$$\psi(t)=\psi_1(t), \quad \text{and} \quad \zeta(t)=\zeta_1(t). \tag{4.31}$$

Therefore, we verified the uniqueness of solutions of Eq. (4.3). □

### 4.2.2 The TFCM in the Mittag-Leffler kernel sense

The fractional-order TFCM with the Mittag-Leffler kernel is given by

$$\begin{cases} {}^{ABC}_{\ \ \ 0}D_t^{\alpha}\xi(t)=\xi(t)\left(a_1-b_1\xi(t)-c_1\psi(t)\right), \\ {}^{ABC}_{\ \ \ 0}D_t^{\alpha}\psi(t)=\psi(t)\left(-a_2+b_2\xi(t)-d_1\zeta(t)\right), \\ {}^{ABC}_{\ \ \ 0}D_t^{\alpha}\zeta(t)=\zeta(t)\left(-a_3+c_2\psi(t)\right), \end{cases} \tag{4.32}$$

with the initial condition Eq. (4.2), where ${}^{ABC}_{\ \ \ 0}D_t^{\alpha}$ is the ABC type fractional operator, and $\alpha$ with $0<\alpha\leq 1$ is the fractional order. Now, by using Eq. (4.7), Eq. (4.32) may be converted into the following expressions:

$$\begin{aligned} \xi(t)-\xi(0)&=\frac{(1-\alpha)}{A(\alpha)}\left\{a_1\xi(t)-b_1(\xi(t))^2-c_1\xi(t)\psi(t)\right\} \\ &\quad+\frac{\alpha}{\Gamma(\alpha)A(\alpha)}\int_0^t(t-\tau)^{\alpha-1}\left\{a_1\xi(\tau)-b_1(\xi(\tau))^2-c_1\xi(\tau)\psi(\tau)\right\}d\tau, \\ \psi(t)-\psi(0)&=\frac{(1-\alpha)}{A(\alpha)}\left\{-a_2\psi(t)+b_2\xi(t)\psi(t)-d_1\zeta(t)\psi(t)\right\}+ \\ \zeta(t)-\zeta(0)&=\frac{(1-\alpha)}{A(\alpha)}\left\{-a_3\zeta(t)+c_2\psi(t)\zeta(t)\right\} \\ &\quad+\frac{\alpha}{\Gamma(\alpha)A(\alpha)}\int_0^t(t-\tau)^{\alpha-1}\left\{-a_3\zeta(\tau)+c_2\psi(\tau)\zeta(\tau)\right\}d\tau. \end{aligned} \tag{4.33}$$

Let us assume

$$\begin{aligned} \varpi_1(t,\xi)&=a_1\xi-b_1\xi^2-c_1\xi\psi, \\ \varpi_2(t,\psi)&=-a_2\psi+b_2\xi\psi-d_1\zeta\psi, \\ \varpi_3(t,\zeta)&=-a_3\zeta+c_2\psi\zeta \end{aligned} \tag{4.34}$$

**Theorem 4.2.4.** *The kernels $\varpi_1$, $\varpi_2$, and $\varpi_3$ satisfy the Lipschitz condition and contraction if$0\leq\chi_1,\chi_2,\chi_3<1$ holds.*

*Proof.* Let us first start with the kernel $\varpi_1$. Suppose $\xi$ and $\xi_1$ are two functions, then we obtain the following expression:

$$\left\|\varpi_1(t,\xi)-\varpi_1(t,\xi_1)\right\|=\left\|a_1\xi(t)-b_1\xi^2(t)-c_1\xi(t)\psi(t)-a_1\xi_1(t)+b_1\xi_1^2(t)+c_1\xi_1(t)\psi(t)\right\|. \tag{4.35}$$

Applying Cauchy's inequality on Eq. (4.35), we get

$$\begin{aligned}
\left\|\varpi_1(t,\xi)-\varpi_1(t,\xi_1)\right\| &\leq \left\|a_1\{\xi(t)-\xi_1(t)\}\right\|+\left\|b_1\left\{\xi^2(t)-\xi_1^2(t)\right\}\right\|+\left\|c_1\psi(t)\{\xi(t)-\xi_1(t)\}\right\|,\\
&\leq a_1\left\|\xi(t)-\xi_1(t)\right\|+b_1\left\|\xi(t)-\xi_1(t)\right\|\left\|\xi(t)+\xi_1(t)\right\|+c_1\psi(t)\left\|\xi(t)-\xi_1(t)\right\|,\\
&\leq a_1\left\|\xi(t)-\xi_1(t)\right\|+b_1\left\|\xi(t)-\xi_1(t)\right\|\left\|\xi(t)+\xi_1(t)\right\|+c_1\left\|\psi(t)\right\|\left\|\xi(t)-\xi_1(t)\right\|,\\
&\leq a_1\left\|\xi(t)-\xi_1(t)\right\|+b_1\left\|\xi(t)-\xi_1(t)\right\|\left(\left\|\xi(t)\right\|+\left\|\xi_1(t)\right\|\right)+c_1\left\|\psi(t)\right\|\left\|\xi(t)-\xi_1(t)\right\|.
\end{aligned} \tag{4.36}$$

Assuming $\|\xi(t)\|\leq\theta_1$, $\|\xi_1(t)\|\leq\theta_1$, $\|\psi(t)\|\leq\theta_2$, and $\|\zeta(t)\|\leq\theta_3$ are bounded functions, we have

$$\left\|\varpi_1(t,\xi)-\varpi_1(t,\xi_1)\right\|\leq\{a_1+2\theta_1b_1+\theta_2c_1\}\left\|\xi(t)-\xi_1(t)\right\|=\chi_1\left\|\xi(t)-\xi_1(t)\right\|, \tag{4.37}$$

where $\chi_1=a_1+2\theta_1b_1+\theta_2c_1$. So $\varpi_1$ satisfies the Lipschitz condition. If we restrict $0\leq\chi_1<1$, then it will also be a contraction. Similarly, $\varpi_2$ and $\varpi_3$ will satisfy the Lipschitz condition, which is

$$\begin{aligned}
\left\|\varpi_2(t,\psi)-\varpi_2(t,\psi_1)\right\| &\leq \chi_2\left\|\psi(t)-\psi_1(t)\right\|,\\
\left\|\varpi_3(t,\zeta)-\varpi_3(t,\zeta_1)\right\| &\leq \chi_3\left\|\zeta(t)-\zeta_1(t)\right\|.
\end{aligned} \tag{4.38}$$

Substituting Eq. (4.34) into Eq. (4.33), we obtain

$$\begin{aligned}
\xi(t) &= \xi_0+\frac{(1-\alpha)}{A(\alpha)}\varpi_1(t,\xi)+\frac{\alpha}{\Gamma(\alpha)A(\alpha)}\int_0^t(t-\tau)^{\alpha-1}\varpi_1(\tau,\xi)\,d\tau,\\
\psi(t) &= \psi_0+\frac{(1-\alpha)}{A(\alpha)}\varpi_2(t,\psi)+\frac{\alpha}{\Gamma(\alpha)A(\alpha)}\int_0^t(t-\tau)^{\alpha-1}\varpi_2(\tau,\psi)\,d\tau,\\
\zeta(t) &= \zeta_0+\frac{(1-\alpha)}{A(\alpha)}\varpi_3(t,\zeta)+\frac{\alpha}{\Gamma(\alpha)A(\alpha)}\int_0^t(t-\tau)^{\alpha-1}\varpi_3(\tau,\zeta)\,d\tau.
\end{aligned} \tag{4.39}$$

Now we express the following recursive formula:

$$\begin{aligned}
\xi_n(t) &= \frac{(1-\alpha)}{A(\alpha)}\varpi_1(t,\xi_{n-1})+\frac{\alpha}{\Gamma(\alpha)A(\alpha)}\int_0^t(t-\tau)^{\alpha-1}\varpi_1(\tau,\xi_{n-1})\,d\tau,\\
\psi_n(t) &= \frac{(1-\alpha)}{A(\alpha)}\varpi_2(t,\psi_{n-1})+\frac{\alpha}{\Gamma(\alpha)A(\alpha)}\int_0^t(t-\tau)^{\alpha-1}\varpi_2(\tau,\psi_{n-1})\,d\tau,\\
\zeta(t) &= \zeta_0+\frac{(1-\alpha)}{A(\alpha)}\varpi_3(t,\zeta)+\frac{\alpha}{\Gamma(\alpha)A(\alpha)}\int_0^t(t-\tau)^{\alpha-1}\varpi_3(\tau,\zeta)\,d\tau,
\end{aligned} \tag{4.40}$$

with the initial conditions Eq. (4.2).

The difference between the two consecutive terms may be constructed as

$$\begin{aligned}
\Theta_{1(n)}(t) &= \xi_n(t) - \xi_{n-1}(t) \\
&= \frac{(1-\alpha)}{A(\alpha)}\{\varpi_1(t,\xi_{n-1}) - \varpi_1(t,\xi_{n-2})\} \\
&\quad + \frac{\alpha}{\Gamma(\alpha)A(\alpha)}\int_0^t (t-\tau)^{\alpha-1}\{\varpi_1(\tau,\xi_{n-1}) - \varpi_1(\tau,\xi_{n-2})\}d\tau, \\
\Theta_{2(n)}(t) &= \psi_n(t) - \psi_{n-1}(t) = \frac{(1-\alpha)}{A(\alpha)}\{\varpi_2(t,\psi_{n-1}) - \varpi_2(t,\psi_{n-2})\}+, \\
\Theta_{3(n)}(t) &= \zeta_n(t) - \zeta_{n-1}(t) \\
&= \frac{(1-\alpha)}{A(\alpha)}\{\varpi_3(t,\zeta_{n-1}) - \varpi_3(t,\zeta_{n-2})\} \\
&\quad + \frac{\alpha}{\Gamma(\alpha)A(\alpha)}\int_0^t (t-\tau)^{\alpha-1}\{\varpi_3(\tau,\zeta_{n-1}) - \varpi_3(\tau,\zeta_{n-2})\}d\tau,
\end{aligned} \tag{4.41}$$

where

$$\xi_n(t) = \sum_{i=0}^{\infty}\Theta_{1(i)},\ \psi_n(t) = \sum_{i=0}^{\infty}\Theta_{2(i)},\quad \text{and}\quad \zeta_n(t) = \sum_{i=0}^{\infty}\Phi_{3(i)}. \tag{4.42}$$

Now from Eq. (4.41), we have

$$\left\|\Theta_{1(n)}(t)\right\| = \|\xi_n(t) - \xi_{n-1}(t)\| = \left\| \begin{array}{l} \frac{(1-\alpha)}{A(\alpha)}\{\varpi_1(t,\xi_{n-1}) - \varpi_1(t,\xi_{n-2})\} + \\ \frac{\alpha}{\Gamma(\alpha)A(\alpha)}\int_0^t (t-\tau)^{\alpha-1}\{\varpi_1(\tau,\xi_{n-1}) - \varpi_1(\tau,\xi_{n-2})\}d\tau \end{array} \right\|. \tag{4.43}$$

With the help of triangular inequality, Eq. (4.43) reduces to

$$\begin{aligned}
\|\xi_n(t) - \xi_{n-1}(t)\| &\le \frac{(1-\alpha)}{A(\alpha)}\|\varpi_1(t,\xi_{n-1}) - \varpi_1(t,\xi_{n-2})\| \\
&\quad + \frac{\alpha}{\Gamma(\alpha)A(\alpha)}\left\|\int_0^t (t-\tau)^{\alpha-1}\{\varpi_1(\tau,\xi_{n-1}) - \varpi_1(\tau,\xi_{n-2})\}d\tau\right\|.
\end{aligned} \tag{4.44}$$

As the kernel $\varpi_1$ satisfies the Lipschitz condition, so

$$\|\xi_n(t) - \xi_{n-1}(t)\| \le \frac{(1-\alpha)}{A(\alpha)}\chi_1\|\xi_{n-1} - \xi_{n-2}\| + \frac{t^\alpha}{\Gamma(\alpha)A(\alpha)}\chi_1. \tag{4.45}$$

Now, from Eq. (4.45), we have

$$\left\|\Theta_{1(n)}(t)\right\| \le \frac{(1-\alpha)}{A(\alpha)}\chi_1\left\|\Theta_{1(n-1)}(t)\right\| + \frac{t^\alpha}{\Gamma(\alpha)A(\alpha)}\chi_1\int_0^t \left\|\Theta_{1(n-1)}(t)\right\|d\tau. \tag{4.46}$$

Similarly, the following results may be obtained:

$$\left\|\Theta_{2(n)}(t)\right\| \leq \frac{(1-\alpha)}{A(\alpha)}\chi_2\left\|\Theta_{2(n-1)}(t)\right\| + \frac{t^{\alpha}}{\Gamma(\alpha)A(\alpha)}\chi_2\int_0^t\left\|\Theta_{2(n-1)}(t)\right\|d\tau,$$

$$\left\|\Theta_{3(n)}(t)\right\| \leq \frac{(1-\alpha)}{A(\alpha)}\chi_3\left\|\Theta_{3(n-1)}(t)\right\| + \frac{t^{\alpha}}{\Gamma(\alpha)A(\alpha)}\chi_3\int_0^t\left\|\Theta_{3(n-1)}(t)\right\|d\tau. \tag{4.47}$$

□

Considering the above results, we have derived the following theorem:

**Theorem 4.2.5.** *The fractional TFCM Eq. (4.32) has a solution if there exists a* $t_0$ *such that*

$$\left(\frac{(1-\alpha)}{A(\alpha)}\chi_i + \frac{t_0^{\alpha}}{\Gamma(\alpha)A(\alpha)}\chi_i\right) < 1, \quad fori = 1, 2, 3$$

*Proof.* Let $\xi(t)$, $\psi(t)$, and $\zeta(t)$ are bounded functions that satisfy the Lipschitz condition. Now, from Eqs. (4.46) and (4.47), we have

$$\left\|\Theta_{1(n)}(t)\right\| \leq \|\xi_n(0)\|\left[\frac{(1-\alpha)}{A(\alpha)}\chi_1 + \frac{t^{\alpha}}{\Gamma(\alpha)A(\alpha)}\chi_1\right]^n,$$

$$\left\|\Theta_{2(n)}(t)\right\| \leq \|\psi_n(0)\|\left[\frac{(1-\alpha)}{A(\alpha)}\chi_2 + \frac{t^{\alpha}}{\Gamma(\alpha)A(\alpha)}\chi_2\right]^n, \tag{4.48}$$

$$\left\|\Theta_{3(n)}(t)\right\| \leq \|\zeta_n(0)\|\left[\frac{(1-\alpha)}{A(\alpha)}\chi_3 + \frac{t^{\alpha}}{\Gamma(\alpha)A(\alpha)}\chi_3\right]^n$$

The solution, as mentioned above, therefore exists and is continuous. So to show that the system Eq. (4.48) is a solution to Eq. (4.32), we consider

$$\xi(t) - \xi(0) = \xi_n(t) - \Omega_n(t),$$

$$\psi(t) - \psi(0) = \psi_n(t) - \Delta_n(t), \tag{4.49}$$

$$\zeta(t) - \zeta(0) = \zeta_n(t) - \vartheta_n(t).$$

Thus we obtain

$$\|\Omega_n(t)\| = \left\|\frac{(1-\alpha)}{A(\alpha)}\{\varpi_1(t,\xi) - \varpi_1(t,\xi_{n-1})\} + \frac{\alpha}{\Gamma(\alpha)A(\alpha)}\int_0^t\{\varpi_1(\tau,\xi) - \varpi_1(\tau,\xi_{n-1})\}(t-\tau)^{\alpha-1}d\tau\right\|,$$

$$\leq \frac{(1-\alpha)}{A(\alpha)}\|\varpi_1(t,\xi) - \varpi_1(t,\xi_{n-1})\| + \frac{t^{\alpha}}{\Gamma(\alpha)A(\alpha)}\int_0^t\|\varpi_1(\tau,\xi) - \varpi_1(\tau,\xi_{n-1})\|d\tau,$$

$$\leq \frac{(1-\alpha)}{A(\alpha)}\chi_1\|\xi - \xi_{n-1}\| + \frac{t^{\alpha}}{\Gamma(\alpha)A(\alpha)}\chi_1\|\xi - \xi_{n-1}\|. \tag{4.50}$$

Following the above process recursively, we have

$$\|\Omega_n(t)\| \leq \left(\frac{(1-\alpha)}{A(\alpha)} + \frac{t^\alpha}{\Gamma(\alpha)A(\alpha)}\right)^{n+1} \chi_1^{n+1}\theta_1. \tag{4.51}$$

Then at $t = t_0$, we get

$$\|\Omega_n(t)\| \leq \left(\frac{(1-\alpha)}{A(\alpha)(\alpha)} + \frac{t_0^\alpha}{\Gamma(\alpha)A(\alpha)}\right)^{n+1} \chi_1^{n+1}\theta_1. \tag{4.52}$$

Now, taking a limit $n \to \infty$ in Eq. (4.52), we have

$$\|\Omega_n(t)\| \to 0$$

Similarly, we may get

$$\|\Delta_n(t)\| \to 0 \quad \text{and} \quad \|\vartheta_n(t)\| \to 0.$$

Hence the existence of the solution is proved. □

Now, we need to show the uniqueness of the solution of the considered model Eq. (4.32).

Let us assume the presence of another system of the solution of Eq. (4.32) viz. $\xi^*(t)$, $\psi^*(t)$, and $\zeta^*(t)$, then

$$\begin{aligned}\xi(t) - \xi^*(t) = {} & \frac{(1-\alpha)}{A(\alpha)}\left\{\varpi_1(t,\xi) - \varpi_1\left(t,\xi^*\right)\right\} \\ & + \frac{\alpha}{\Gamma(\alpha)A(\alpha)}\int_0^t \left\{\varpi_1(t,\xi) - \varpi_1\left(t,\xi^*\right)\right\}(t-\tau)^{\alpha-1}\,d\tau.\end{aligned} \tag{4.53}$$

Applying the norm on Eq. (4.53), we get

$$\left\|\xi(t) - \xi^*(t)\right\| \leq \frac{(1-\alpha)}{A(\alpha)}\left\|\varpi_1(t,\xi) - \varpi_1\left(t,\xi^*\right)\right\| + \frac{\alpha}{\Gamma(\alpha)A(\alpha)}, \tag{4.54}$$

$$\left\|\xi(t) - \xi^*(t)\right\| \leq \frac{(1-\alpha)}{A(\alpha)}\left\|\varpi_1(t,\xi) - \varpi_1\left(t,\xi^*\right)\right\| + \frac{t^\alpha}{\Gamma(\alpha)A(\alpha)}\int_0^t \left\|\varpi_1(t,\xi) - \varpi_1\left(t,\xi^*\right)\right\| d\tau. \tag{4.55}$$

It gives

$$\left\|\xi(t) - \xi^*(t)\right\| \leq \frac{\chi_1}{A(\alpha)}\left\{(1-\alpha) + \frac{t^\alpha}{\Gamma(\alpha)}\right\}\left\|\xi(t) - \xi^*(t)\right\|, \tag{4.56}$$

which leads to

$$\left\|\xi(t) - \xi^*(t)\right\|\left(1 - \frac{\chi_1}{A(\alpha)}\left((1-\alpha) + \frac{t^\alpha}{\Gamma(\alpha)}\right)\right) \leq 0. \tag{4.57}$$

**Theorem 4.2.6.** *Eq. (4.32) gives a unique solution if the following inequality holds*

$$\left(1 - \frac{\chi_1}{A(\alpha)}\left((1-\alpha) + \frac{t^\alpha}{\Gamma(\alpha)}\right)\right) > 0. \tag{4.58}$$

*Proof.* Assume that the condition Eq. (4.50) holds, then from Eq. (4.57), we obtain

$$\left\|\xi(t)-\xi^*(t)\right\|\left(1-\frac{\chi_1}{A(\alpha)}\left((1-\alpha)+\frac{t^\alpha}{\Gamma(\alpha)}\right)\right)\le 0, \quad \text{that implies}$$
$$\left\|\xi(t)-\xi^*(t)\right\|=0 \quad \text{or} \quad \xi(t)=\xi^*(t). \tag{4.59}$$

Similarly, we may obtain

$$\psi(t)=\psi^*(t), \quad \text{and} \quad \zeta(t)=\zeta^*(t). \tag{4.60}$$

Thus the uniqueness of the system of solutions of Eq. (4.32) is verified. □

## 4.3 Equilibria of time-fractional order TFCM and their stability

To calculate the equilibria of system Eq. (4.1), we have to compute

$$D_t^\alpha \xi(t)=0,\ D_t^\alpha \psi(t)=0, \quad \text{and} \quad D_t^\alpha \zeta(t)=0.$$

Thus the equilibrium points are

$$E_0(0,0,0),\ E_1\left(\frac{a_1}{b_1},0,0\right),\ E_2\left(0,\frac{a_3}{c_2},\frac{a_2}{d_1}\right),\ E_3\left(\frac{a_2}{b_2},\frac{a_1b_2-a_2b_1}{b_2c_1},0\right), \quad \text{and} \quad E_4\left(\xi^*,\psi^*,\zeta^*\right),$$

where $\xi^*=\frac{a_1c_2-a_3c_1}{b_1c_2}$, $\psi^*=\frac{a_3}{c_2}$, and $\zeta^*=\frac{a_1b_2c_2-a_2b_1c_2-a_3b_2c_1}{b_1c_2d_1}$, $\zeta^*$ is feasible if $c_2>\frac{a_3b_2c_1}{a_1b_2-a_2b_1}$.
Now the Jacobian matrix of Eq. (4.1) at any point $(\xi,\psi,\zeta)$ is written as

$$J(\xi,\psi,\zeta)=\begin{bmatrix} -c_1\psi-2b_1\xi+a_1 & -c_1\xi & 0 \\ b_2\psi & -d_1\zeta+b_2\xi-a_2 & -d_1\psi \\ 0 & c_2\zeta & c_2\psi-a_3 \end{bmatrix}$$

**Theorem 4.3.1.** *The equilibrium point $E_0$ of Eq. (4.1) is always a saddle point.*

*Proof.* The Jacobian matrix at $E_0(0,0,0)$ is given by

$$J(E_0)=\begin{bmatrix} a_1 & 0 & 0 \\ 0 & -a_2 & 0 \\ 0 & 0 & -a_3 \end{bmatrix}$$

Eigenvalues of the above matrix are obtained by solving the characteristic equation.

$$P(\lambda)=-(\lambda+a_3)(\lambda+a_2)(a_1-\lambda)$$

Eigenvalues of the Jacobian matrix are given as $\lambda_1=a_1>0$, $\lambda_2=-a_2<0$, and $\lambda_3=-a_3<0$. Thus $|arg(\lambda_1)|=0<\frac{\alpha\pi}{2}$, $|arg(\lambda_2)|=\pi>\frac{\alpha\pi}{2}$, $|arg(\lambda_3)|=\pi>\frac{\alpha\pi}{2}$. The convergence of the Mittag-Leffler function results the equilibrium point $E_0$ as a saddle point. □

**Theorem 4.3.2.** *The equilibrium point $E_1$ of Eq. (4.1) is locally asymptotically stable if $b_2 < \frac{a_2 b_1}{a_1}$.*

*Proof.* The Jacobian matrix of system (4.1) at the equilibrium point $E_1$ is written as

$$J(E_1) = \begin{bmatrix} -a_1 & -\dfrac{a_1 c_1}{b_1} & 0 \\ 0 & \dfrac{a_1 b_2 - a_2 b_1}{b_1} & 0 \\ 0 & 0 & -a_3 \end{bmatrix}$$

$E_1$ is locally asymptotically stable if $|arg(\lambda_i)| > \frac{\alpha\pi}{2}$, for $i = 1, 2, 3$. So the eigenvalues of the Jacobian matrix are $\lambda_1 = -a_3 < 0$, $\lambda_2 = -a_1 < 0$, and $\lambda_3 = \frac{a_1 b_2 - a_2 b_1}{b_1}$.

Hence $|arg(\lambda_1)| = \pi > \frac{\alpha\pi}{2}$, $|arg(\lambda_2)| = \pi > \frac{\alpha\pi}{2}$. If $b_2 < \frac{a_2 b_1}{a_1}$, then $\lambda_3 < 0$ such that $|arg(\lambda_3)| = \pi > \frac{\alpha\pi}{2}$. This completes the proof. □

Furthermore, to analyze the stability of the equilibrium point $E_4(\xi^*, \psi^*, \zeta^*)$, we have to calculate $J(E_4)$ which is as follows:

$$J(E_4) = \begin{bmatrix} \dfrac{c_1 a_3 - a_1 c_2}{c_2} & \dfrac{-(a_1 c_2 - c_1 a_3) c_1}{b_1 c_2} & 0 \\ \dfrac{a_3 b_2}{c_2} & 0 & \dfrac{-a_3 d_1}{c_2} \\ 0 & \dfrac{a_1 b_2 c_2 - a_2 b_1 c_2 - a_3 b_2 c_1}{b_1 d_1} & 0 \end{bmatrix}$$

One may obtain eigenvalues of the matrix by solving the following characteristic equation

$$P(\lambda) = \lambda^3 + k_1 \lambda^2 + k_2 \lambda + k_3, \quad \text{where}$$

$$k_1 = \frac{a_1 c_2 - c_1 a_3}{c_2}, \; k_2 = \frac{a_1 b_2 c_1 c_2 + a_1 b_2 c_2^2 - a_2 b_1 c_2^2 - a_3 b_2 c_1^2 - a_3 b_2 c_1 c_2}{b_1 c_2^2},$$

$$k_3 = \frac{(a_1 b_2 c_2 - a_2 b_1 c_2 - a_3 b_2 c_1) a_3 (a_1 c_2 - c_1 a_3)}{b_1 c_2^2}$$

Let $\Lambda(p)$ is the discriminant of the polynomial $P(\lambda)$, then we can write discriminant as

$$\Lambda(p) = 18 k_1 k_2 k_3 - 27 k_3^2 - 4 a_1^3 a_3 + a_1^2 a_2^2 - 4 a_2^3$$

**Proposition 4.3.3.** *The equilibrium point $E_4$ is asymptotically stable if one of the following conditions (Ahmed et al., 2007) holds good:*

1. $\Lambda(p) > 0$, $k_1 > 0$, $k_3 > 0$, *and* $k_1 k_2 > k_3$;
2. $\Lambda(p) < 0$, $k_1 \geq 0$, $k_2 \geq 0$, $k_3 > 0$, *and* $\alpha < \frac{2}{3}$;
3. $\Lambda(p) < 0$, $k_1 > 0$, $k_2 > 0$, $k_1 k_2 = k_3$, *and* $\forall \alpha \in (0, 1)$.

## 4.4 Algorithm for HPETM with CFC fractional-order derivative

To explain the detailed procedure of this method, let us consider the following expression:

$$ {}^{CFC}_{0}D_t^{\alpha}\xi(x,t) + R\xi(x,t) + N\xi(x,t) = \eta(x,t), \qquad 0 < \alpha \leq 1, \tag{4.1} $$

with initial condition (IC):

$$ \xi(x,0) = g(x), \tag{4.2} $$

where ${}^{CFC}_{0}D_t^{\alpha} = \frac{\partial^{\alpha}}{\partial t^{\alpha}}$ is the differential operator, $R$, $N$ are the linear and nonlinear differential operators, $\eta(x,t)$ is known function, and $\xi(x,t)$ is an unknown function. The HPETM approach involves mainly the following steps.

**Step 1.** Applying ET on both sides of Eq. (4.1), we get

$$ C(\alpha)\left(\frac{E(\xi(x,t)) - s^2\xi(x,0)}{1+\alpha(s-1)}\right) = E[\eta(x,t) - R\xi(x,t) - N\xi(x,t)], \tag{4.3} $$

$$ E(\xi(x,t)) = s^2\xi(x,0) + \left(\frac{1+\alpha(s-1)}{C(\alpha)}\right)\{E[\eta(x,t)] - E[R\xi(x,t) + N\xi(x,t)]\}. \tag{4.4} $$

Using IC Eq. (4.2), Eq. (4.4) reduces to

$$ E(\xi(x,t)) = s^2 g(x) + \left(\frac{1+\alpha(s-1)}{C(\alpha)}\right)\{E[\eta(x,t)] - E[R\xi(x,t) + N\xi(x,t)]\}. \tag{4.5} $$

**Step 2.** Taking the inverse ET operator on Eq. (4.5), we obtain

$$ \xi(x,t) = H(x,t) - E^{-1}\left[\left(\frac{1+\alpha(s-1)}{C(\alpha)}\right)\{E(R\xi(x,t) + N\xi(x,t))\}\right], \tag{4.6} $$

where $H(x,t)$ denotes the term obtained from the known function and IC by taking inverse ET.

**Step 3.** To apply HPM, we first express the solution as powers series in terms of homotopy parameter $p \in [0, 1]$ as

$$ \boldsymbol{\xi}(\boldsymbol{x},\boldsymbol{t}) = \sum_{n=0}^{\infty} \boldsymbol{p}^n \boldsymbol{\xi}_n(\boldsymbol{x},\boldsymbol{t}) \tag{4.7} $$

and the nonlinear term is decomposed as

$$ N\xi(x,t) = \sum_{n=0}^{\infty} p^n H_n(\xi), \tag{4.8} $$

where $H_n(\xi)$ is the He's polynomials (R.M. Jena and Chakraverty, 2019d), and their values can be computed by using the following expression:

$$H_n(\xi_0, \xi_1, \xi_2, .., \xi_n) = \frac{1}{n!}\frac{\partial^n}{\partial p^n}\left[N\sum_{j=0}^{n} p^j \xi_j\right]_{p=0}, \qquad n = 0, 1, 2. \tag{4.9}$$

**Step 4.** Substituting Eqs. (4.7) and (4.8) into Eq. (4.6) and using HPM, the following expression is obtained:

$$\sum_{n=0}^{\infty} p^n \xi_n(x,t) = H(x,t) - p\left[E^{-1}\left[\left(\frac{1+\alpha(s-1)}{C(\alpha)}\right)\left\{E\left(R\sum_{n=0}^{\infty} p^n \xi_n(x,t) + \sum_{n=0}^{\infty} p^n H_n(\xi)\right)\right\}\right]\right]. \tag{4.10}$$

Comparing the coefficient of the same power of p of Eq. (4.10) gives the following expressions:

$$p^0 : \xi_0(x,t) = H(x,t),$$

$$p^1 : \xi_1(x,t) = -E^{-1}\left[\left(\frac{1+\alpha(s-1)}{C(\alpha)}\right)E\{R\xi_0(x,t) + H_0(\xi)\}\right],$$

$$p^2 : \xi_2(x,t) = -E^{-1}\left[\left(\frac{1+\alpha(s-1)}{C(\alpha)}\right)E\{R\xi_1(x,t) + H_1(\xi)\}\right],$$

$$\vdots$$

Continuing the procedure, likewise, all the values $\{\xi_n(x,t)\}_{n=0}^{\infty}$ may be calculated. So the nth-~order approximate solution of Eqs. (4.1) and (4.2) may be written as

$$\tilde{\xi}_n(x,t) = \sum_{k=0}^{n} \xi_k(x,t). \tag{4.11}$$

Suppose taking $n \to \infty$, if we observe $\tilde{\xi}_n(x,t)$ converges to a particular solution, then that solution will be the exact solution of Eqs. (4.1) and (4.2).

## 4.5 Algorithm for HPETM with ABC fractional-order derivative

Similar to Section 4.4, again, let us consider the following expression:

$${}^{ABC}_{\ \ 0}D_t^{\alpha}\xi(x,t) + R\xi(x,t) + N\xi(x,t) = \eta(x,t), \qquad 0 < \alpha \le 1, \tag{4.1}$$

subject to IC:

$$\xi(x,0) = g(x). \tag{4.2}$$

**Step 1.** Applying ET on Eq. (4.1) and using IC, we get

$$A(\alpha)\left(\frac{E(\xi(x,t)) - s^2\xi(x,0)}{1+\alpha(s^\alpha - 1)}\right) = E[\eta(x,t) - R\xi(x,t) - N\xi(x,t)], \tag{4.3}$$

$$E(\xi(x,t)) = s^2 g(x) + \left(\frac{1+\alpha(s^\alpha - 1)}{A(\alpha)}\right)\{E[\eta(x,t)] - E[R\xi(x,t) + N\xi(x,t)]\}. \tag{4.4}$$

**Step 2.** Applying the inverse ET operator on Eq. (4.4), it gives

$$\xi(x,t) = H(x,t) - E^{-1}\left[\left(\frac{1+\alpha(s^\alpha - 1)}{A(\alpha)}\right)\{E(R\xi(x,t) + N\xi(x,t))\}\right], \tag{4.5}$$

where $H(x,t)$ denotes the term obtained from the known function and IC by taking inverse ET.

**Step 3.** Let us assume $\boldsymbol{\xi(x,t)} = \sum_{n=0}^{\infty} \boldsymbol{p^n \xi_n(x,t)}$, and $N\xi(x,t) = \sum_{n=0}^{\infty} p^n H_n(\xi)$, where $H_n(\xi)$ are the He's polynomials.

**Step 4.** Using HPM combined with ET, the following expression is obtained:

$$\sum_{n=0}^{\infty} p^n \xi_n(x,t) = H(x,t) - p\left[E^{-1}\left[\left(\frac{1+\alpha(s^\alpha - 1)}{A(\alpha)}\right)\left\{E\left(R\sum_{n=0}^{\infty} p^n \xi_n(x,t) + \sum_{n=0}^{\infty} p^n H_n(\xi)\right)\right\}\right]\right]. \tag{4.6}$$

Equating the terms with the like power of p, we get

$$p^0 : \xi_0(x,t) = H(x,t),$$

$$p^1 : \xi_1(x,t) = -E^{-1}\left[\left(\frac{1+\alpha(s^\alpha - 1)}{A(\alpha)}\right) E\{R\xi_0(x,t) + H_0(\xi)\}\right],$$

$$p^2 : \xi_2(x,t) = -E^{-1}\left[\left(\frac{1+\alpha(s^\alpha - 1)}{A(\alpha)}\right) E\{R\xi_1(x,t) + H_1(\xi)\}\right],$$

$$\vdots$$

Likewise, all the values $\{\xi_n(x,t)\}_{n=0}^{\infty}$ can be calculated. So the approximate solution is written as

$$\tilde{\xi}_n(x,t) = \sum_{k=0}^{n} \xi_k(x,t). \tag{4.7}$$

## 4.6 Convergence analysis

**Theorem 4.6.1.** *The series solution* $\sum_{k=0}^{\infty} \xi_k(x,t)$ *converges if for each* $\varepsilon \geq 0$ *there exist a number* $n \in N$ *such that* $|\xi_{m+p}| < \varepsilon, \forall m \geq n$ *and* $p = 1, 2, 3$.

*Proof.* Let us take the sequence of functions $\{F_m\}_{m=0}^{\infty}$, which is defined as follows:

$$F_0 = \xi_0,$$
$$F_1 = \xi_0 + \xi_1,$$
$$F_2 = \xi_0 + \xi_1 + \xi_2,$$
$$F_m = \xi_0 + \xi_1 + \xi_2 + . + \xi_m$$

It is noted that $\{F_m\}_{m=0}^{\infty}$ converges uniformly on $\mathfrak{R}$. From Theorem 4.6.1, we obtain

$$\left|F_{m+p} - F_m\right| = \left|\xi_{m+p}\right| < \varepsilon, \qquad \forall m \geq n \quad \text{and} \quad p = 1, 2, 3. \tag{4.1}$$

Hence by Cauchy criteria, $\{F_m\}_{m=0}^{\infty}$ converges uniformly on $\mathfrak{R}$, and thus the series solution $\sum_{k=0}^{\infty} \xi_k (x, t)$ converges. □

## 4.7 Numerical results and discussion

Numerical solutions have been addressed to check the reliability and efficiency of the present technique. All the numerical calculations for both the Caputo–Fabrizio and Atangana–Baleanu fractional operators are computed by truncating the HPETM infinite series solutions to a finite number (n = 3) of terms of solutions. Here we have considered the initial condition involved in the time-fractional three-species food chain model as $\xi(0) = 0.05$, $\psi(0) = 0.1$, $\zeta(0) = 0.5$, and the parameters as $a_1 = 4$, $b_1 = 0.01$, $c_1 = 14$, $a_2 = 0.3$, $c_2 = 0.05$, $a_3 = 0.15$, $d_1 = 0.1$, and $b_2 = 0.3$. The values of the initial conditions and the parameters involved in this model are taken from reference (Merdan and Khaniyev, 2008). Fig. 4.1 and Fig. 4.2 examine the effects of fractional-order derivatives $\alpha$ on the population density of $\xi$ (prey), $\psi$ (mid-predator), and $\zeta$ (top-predator) defined in the ABC and CFC sense. The numerical results of $\xi$, $\psi$ and $\zeta$ for different fractional Brownian motions $\alpha = 0.2, 0.4, 0.6, 0.8$ and standard motion $\alpha = 1$ have been given graphically. The variation of results of $\xi$, $\psi$ and $\zeta$ with respect to time $(t)$ for various values of $\alpha$ are depicted in Fig. 4.1 and Fig. 4.2. Fig. 4.1 and Fig. 4.2 clearly show the time history of the densities of the prey and the predator population. The graphical representations illustrate that the obtained solutions of this model depend on the fractional-order derivatives. Fig. 4.1 and Fig. 4.2 show that the density of the prey population $\xi$ decreases, whereas the population densities $\psi$ and $\zeta$ increase with the increasing values of $t$ and $\alpha$. Tables 4.1 and 4.2 and show the comparison results of the method solution defined in two different kernels with the existing solution. Both tables clearly show that the obtained results are in good agreement with the other method solution up to $3^{\text{rd}}$-order approximate solution at $\alpha = 1$. The graphical analysis indicates the precision of the results obtained by HPETM by considering only three iterative terms. However, by increasing the number of terms of solution, the accuracy of the approximation of the solutions may be improved.

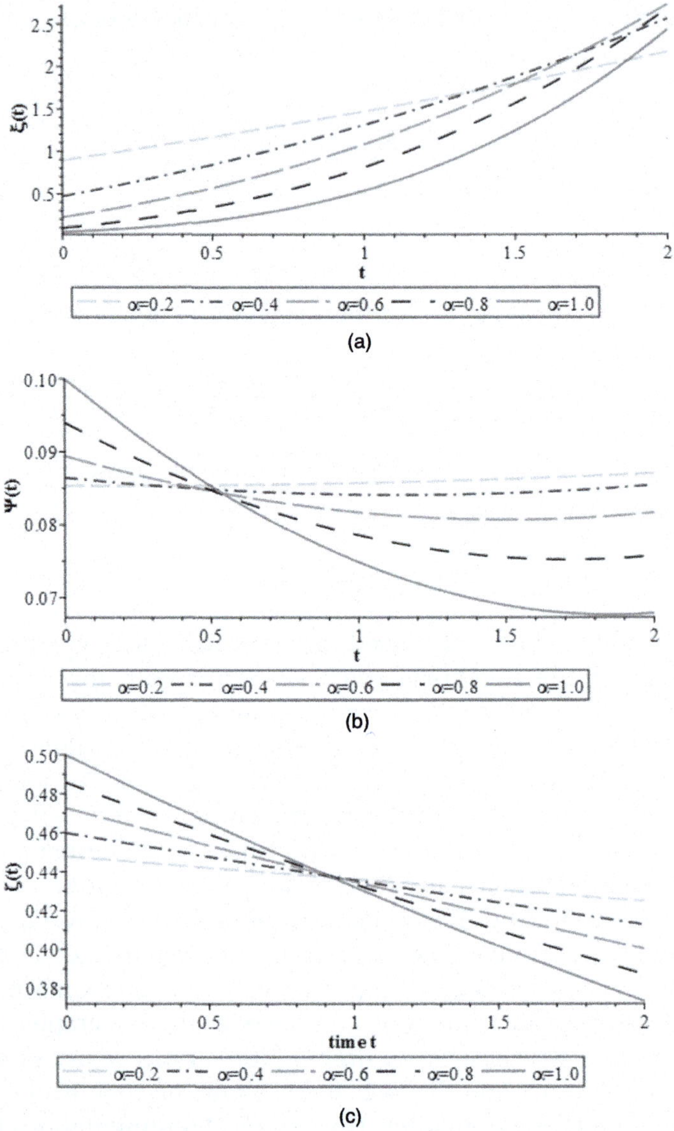


**FIGURE 4.1** Plots of (a) $\xi(t)$ (b) $\psi(t)$ and (c) $\zeta(t)$ vs. t defined in the CFC sense for various values of $\alpha$. Solution plots of the titled model defined in CFC sense for different values of fractional order.

## 4.8 Conclusion

In this paper, the Caputo–Fabrizio and Atangana–Baleanu fractional derivatives are used to analyze the three-species food chain model. The fixed-point theorem is used to demonstrate the existence and uniqueness of the solution of the titled model. The solutions for the fractional-order models are obtained by using the perturbation transform

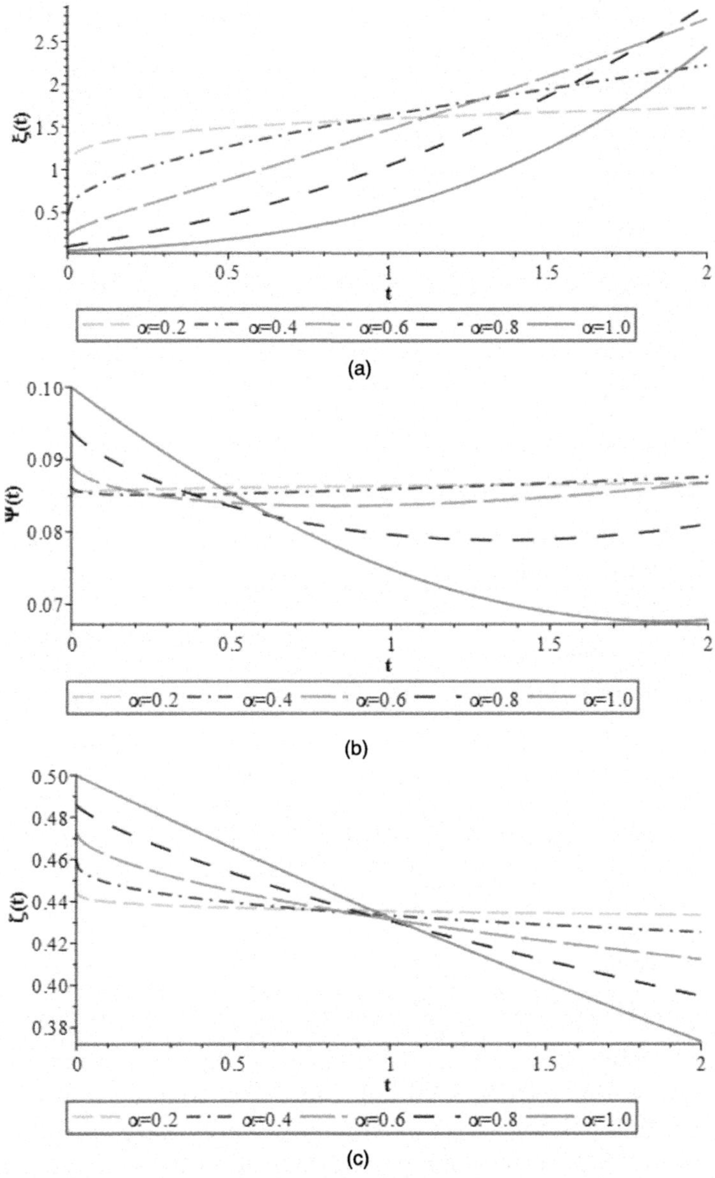


**FIGURE 4.2** Plots of (a) $\xi(t)$ (b) $\psi(t)$ and (c) $\zeta(t)$ vs. t defined in the ABC sense for various values of $\alpha$. Solution plots of the titled model defined in ABC sense for different values of fractional order.

method, namely HPETM. Related simulations are performed to demonstrate the effect of fractional order. By simulation, it is clear that the Atangana–Baleanu and Caputo–Fabrizio fractional derivatives exhibit interesting behaviors for different values $\alpha$. We have also compared the present solution with the existing one, which agrees well. The newly defined fractional derivatives may be helpful in modeling other real-life problems. The

**Table 4.1** Comparison of the present solution defined in CFC sense with the existing solution at $\alpha = 1$.

| $t$ | Present solution | | | Solution (Merdan and Khaniyev, 2008) | | |
|---|---|---|---|---|---|---|
| | $\xi(t)$ | $\psi(t)$ | $\zeta(t)$ | $\xi(t)$ | $\psi(t)$ | $\zeta(t)$ |
| 0.2 | 0.084619 | 0.093620 | 0.485692 | 0.084619 | 0.093620 | 0.485692 |
| 0.4 | 0.142083 | 0.087892 | 0.471765 | 0.142083 | 0.087892 | 0.471765 |
| 0.6 | 0.230788 | 0.082836 | 0.458214 | 0.230788 | 0.082836 | 0.458214 |
| 0.8 | 0.359129 | 0.078470 | 0.445030 | 0.359129 | 0.078470 | 0.445030 |
| 1.0 | 0.535500 | 0.074812 | 0.432210 | 0.535500 | 0.074812 | 0.432210 |
| 1.2 | 0.768297 | 0.071881 | 0.419746 | 0.768297 | 0.071881 | 0.419746 |
| 1.4 | 1.065914 | 0.069697 | 0.407632 | 1.065914 | 0.069697 | 0.407632 |
| 1.6 | 1.436748 | 0.068277 | 0.395862 | 1.436748 | 0.068277 | 0.395862 |
| 1.8 | 1.889193 | 0.067640 | 0.384431 | 1.889193 | 0.067640 | 0.384431 |
| 2.0 | 2.431643 | 0.067805 | 0.373331 | 2.431643 | 0.067805 | 0.373331 |

**Table 4.2** Comparison of the present solution defined in ABC sense with the existing solution at $\alpha = 1$.

| $t$ | Present solution | | | Solution (Merdan and Khaniyev, 2008) | | |
|---|---|---|---|---|---|---|
| | $\xi(t)$ | $\psi(t)$ | $\zeta(t)$ | $\xi(t)$ | $\psi(t)$ | $\zeta(t)$ |
| 0.2 | 0.084619 | 0.093620 | 0.485692 | 0.084619 | 0.093620 | 0.485692 |
| 0.4 | 0.142083 | 0.087892 | 0.471765 | 0.142083 | 0.087892 | 0.471765 |
| 0.6 | 0.230788 | 0.082836 | 0.458214 | 0.230788 | 0.082836 | 0.458214 |
| 0.8 | 0.359129 | 0.078470 | 0.445030 | 0.359129 | 0.078470 | 0.445030 |
| 1.0 | 0.535500 | 0.074812 | 0.432210 | 0.535500 | 0.074812 | 0.432210 |
| 1.2 | 0.768297 | 0.071881 | 0.419746 | 0.768297 | 0.071881 | 0.419746 |
| 1.4 | 1.065914 | 0.069697 | 0.407632 | 1.065914 | 0.069697 | 0.407632 |
| 1.6 | 1.436748 | 0.068277 | 0.395862 | 1.436748 | 0.068277 | 0.395862 |
| 1.8 | 1.889193 | 0.067640 | 0.384431 | 1.889193 | 0.067640 | 0.384431 |
| 2.0 | 2.431643 | 0.067805 | 0.373331 | 2.431643 | 0.067805 | 0.373331 |

higher-order solution may be considered by increasing the number of terms of the HPETM solution to achieve better results. Although the Elzaki transform is efficient in solving FDEs, it sometimes fails to handle nonlinear representations arising from the FDEs. These difficulties may be overcome by coupling this transform with that of HPM. Lastly, this study ensures that the HPETM is precise, converges rapidly, and is a useful technique for dealing with the nonlinear FDEs defined in Mittag-Leffler and exponential law kernels. It is also a very easy-to-use mathematical method that can be applied to solve real-life problems in various fields of science and engineering.

## Declaration of competing interest

There are no conflicts of interest in this work.

# Acknowledgment

The first author acknowledges the Department of Science and Technology of the Government of India for providing financial assistance under the INSPIRE Fellowship scheme (IF170207).

# References

Ahmed, E., El-Sayed, A.M.A., El-Saka, H.A.A., 2007. Equilibrium points, stability and numerical solutions of fractional-order predator–prey and rabies models. Journal of Mathematical Analysis and Applications (ISSN 0022-247X) 325 (1), 542–553. https://doi.org/10.1016/j.jmaa.2006.01.087.

Atangana, A., Baleanu, D., 2016. New fractional derivatives with non-local and non-singular kernel: theory and application to heat transfer model. Thermal Science (ISSN 0354-9836) 20 (2), 763–769. https://doi.org/10.2298/TSCI160111018A. Serbian Society of Heat Transfer Engineers, South Africa, Available from: http://www.doiserbia.nb.rs/img/doi/0354-9836/2016/0354-98361600018A.pdf.

Atangana, A., Gómez-Aguilar, J.F., 2018. Decolonisation of fractional calculus rules: breaking commutativity and associativity to capture more natural phenomena. The European Physical Journal Plus (ISSN 2190-5444) 133 (4). https://doi.org/10.1140/epjp/i2018-12021-3. Springer Verlag, South Africa, Available from: https://www.springer.com/journal/13360.

Baek, H., Do, Y., Lim, Y., Lim, D., 2011. A three-species food chain system with two types of functional responses. Abstract and Applied Analysis (ISSN 1687-0409) 2011. https://doi.org/10.1155/2011/934569.

Baleanu, D., Diethelm, K., Scalas, E., Trujillo, J.J., 2012. Fractional Calculus: Models and Numerical Methods. World Scientific, Singapore.

Caputo, M., Fabrizio, M., 2015. A new definition of fractional derivative without singular kernel. Progress in Fractional Differentiation and Applications (ISSN 2356-9344) 1 (2), 73–85. https://doi.org/10.12785/pfda/010201. Natural Sciences Publishing, United States, Available from: http://naturalspublishing.com/files/published/0gb83k287mo759.pdf.

Chauvet, E., Paulett, J.E., Previte, J.P., et al., 2002. A Lotka-Volterra three species food chain. Mathematics Magazine 75 (4), 243–255.

Dubey, B., Upadhyay, R.K., 2004. Persistence and extinction of one-prey and two-predators system. Nonlinear Analysis: Modeling & Control (ISSN 2335-8963) 9 (4), 307–329. https://doi.org/10.15388/na.2004.9.4.15147.

Elzaki, T.M., Elzaki, S.M., Hilal, E.M.A., 2012. Elzaki and Sumudu transforms for solving some differential equations. Global Journal of Pure and Applied Mathematics (ISSN 0973-9750) 8 (2), 167–173. Saudi Arabia, Available from. http://www.ripublication.com/gjpamv7/gjpamv8n2_7.pdf.

Guo, C., Fang, S., 2016. Stability and approximate analytic solutions of the fractional Lotka-Volterra equations for three competitors. Advances in Difference Equations (ISSN 1687-1847) 2016 (1). https://doi.org/10.1186/s13662-016-0943-y. Available from: http://www.springerlink.com/content/1687-1847/.

Hastings, A., Powell, T., 1991. Chaos in a three-species food chain. Ecology (ISSN 0012-9658) 72 (3), 896–903. https://doi.org/10.2307/1940591.

Jena, R.M., Chakraverty, S., 2019a. A new iterative method based solution for fractional Black–Scholes option pricing equations (BSOPE). SN Applied Sciences (ISSN 2523-3971) 1 (1). https://doi.org/10.1007/s42452-018-0106-8. Springer Nature, India, Available from: springer.com/snas.

Jena, R.M., Chakraverty, S., 2019b. Analytical solution of Bagley-Torvik equations using Sumudu transformation method. SN Applied Sciences (ISSN 2523-3971) 1 (3). https://doi.org/10.1007/s42452-019-0259-0. Springer Nature, India, Available from: springer.com/snas.

Jena, R.M., Chakraverty, S., 2019c. Residual power series method for solving time-fractional model of vibration equation of large membranes. Journal of Applied and Computational Mechanics (ISSN 2383-4536) 5 (4), 603–615. https://doi.org/10.22055/JACM.2018.26668.1347. Shahid Chamran University of Ahvaz, India, Available from: http://jacm.scu.ac.ir/article_13838_70ed6fb779bee4d57b2d80cca3bc2ad2.pdf.

Jena, R.M., Chakraverty, S., 2019d. Solving time-fractional Navier–Stokes equations using homotopy perturbation Elzaki transform. SN Applied Sciences (ISSN 2523-3971) 1 (1). https://doi.org/10.1007/s42452-018-0016-9. Springer Nature, India, Available from: https://www.springer.com/journal/42452.

Jena, R.M., Chakraverty, S., Baleanu, D., 2019a. On new solutions of time-fractional wave equations arising in shallow water wave propagation. Mathematics (ISSN 2227-7390) 7 (8). https://doi.org/10.3390/

math7080722. MDPI AG, India, Available from: https://res.mdpi.com/mathematics/mathematics-07-00722/article_deploy/mathematics-07-00722.pdf?filename=&attachment=1.

Jena, R.M., Chakraverty, S., Baleanu, D., 2019b. On the solution of imprecisely defined nonlinear time-fractional dynamical model of marriage. Mathematics 7, 689–704.

Jena, R.M., Chakraverty, S., Jena, S.K., 2019c. Dynamic response analysis of fractionally damped beams subjected to external loads using homotopy analysis method. Journal of Applied and Computational Mechanics (ISSN 2383-4536) 5 (2), 355–366. https://doi.org/10.22055/jacm.2019.27592.1419. Shahid Chamran University of Ahvaz, India, Available from: http://jacm.scu.ac.ir/article_14072_d4c690cf2cf2984ab3bb8e1450798d4f.pdf.

Jena, R.M., Chakraverty, S., Baleanu, D., Jena, S.K., 2021. Analysis of time-fractional dynamical model of romantic and interpersonal relationships with non-singular kernels: a comparative study. Mathematical Methods in the Applied Sciences (ISSN 1099-1476) 44 (2), 2183–2199. https://doi.org/10.1002/mma.6929. John Wiley and Sons Ltd, India, Available from: http://onlinelibrary.wiley.com/journal/10.1002/(ISSN)1099-1476.

Jena, Rajarama Mohan, Chakraverty, S., 2019e. Boundary Characteristic Orthogonal Polynomials-Based Galerkin and Least Square Methods for Solving Bagley–Torvik Equations. Springer Science and Business Media LLC. Springer Science and Business Media LLC, pp. 327–342.

Klebanoff, A., Hastings, A., 1994. Chaos in three species food chains. Journal of Mathematical Biology (ISSN 1432-1416) 32 (5), 427–451. https://doi.org/10.1007/BF00160167.

Losada, J., Nieto, J.J., 2015. Properties of a new fractional derivative without singular kernel. Progress in Fractional Differentiation and Applications (ISSN 2356-9344) 1 (2), 87–92. https://doi.org/10.12785/pfda/010202. Natural Sciences Publishing, Spain, Available from: http://naturalspublishing.com/files/published/2j1ns3h8o2s789.pdf.

Lotka, A.J., 1925. Elements of Physical Biology. Williams & Wilkins Co., Baltimore.

Maiti, A., Patra, B., Samanta, G.P., 2006. Persistence and stability of a food chain model with mixed selection of functional responses. Nonlinear Analysis: Modeling & Control (ISSN 2335-8963) 11 (2), 171–185. https://doi.org/10.15388/na.2006.11.2.14757.

Merdan, M., 2010. Homotopy perturbation method for solving a three-species food chain model. Selçuk Journal of Applied Mathematics 11 (1), 3–13.

Merdan, M., Khaniyev, T., 2008. Variational iteration and homotopy perturbation method for solving a three-species food chain model. AKÜ J. Sci. 8 (2), 27–36.

Merdan, M., Gokdogan, A., Erturk, V.S., 2011. A numeric-analytic method for approximating three-species food chain models. International Journal of Physical Sciences 6 (7), 1822–1833.

Podlubny, I., 1999. Fractional Differential Equations: An Introduction to Fractional Derivatives, Fractional Differential Equations, Methods of Their Solution and Some of Their Applications. Elsevier, Academic Press.

Rao, K.K., Narayan, K.L., 2017. Stability analysis of a three species food chain model with harvesting. Journal of Mathematical and Statistical Sciences 9, 285–294.

Saad, K.M., Baleanu, D., Atangana, A., 2018. New fractional derivatives applied to the Korteweg–de Vries and Korteweg–de Vries–Burger's equations. Computational & Applied Mathematics (ISSN 1807-0302) 37 (4), 5203–5216. https://doi.org/10.1007/s40314-018-0627-1. Springer Science and Business Media, LLC, Saudi Arabia, Available from: https://www.springer.com/journal/40314.

Sanjaya, W.S.M., Mohammad, I.B., Mahmat, M., Salleh, Z., 2012. Mathematical model of three-species food chain interaction with mixed functional response. International Journal of Modern Physics: Conference Series 9, 334–340.

Sanjaya, W.S.M., Salleh, Z., Mahmat, M., 2013. Mathematical models of three species food chain with Holling type-III functional response. International Journal of Pure and Applied Mathematics 89 (5), 647–657.

Volterra, V., 1926. Variazioni e fluttuazioni del numero d'individui in specie animali conviventi. Memorie Della Reale Accademia Nazionale dei Lincei. Ser. VI 2 (1).

Wu, X.J., Li, J., Upadhyay, R.K., 2010. Chaos control and synchronization of a three-species food chain model via Holling functional response. International Journal of Computer Mathematics (ISSN 1029-0265) 87 (1), 199–214. https://doi.org/10.1080/00207160801993232.

Zibaei, S., Namjoo, M., 2016. A nonstandard finite difference scheme for solving three species food chain with fractional-order Lotka-Volterra model. Iranian Journal of Numerical Analysis and Optimization 6 (1), 53–78.

# 5

# Uncertain dynamic responses of imprecisely defined beam fractionally damped beam subject to external loads using double parametric form-based homotopy analysis method

Smita Tapaswini[a], Jashobanta Sa[a], Basanta Kumar Sahoo[b], and Anasuya Nath[c]
[a]PG Department of Mathematics, Government (Autonomous) College, Angul, Odisha, India [b]PG Department of Physics, Government (Autonomous) College, Angul, Odisha, India [c]PG Department of Mathematics, Utkal University, Vani Vihar, Bhubaneswar, India

## 5.1 Introduction

Fractional calculus has acquired great importance in the past few years, because it can model complex phenomena more accurately. It has been utilized in diverse fields, from science and engineering to finance and social sciences. It is a challenging task to obtain analytical solutions to fractional-order differential equations; a simple, robust and efficient computational method to solve these differential equations may be required. Numerous notable studies have been published on fractional calculus in recent decades in the form of some outstanding books by Ullah et al. (2020); Kilbas et al. (2006); Kiryakova (1993); Miller & Ross (1993); Oldham and Spanier (1974); Podlubny (1999); Samko et al. (1993). The books listed here, provide a detailed analysis of the field, which assist readers in comprehending the fundamental ideas of fractional calculus. In this context, several authors who have proposed diverse methodologies for solving fractional-order differential equations have been cited by Iyiola and Zaman (2014); Ucar et al. (2019); Abaid Ur Rehman et al. (2022); Muhammad and Akram (2024); Kumar et al. (2018); Sartanpara et al. (2022); Qayyum and Ahmad (2024); Guo (2019); Alderremy et al. (2020); Liang and Tang (2007); Zhang et al. (2008).

In most cases, parameters and conditions in modeling are treated as deterministic or exactly defined. Nevertheless, in place of precise fixed values, we often have ambigu-

Advanced Computational and Mathematical Approaches in Applied Differential Equations
https://doi.org/10.1016/B978-0-44-345743-2.00016-5

ous, undetermined, and partial details about variables and parameters because of measurement faults, observational uncertainties, experimental inaccuracies, varying operating conditions, or maintenance-related mistakes. The foregoing lead to uncertainties. In general, these uncertainties are well described by employing probabilistic, interval, and fuzzy approaches. In probabilistic techniques, uncertain variables are modeled as random variables defined by joint probability distributions or joint probability density functions. However, probabilistic approaches may struggle to give accurate conclusions with the appropriate level of accuracy in the absence of adequate experimental data. This limitation may be attributed to the underlying distributions of probability. Over the past few years, interval analysis approaches, including fuzzy theory, have evolved as solid tools in modeling uncertainty in many real-world situations. These approaches express uncertain parameters and variables using interval and fuzzy numbers. Notably, intervals are a special case of fuzzy numbers. Therefore we have employed fuzzy set theory, where interval solutions have been defined as special cases.

The concept of fuzzy sets, introduced by Zadeh (1965), has been widely studied and applied to analyze uncertainty in various contexts (Tapaswini et al., 2015, 2017; Jamal et al., 2023; Zureigat et al., 2023b,a; Georgieva and Spasova, 2022; Ullah et al., 2020). The importance of fractional and fuzzy methodologies in modeling and design has led to attempts to merge these two concepts. Recently, there have been some significant developments in the theory of fuzzy fractional modeling as mentioned below. The innovative concept of FFDEs is introduced by Agrawal et al. (2010) in their research. A study in Arshad and Lupulescu (2011) investigated fractional differential equations featuring fuzzy initial value problems. Various contributions can be seen to the theory of FFDEs have been described in Tapaswini and Behera (2020, 2021); Chakraverty and Behera (2013, 2015); Chakraverty et al. (2016a); Ullah et al. (2023); Ahmad et al. (2013). Tapaswini and Behera (2020) adopted the Adomian Decomposition method for solving imprecisely defined fuzzy space-fractional telegraph equations. The imprecisely defined fractional-order Fokker–Planck equation subjected to fuzzy uncertainty was studied in Tapaswini and Behera (2021). However, Zadeh's extension principle was implemented in Ahmad et al. (2013) for solving FFDEs.

The homotopy analysis method (HAM) has attracted substantial attention in the past few years being an adaptable method for studying both linear and nonlinear physical issues. The HAM approach, first proposed in Liao (1992), has become a valuable tool in various fields. The convergence analysis of the HAM is discussed in Liao (2003). The technique has been successfully studied and utilized to address diverse equations emerging in science and engineering applications (see Qu et al., 2020; Odibat and Baleanu, 2019; Saratha et al., 2020; Al-Qudah et al., 2022; Jena et al., 2019).

The current study employs HAM to address the uncertain dynamic behavior of a fuzzy fractionally damped beam with fuzzy initial conditions. The uncertainty in the initial condition is characterized by fuzzy numbers. The method involves a newly developed DPF of fuzzy numbers. Both loads are analyzed in this research. A review of the literature shows that fuzzy differential equations are typically converted into two crisp differential equations to establish solution bounds. In contrast, the proposed method converts the fuzzy differential equation into a single crisp differential equation using the DPF of fuzzy numbers, and then solves this crisp differential equation to obtain the final fuzzy solution by substituting parametric values.

This paper is structured as follows. Section 5.2 presents the essential preliminaries connected to the present inquiry. HAM is applied in conjunction with the proposed approach to solve the titled problem in Section 5.3. The uncertain responses for both the loads are examined in Section 5.4. Section 5.5 gives numerical results and discusses different parameters, followed by the conclusion within the last section.

## 5.2 Preliminaries

We introduce below the notations and definitions that will be employed throughout this paper.

**Definition 5.2.1** (Fuzzy set). (Chakraverty et al., 2016b) A fuzzy set $\tilde{B}$ on the real line $R$ is defined as the set of ordered pairs such that

$$\tilde{B} = \left\{ \left(x, \mu_{\tilde{B}}(x)\right) \mid x \in R, \mu_{\tilde{B}}(x) \in [0, 1] \right\},$$

where $\mu_{\tilde{B}}(x)$ is called the membership function.

**Definition 5.2.2** (Fuzzy number). (Chakraverty et al., 2016b) A fuzzy number $\tilde{B}$ is convex normalized fuzzy set $\tilde{B}$ of the real line $R$ such that

$$\{\mu_{\tilde{B}}(x) : R \to [0, 1], \forall x \in R\};$$

$\mu_{\tilde{B}}$ is called the membership function of the fuzzy set such that it is piecewise continuous.

**Definition 5.2.3** (TFN). (Chakraverty et al., 2016b) A TFN $\tilde{B}$ is a convex normalized fuzzy set $\tilde{B}$ of the real line $R$ such that

- There exists exactly one $x_0 \in R$ with $\mu_{\tilde{A}}(x_0) = 1$ ($x_0$ is called the mean value of $\tilde{B}$), where $\mu_{\tilde{B}}$ is called the membership function of the fuzzy set.
- $\mu_{\tilde{B}}(x)\mu_{\mathrm{U}}$ (x) is piecewise continuous.

Let us consider an arbitrary triangular fuzzy number $\tilde{B} = (a, b, c)$. The membership function $\mu_{\tilde{B}}$ of $\tilde{B}$ will be define as follows:

$$\mu_{\tilde{B}}(x) = \begin{cases} 0, x \le a \\ \dfrac{x-a}{b-a}, a \le x \le b \\ \dfrac{c-x}{c-b}, b \le x \le c \\ 0, x \ge c \end{cases}.$$

**Definition 5.2.4** (SPF of fuzzy number). (Chakraverty et al., 2016b) The TFN $\tilde{B} = (a, b, c)\tilde{\mathrm{U}} = (\mathrm{a}, \mathrm{b}, \mathrm{c})$ can be expressed as an ordered pair of functions through the application of the $\gamma -$ cut method, as demonstrated below:

$$\tilde{B}(\gamma) = [\underline{a}(\gamma), \overline{a}(\gamma)] = [(b-a)\gamma + a, -(c-b)\gamma + c] \quad \text{where,} \quad \gamma \in [0, 1].$$

The $\gamma -$ cut form is known as parametric form or SPF of fuzzy numbers.

It is important to note that the bounds of the fuzzy numbers meet the following requirements:

1. $\underline{a}(\gamma)$ is a bounded function that is left continuous and nondecreasing over [0, 1].
2. $\overline{a}(\gamma)$ is a bounded function that is right continuous and nonincreasing over [0, 1]
3. $\underline{a}(\gamma) \leq \overline{a}(\gamma), 0 \leq \gamma \leq 1$.

**Definition 5.2.5** (DPF of fuzzy number). Referring to the parametric form as outlined in Definition 5.2.3, we find that $\tilde{A}(\gamma) = [\underline{a}(\gamma), \overline{a}(\gamma)]$. Now this can be expressed as a crisp number with DPF

$$\tilde{A}(\gamma, \beta) = \beta(\overline{a}(\gamma) - \underline{a}(\gamma)) + \underline{a}(\gamma) \quad \text{where} \quad \beta \epsilon [0, 1].$$

It is worth mentioning that the lower and upper bounds in SPF may be obtained if we put $\beta = 0$ and 1, respectively, in the above DPF. This may be represented as

$$\tilde{A}(\gamma, 0) = \underline{a}(\gamma) \quad \text{and} \quad \tilde{A}(\gamma, 1) = \overline{a}(\gamma).$$

**Definition 5.2.6** (Fuzzy arithmetic). For any two arbitrary fuzzy number $\tilde{x} = [\underline{x}(\gamma), \overline{x}(\gamma)]$, $\tilde{y} = [\underline{y}(\gamma), \overline{y}(\gamma)]$, and scalar k, the fuzzy arithmetic is defined as follows:

1. $\tilde{x} = \tilde{y}$ if and only if $\underline{x}(\gamma) = \underline{y}(\gamma)$ and $\overline{x}(\gamma) = \overline{y}(\gamma)\mathrm{x} = \mathrm{y}$ if and only if $\underline{\mathrm{x}}(\gamma) = \underline{\mathrm{y}}(\gamma)$ and $\overline{\mathrm{x}}(\gamma) = \overline{\mathrm{y}}(\gamma)$;
2. $\mathrm{x} + \mathrm{y} = (\underline{\mathrm{x}}(\gamma) + \underline{\mathrm{y}}(\gamma), \overline{\mathrm{x}}(\gamma) + \overline{\mathrm{y}}(\gamma))\tilde{x} + \tilde{y} = [\underline{x}(\gamma) + \underline{y}(\gamma), \overline{x}(\gamma) + \overline{y}(\gamma)]$;
3. $\tilde{x} \times \tilde{y} = [\min(M), \max(M)]$, where $M = \{\underline{x}(\gamma) \times \underline{y}(\gamma), \underline{x}(\gamma) \times \overline{y}(\gamma), \overline{x}(\gamma) \times \underline{y}(\gamma), \overline{x}(\gamma) \times \overline{y}(\gamma)\}$;
4. $k\tilde{x} = \begin{cases} [k\overline{x}(\gamma), k\underline{x}(\gamma)], k < 0 \\ [k\underline{x}(\gamma), k\overline{x}(\gamma)], k \geq 0. \end{cases}$

## 5.3 Implementation of proposed method on imprecisely defined fractionally damped beam

**To construct simulation techniques** for an imprecisely fractionally damped beam. The initial step in the solution procedure involves transforming the given equation into an interval-based fuzzy fractional differential equation using a SPF. **Following this,** the interval-based differential equation is reduced to a crisp differential equation using the

DPF. Thereafter the HAM is employed to address the differential equation associated with the titled problem.

We **now consider** the fuzzy fractional beam equation:

$$\rho A\frac{\partial^2 \tilde{u}}{\partial t^2}+c\frac{\partial^p \tilde{u}}{\partial t^p}+EI\frac{\partial^4 \tilde{u}}{\partial x^4}=F(x,t). \tag{5.1}$$

The variables $\rho$, $A$, $c$, $E$ and $I$ denote the mass density, cross-sectional area, damping coefficients per unit length, Young's modulus of elasticity, and the beam's moment of inertia, respectively. $F(x,t)$ represents the external force applied, while $\tilde{v}(x,t)$ indicates the transverse fuzzy displacement. $\frac{\partial^p}{\partial t^p}$ is the fractional derivative of order $p\in(0,1)$ of the fuzzy displacement function U. The initial conditions are considered to be fuzzy, specifically $\tilde{u}(0)=\tilde{u}'(0)=(-0.1,0,0.1)$.

Eq. (5.1) is equivalent to

$$\frac{\partial^2 \tilde{u}}{\partial t^2}+\frac{c}{\rho A}\frac{\partial^p \tilde{u}}{\partial t^p}+\frac{EI}{\rho A}\frac{\partial^4 \tilde{u}}{\partial x^4}=\frac{F(x,t)}{\rho A}. \tag{5.2}$$

Eq. (5.2) may be expressed in the SPF as

$$\left[\frac{\partial^2 \underline{u}(x,t;\gamma)}{\partial t^2},\frac{\partial^2 \bar{u}(x,t;\gamma)}{\partial t^2}\right]+\frac{c}{\rho A}\left[\frac{\partial^p \underline{u}(x,t;\gamma)}{\partial t^p},\frac{\partial^p \bar{u}(x,t;\gamma)}{\partial t^p}\right]$$
$$+\frac{EI}{\rho A}\left[\frac{\partial^4 \underline{u}(x,t;\gamma)}{\partial x^4},\frac{\partial^4 \overline{u}(x,t;\gamma)}{\partial x^4}\right]=\frac{F(x,t)}{\rho A}, \tag{5.3}$$

subject to fuzzy initial condition

$$\left[\underline{u}(x,0;\gamma),\bar{u}(x,0;\gamma)\right]=\left[\underline{u}'(x,0;\gamma),\bar{u}'(x,0;\gamma)\right]=[0.1\gamma-0.1,0.1-0.1\gamma],\quad \text{where, } \gamma\in[0,1]. \tag{5.4}$$

Subsequently, utilizing the DPF as outlined in Definition 5.2.4, Eq. (5.3) can be expressed as

$$\left\{\beta\left(\frac{\partial^2 \bar{u}(x,t;\gamma)}{\partial t^2}-\frac{\partial^2 \underline{u}(x,t;\gamma)}{\partial t^2}\right)+\frac{\partial^2 \underline{u}(x,t;\gamma)}{\partial t^2}\right\}$$
$$+\frac{c}{\rho A}\left\{\beta\left(\frac{\partial^p \overline{u}(x,t;\gamma)}{\partial t^p}-\frac{\partial^p \underline{u}(x,t;\gamma)}{\partial t^p}\right)+\frac{\partial^p \underline{u}(x,t;\gamma)}{\partial t^p}\right\}$$
$$+\frac{EI}{\rho A}\left\{\beta\left(\frac{\partial^4 \overline{u}(x,t;\gamma)}{\partial x^4}-\frac{\partial^4 \underline{u}(x,t;\gamma)}{\partial x^4}\right)+\frac{\partial^4 \underline{u}(x,t;\gamma)}{\partial x^4}\right\}=F(x,t), \tag{5.5}$$

subject to fuzzy initial conditions

$$\left\{\beta\left(\bar{u}(x,0;\gamma)-\underline{u}(x,0;\gamma)\right)+\underline{u}(x,0;\gamma)\right\}=\{\beta(0.2-0.2\gamma)+(0.1\gamma-0.1)\} \tag{5.6}$$

and

$$\left\{\beta\left(\bar{u}'(x,0;\gamma)-\underline{u}'(x,0;\gamma)\right)+\underline{u}'(x,0;\gamma)\right\}=\{\beta(0.2-0.2\gamma)+(0.1\gamma-0.1)\},$$

$$\text{where} \quad \gamma, \beta \in [0, 1]. \tag{5.7}$$

We now denote

$$\left\{\beta\left(\frac{\partial^2 \bar{u}(x,t;\gamma)}{\partial t^2} - \frac{\partial^2 \underline{u}(x,t;\gamma)}{\partial t^2}\right) + \frac{\partial^2 \underline{u}(x,t;\gamma)}{\partial t^2}\right\} = \frac{\partial^2 \tilde{u}(x,t;\gamma,\beta)}{\partial t^2}, \tag{5.8}$$

$$\left\{\beta\left(\frac{\partial^p \bar{u}(x,t;\gamma)}{\partial t^p} - \frac{\partial^p \underline{u}(x,t;\gamma)}{\partial t^p}\right) + \frac{\partial^p \underline{u}(x,t;\gamma)}{\partial t^p}\right\} = \frac{\partial^p \tilde{u}(x,t;\gamma,\beta)}{\partial t^p}, \tag{5.9}$$

$$\left\{\beta\left(\frac{\partial^4 \bar{u}(x,t;\gamma)}{\partial x^4} - \frac{\partial^4 \underline{u}(x,t;\gamma)}{\partial x^4}\right) + \frac{\partial^4 \underline{u}(x,t;\gamma)}{\partial x^4}\right\} = \frac{\partial^4 \underline{u}(x,t;\gamma,\beta)}{\partial x^4}, \tag{5.10}$$

$$\{\beta\left(\bar{u}(x,0;\gamma) - \underline{u}(x,0;\gamma)\right) + \underline{u}(x,0;\gamma)\} = \tilde{u}(x,0;\gamma,\beta) \tag{5.11}$$

and

$$\{\beta\left(\bar{u}'(x,0;\gamma) - \underline{u}'(x,0;\gamma)\right) + \underline{u}'(x,0;\gamma)\} = \tilde{u}'(x,0;\gamma,\beta). \tag{5.12}$$

By substituting these values in Eq. (5.5), we obtain

$$\frac{\partial^2 \tilde{u}(x,t;\gamma,\beta)}{\partial t^2} + \frac{c}{\rho A}\frac{\partial^\gamma \tilde{u}(x,t;\gamma,\beta)}{\partial t^\gamma} + \frac{EI}{\rho A}\frac{\partial^4 \tilde{u}(x,t;\gamma,\beta)}{\partial x^4} = \frac{F(x,t)}{\rho A}, \tag{5.13}$$

with initial conditions

$$\tilde{u}(x,0;\gamma,\beta) = \tilde{u}'(x,0;\gamma,\beta) = \{\beta(0.2 - 0.2\gamma) + (0.1\gamma - 0.1)\}. \tag{5.14}$$

Solving Eq. (5.13) yields the uncertain transverse fuzzy displacement $\tilde{u}(x,t;\gamma,\beta)$.
Using HAM on Eq. (5.13) and choosing the auxiliary linear operator

$$L\left(\varphi(x,t;q)\right) = \frac{\partial^2 \tilde{u}(x,t;\gamma,\beta)}{\partial t^2}, \tag{5.15}$$

let the operator be defined as

$$NFD\left[\tilde{u}\left(x,t;q,\gamma,\beta\right)\right] = D_t^2 \tilde{u}\left(x,t;\gamma,\beta\right) + \frac{c}{\rho A} D_t^\gamma \tilde{u}\left(x,t;\gamma,\beta\right) + \frac{EI}{\rho A} D_x^4 \tilde{u}\left(x,t;\gamma,\beta\right) - \frac{F\left(x,t\right)}{\rho A}. \tag{5.16}$$

Using HAM, the zeroth-order differential equation is formulated as

$$(1-q)L\left[v(x,t;q) - u_0(x,t)\right] = q\hbar NFD(v(x,t;q)). \tag{5.17}$$

Next by HAM nth-order, DE is written as

$$L\left[\tilde{u}_n(x,t;\gamma,\beta) - \chi_n \tilde{u}_{n-1}(x,t;\gamma,\beta)\right] = \hbar FR\left(\vec{\tilde{u}}_{n-1}(x,t;\gamma,\beta)\right). \tag{5.18}$$

Solving Eq. (5.18) for $n \geq 1$, we have

$$\tilde{u}_n(x,t;\gamma,\beta) = \chi_n \tilde{u}_{n-1}(x,t;\gamma,\beta) + \hbar L^{-1} FR\left(\vec{\tilde{u}}_{n-1}(x,t;\gamma,\beta)\right), \tag{5.19}$$

where

$$FR\left(\vec{\tilde{u}}_{n-1}(x,t;\gamma,\beta)\right) = D_t^2\tilde{u}_{n-1}(x,t;\gamma,\beta) + \frac{c}{\rho A}D_t^{\gamma}\tilde{u}_{n-1}(x,t;\gamma,\beta) + \frac{EI}{\rho A}D_x^4\tilde{u}_{n-1}(x,t;\gamma,\beta) - (1-\chi_n)\frac{F(x,t)}{\rho A}. \tag{5.20}$$

Using Eq. (5.6), (5.7), and (5.19), we obtain

$$\tilde{u}_0(x,t;\gamma,\beta) = \tilde{u}'(x,0;\gamma,\beta)t + \tilde{u}(x,0;\gamma,\beta), \tag{5.21}$$

$$\tilde{u}_1(x,t;\gamma,\beta) = \hbar L^{-1}\begin{bmatrix} D_t^2\tilde{u}_0(x,t;\gamma,\beta) + \frac{c}{\rho A}D_t^{\gamma}\tilde{u}_0(x,t;\gamma,\beta) \\ +\frac{EI}{\rho A}D_x^4\tilde{u}_0(x,t;\gamma,\beta) - \frac{F(x,t)}{\rho A} \end{bmatrix}, \tag{5.22}$$

$$\tilde{u}_2(x,t;\gamma,\beta) = \tilde{u}_1(x,t;\gamma,\beta) + \hbar\tilde{u}_1(x,t;\gamma,\beta) + \hbar L^{-1}\left[\frac{c}{\rho A}D_t^{\gamma}\tilde{u}_1(x,t;\gamma,\beta) + \frac{EI}{\rho A}D_x^4\tilde{u}_0(x,t;\gamma,\beta)\right], \tag{5.23}$$

$$\tilde{u}_3(x,t;\gamma,\beta) = \tilde{u}_2(x,t;\gamma,\beta) + \hbar\tilde{u}_2(x,t;\gamma,\beta) + \hbar L^{-1}\left[\frac{c}{\rho A}D_t^{\gamma}\tilde{u}_2(x,t;\gamma,\beta) + \frac{EI}{\rho A}D_x^4\tilde{u}_2(x,t;\gamma,\beta)\right]. \tag{5.24}$$

Therefore Eq. (5.1) can be represented as a solution in the form of

$$\tilde{u}(x,t;\gamma,\beta) = \sum_{n=0}^{\infty}\tilde{u}_n(x,t;\gamma,\beta). \tag{5.25}$$

To derive both lower and upper bounds within a single parametric framework, one can set $\beta = 0$ and 1, respectively, allowing the bounds to be expressed as $\tilde{u}(x,t;\gamma,0) = \underline{u}(x,t;\gamma)$ and $\tilde{u}(x,t;\gamma,1) = \bar{u}(x,t;\gamma)$.

The convergence of the method, as demonstrated by the previous simulations, depends on the variables $\hbar$, the initial condition, and the auxiliary linear operator $L$. For a detailed analysis of the HAM's convergence, one can refer to Liao (2003).

## 5.4 Uncertain response analysis

Consider the externally applied force $F(x,t)$ given by

$$F(x,t) = f(x)g(t). \tag{5.26}$$

In this context, $f(x)$ is characterized as a deterministic function related to space, while $g(t)$ represents a process that varies with time. The following paragraph delves into the unpredictable responses of the dynamic system described in Eq. (5.1) when subjected to two distinct loading scenarios.

### 5.4.1 Unit step function response

We now examine the fuzzy beam response associated with a unit step load, characterized by the form $g(t) = Bu(t)$, where $u(t)$ represents the Heaviside function, $B$ is a constant, and taking $\hbar = -1$.

By applying HAM, we obtain

$$\tilde{u}_0(x,t;\gamma,\beta) = \{\beta(0.2-0.2\gamma)+(0.1\gamma-0.1)\}(1+t), \tag{5.27}$$

$$\tilde{u}_1(x,t;\gamma,\beta) = \{\beta(0.2-0.2\gamma)+(0.1\gamma-0.1)\}\left(-\frac{c}{\rho A}\frac{t^{3-p}}{\Gamma(4-p)}\right)+\frac{fBt^2}{2\rho A}, \tag{5.28}$$

$$\tilde{u}_2(x,t;\gamma,\beta) = \{\beta(0.2-0.2\gamma)+(0.1\gamma-0.1)\}\left(\frac{c^2}{\rho^2A^2}\frac{t^{5-2p}}{\Gamma(6-2p)}\right) - \frac{fBc}{\rho^2A^2}\frac{t^{4-p}}{\Gamma(5-p)} - \frac{EIBf^4}{\rho^2A^2}\frac{t^4}{\Gamma(5)}, \tag{5.29}$$

$$\tilde{u}_3(x,t;\gamma,\beta) = \{\beta(0.2-0.2\gamma)+(0.1\gamma-0.1)\}\left(-\frac{c^3}{\rho^3A^3}\frac{t^{7-3p}}{\Gamma(8-3p)}\right) + \frac{fBc^2}{\rho^3A^3}\frac{t^{6-2p}}{\Gamma(7-2p)} + \frac{2EIBcf^4}{\rho^3A^3}\frac{t^{6-p}}{\Gamma(7-p)} + \frac{E^2I^2Bf^8}{\rho^3A^3}\frac{t^6}{\Gamma(7)}, \tag{5.30}$$

$$\tilde{u}_4(x,t;\gamma,\beta) = \{\beta(0.2-0.2\gamma)+(0.1\gamma-0.1)\}\left(-\frac{c^4}{\rho^4A^4}\frac{t^{9-4p}}{\Gamma(10-4p)}\right) - \frac{fBc^3}{\rho^4A^4}\frac{t^{8-3p}}{\Gamma(9-3p)} - \frac{3EIBc^2f^4}{\rho^4A^4}\frac{t^{8-2p}}{\Gamma(9-2p)} - \frac{3E^2I^2Bcf^8}{\rho^4A^4}\frac{t^{8-p}}{\Gamma(9-p)} - \frac{3E^3I^3Bf^{12}}{\rho^4A^4}\frac{t^8}{\Gamma(9)}, \tag{5.31}$$

and following this pattern, where $f^{(i)} = \frac{\partial^i f}{\partial x^i}$.

Accordingly, the general form of the solution is given by

$$\tilde{u}(x,t;\gamma,\beta) = \{\beta(0.2-0.2\gamma)+(0.1\gamma-0.1)\}\left(1+\sum_{m=0}^{\infty}\left(\frac{-c}{\rho A}\right)^m\frac{t^{(2-p)m+1}}{\Gamma((2-p)m+2)}\right) + \frac{B}{\rho A}\sum_{i=0}^{\infty}\frac{(-1)^i}{i!}\left(\frac{EI}{\rho A}\right)^i f^{(4i)}t^{2(i+1)}\sum_{j=0}^{\infty}\left(\frac{-c}{\rho A}\right)^j\frac{(j+i)!t^{(2-p)j}}{j!\Gamma((2-p)j+2i+3}. \tag{5.32}$$

As previously elucidated, to represent the solution bounds in a unified parametric form, we assign $\beta = 0$ and 1 to define the lower and upper bounds of the solution, respectively. This can be articulated as

$$\tilde{u}(x,t;\gamma,0) = \underline{u}(x,t;\gamma) \quad \text{and} \quad \tilde{u}(x,t;\gamma,1) = \bar{u}(x,t;\gamma).$$

$$\underline{u}(x,t;\gamma) = (0.1\gamma-0.1)\left(1+\sum_{m=0}^{\infty}\left(\frac{-c}{\rho A}\right)^m\frac{t^{(2-p)m+1}}{\Gamma((2-p)m+2)}\right)$$

$$+\frac{B}{\rho A}\sum_{i=0}^{\infty}\frac{(-1)^i}{i!}\left(\frac{EI}{\rho A}\right)^i f^{(4i)}t^{2(i+1)}\sum_{j=0}^{\infty}\left(\frac{-c}{\rho A}\right)^j\frac{(j+i)!t^{(2-p)j}}{j!\Gamma((2-p)j+2i+3}, \tag{5.33}$$

$$\bar{u}(x,t;\gamma)=(0.1-0.1\gamma)\left(1+\sum_{m=0}^{\infty}\left(\frac{-c}{\rho A}\right)^m\frac{t^{(2-p)m+1}}{\Gamma((2-p)m+2)}\right)$$
$$+\frac{B}{\rho A}\sum_{i=0}^{\infty}\frac{(-1)^i}{i!}\left(\frac{EI}{\rho A}\right)^i f^{(4i)}t^{2(i+1)}\sum_{j=0}^{\infty}\left(\frac{-c}{\rho A}\right)^j\frac{(j+i)!t^{(2-p)j}}{j!\Gamma((2-p)j+2i+3}. \tag{5.34}$$

### 5.4.2 Unit impulse function response

Next we investigate the behavior of the beam when it is subjected to an impulse load characterized by the form $g(t)=\delta(t)$, where $\delta(t)$ represents the unit impulse function. Employing the homotopy analysis method (HAM) once more in this context, we obtain

$$\tilde{u}_0(t;\gamma,\beta)=\{\beta(0.2-0.2\gamma)+(0.1\gamma-0.1)\}(1+t), \tag{5.35}$$

$$\tilde{u}_1(t;\gamma,\beta)=\{\beta(0.2-0.2\gamma)+(0.1\gamma-0.1)\}\left(-\frac{c}{\rho A}\frac{t^{3-p}}{\Gamma(4-p)}\right)+\frac{fBt}{\rho A}, \tag{5.36}$$

$$\tilde{u}_2(t;\gamma,\beta)=\{\beta(0.2-0.2\gamma)+(0.1\gamma-0.1)\}\left(\frac{c^2}{\rho^2A^2}\frac{t^{5-2p}}{\Gamma(6-2p)}\right)-\frac{fBc}{\rho^2A^2}\frac{t^{3-p}}{\Gamma(4-p)}-\frac{EIBf^4}{\rho^2A^2}\frac{t^3}{\Gamma(4)}, \tag{5.37}$$

$$\tilde{u}_3(t;\gamma,\beta)=\{\beta(0.2-0.2\gamma)+(0.1\gamma-0.1)\}\left(-\frac{c^3}{\rho^3A^3}\frac{t^{7-3p}}{\Gamma(8-3p)}\right)+\frac{fBc^2}{\rho^3A^3}\frac{t^{5-2p}}{\Gamma(6-2p)}$$
$$+\frac{2EIBcf^4}{\rho^3A^3}\frac{t^{5-p}}{\Gamma(6-p)}+\frac{E^2I^2Bf^8}{\rho^3A^3}\frac{t^5}{\Gamma(6)}, \tag{5.38}$$

and so on, where $f^{(i)}=\frac{\partial^i f}{\partial x^i}$. Similarly, the other components can be identified. Consequently, the solution can be expressed in a general format as

$$\tilde{u}(x,t;\gamma,\beta)=\{\beta(0.2-0.2\gamma)+(0.1\gamma-0.1)\}\left\{1+\sum_{k=0}^{\infty}\left(\frac{-c}{\rho A}\right)^k\frac{t^{(2-p)k+1}}{\Gamma((2-p)k+2)}\right\}$$
$$+\frac{BI}{\rho A}\sum_{\gamma=0}^{\infty}\frac{(-1)^\gamma}{\gamma!}\left(\frac{EI}{\rho A}\right)^\gamma f^{(4\gamma)}t^{2\gamma+1}\sum_{j=0}^{\infty}\left(\frac{-c}{\rho A}\right)^j\frac{(j+\gamma)!t^{(2-p)j}}{j!\Gamma((2-p)j+2\gamma+2)}. \tag{5.39}$$

To derive the solution in a SPF, we can substitute $\beta=0$ and 1 to determine the solution's lower and upper bounds, respectively. This can also be represented as $u(x,t;\gamma,0)=\underline{u}(x,t,\gamma)$ and $u(x,t,\gamma,1)=\bar{u}(x,t,\gamma)$.

$$\underline{u}(x,t;\gamma,0)=(0.1\gamma-0.1)\left\{1+\sum_{k=0}^{\infty}\left(\frac{-c}{\rho A}\right)^k\frac{t^{(2-p)k+1}}{\Gamma((2-p)k+2)}\right\}$$

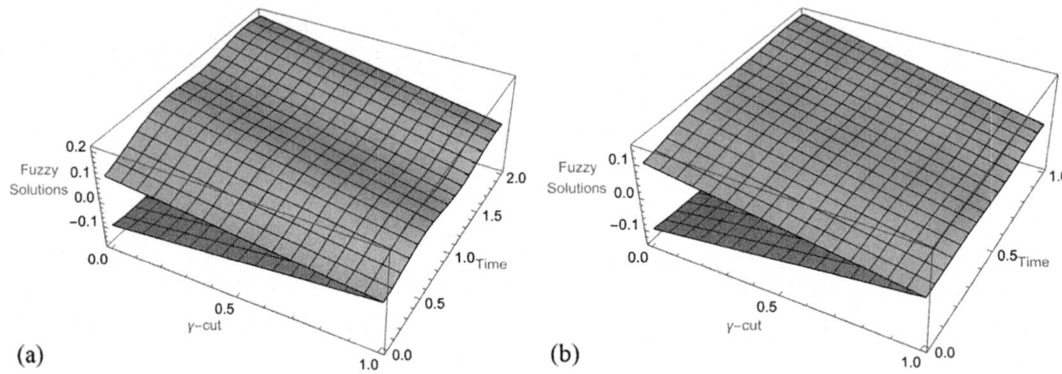


**FIGURE 5.1** Fuzzy step response for $\eta = 0.05$, $p = 0.4$ a) $\omega = 5$ rad/s b) $\omega = 10$ rad/s.

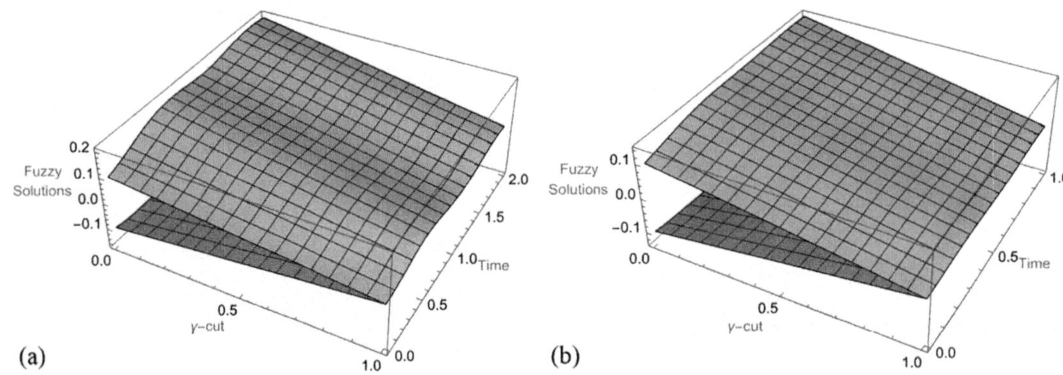


**FIGURE 5.2** Fuzzy step response for $\eta = 0.05$, $p = 0.6$ a) $\omega = 5$ rad/s b) $\omega = 10$ rad/s.

$$+\frac{B}{\rho A}\sum_{\gamma=0}^{\infty}\frac{(-1)^{\gamma}}{\gamma!}\left(\frac{EI}{\rho A}\right)^{\gamma} f^{(4\gamma)}\, t^{2\gamma+1}\sum_{j=0}^{\infty}\left(\frac{-c}{\rho A}\right)^{j}\frac{(j+\gamma)!\,t^{(2-p)j}}{j!\,\Gamma\big((2-p)j+2\gamma+2\big)}, \tag{5.40}$$

$$\begin{aligned}\bar{u}\,(x,t;\gamma,1) = {} & (0.1-0.1\gamma)\left\{1+\sum_{k=0}^{\infty}\left(\frac{-c}{\rho A}\right)^{k}\frac{t^{(2-p)k+1}}{\Gamma\left((2-p)k+2\right)}\right\} \\ & +\frac{B}{\rho A}\sum_{\gamma=0}^{\infty}\frac{(-1)^{\gamma}}{\gamma!}\left(\frac{EI}{\rho A}\right)^{\gamma} f^{(4\gamma)}\, t^{2\gamma+1}\sum_{j=0}^{\infty}\left(\frac{-c}{\rho A}\right)^{j}\frac{(j+\gamma)!\,t^{(2-p)j}}{j!\,\Gamma\big((2-p)j+2\gamma+2\big)}.\end{aligned} \tag{5.41}$$

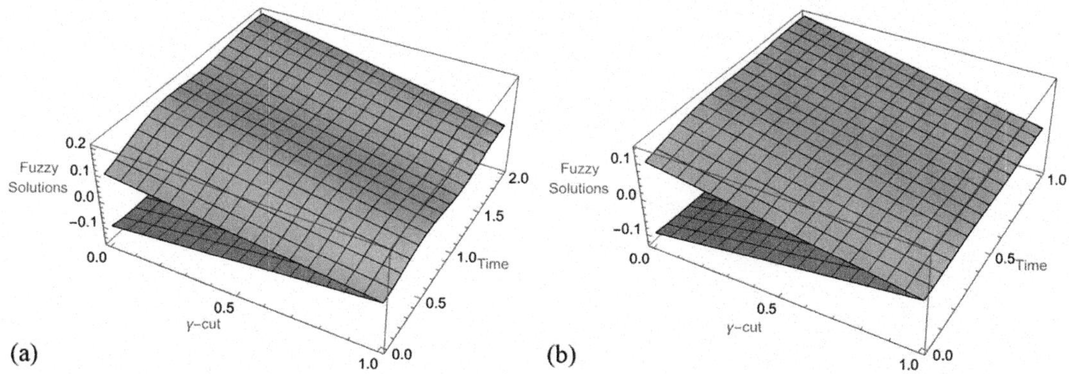


**FIGURE 5.3** Fuzzy step response for $\eta = 0.05$, $p = 0.8$ a) $\omega = 5$ rad/s b) $\omega = 10$ rad/s.

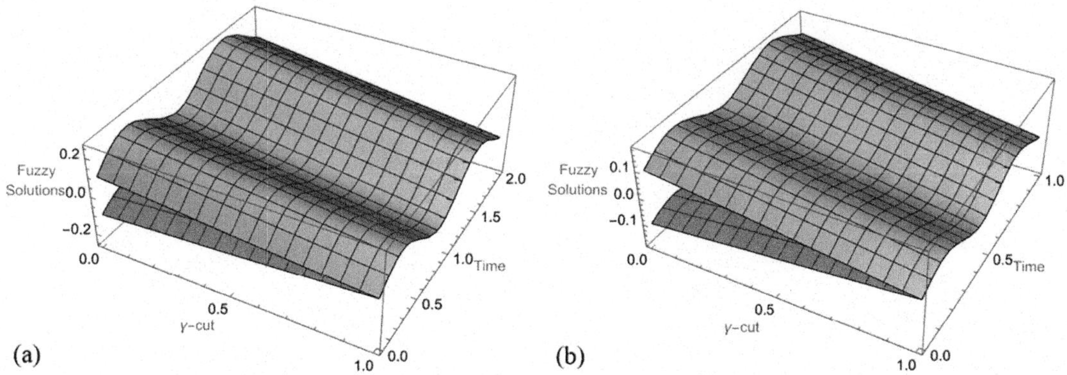


**FIGURE 5.4** Fuzzy impulse response for $\eta = 0.05$, $p = 0.4$ a) $\omega = 5$ rad/s b) $\omega = 10$ rad/ s.

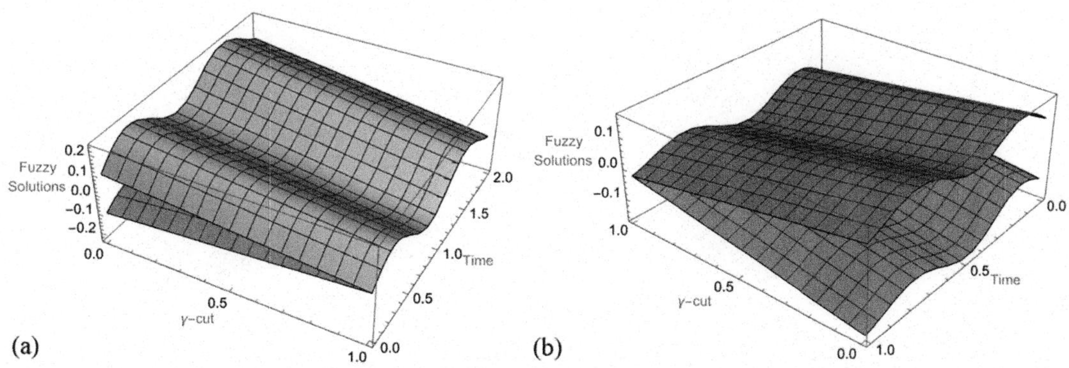


**FIGURE 5.5** Fuzzy impulse response for $\eta = 0.05$, $p = 0.6$ a) $\omega = 5$ rad/s b) $\omega = 10$ rad/s.

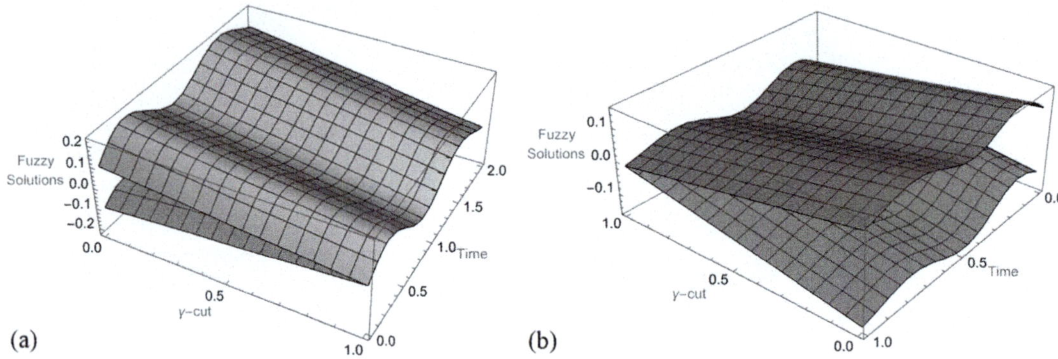


**FIGURE 5.6** Fuzzy unit impulse response for $\eta = 0.05$, $p = 0.8$ a) $\omega = 5$ rad/s b) $\omega = 10$ rad/s.

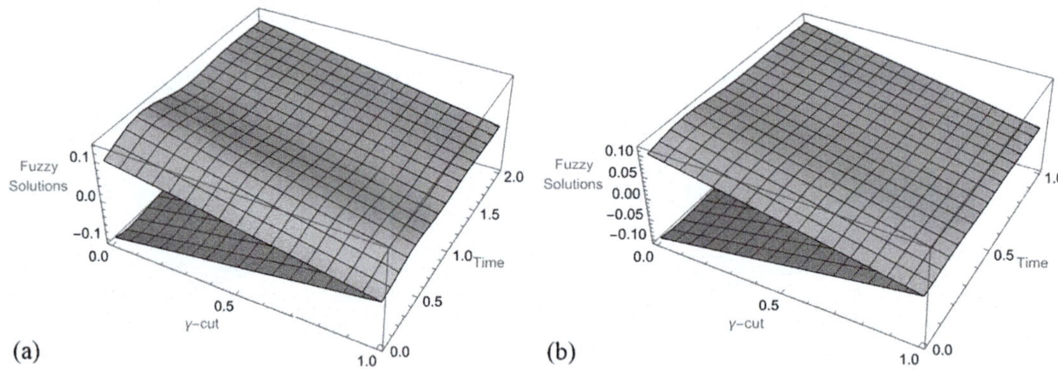


**FIGURE 5.7** Fuzzy step response for $\eta = 0.5$, $p = 0.4$ a) $\omega = 5$ rad/s b) $\omega = 10$ rad/s.

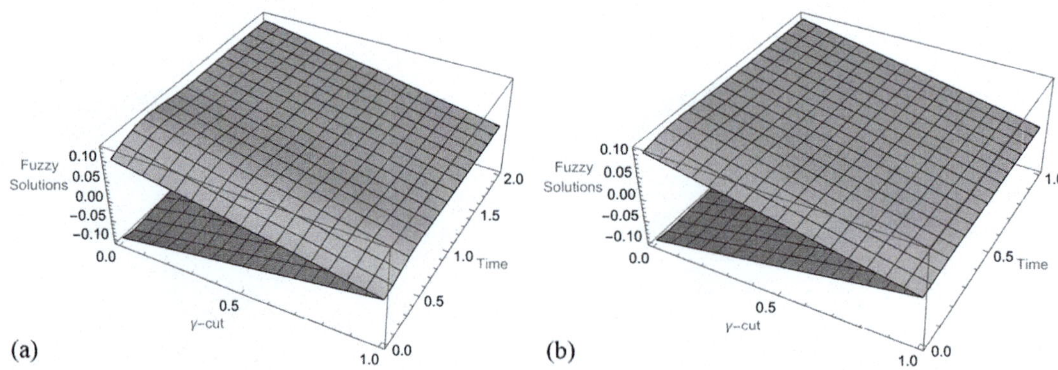


**FIGURE 5.8** Fuzzy step response for $\eta = 0.5$, $p = 0.6$ a) $\omega = 5$ rad/s b) $\omega = 10$ rad/s.

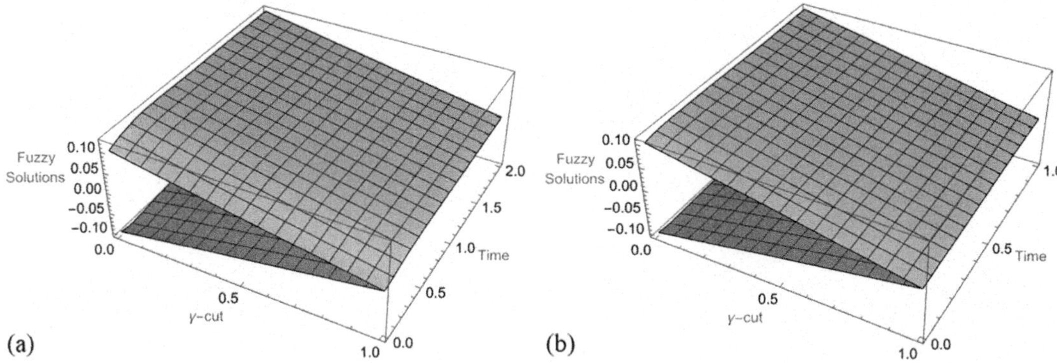


**FIGURE 5.9** Fuzzy step response for $\eta = 0.5$, $p = 0.8$ a) $\omega = 5$ rad/s b) $\omega = 10$ rad/s.

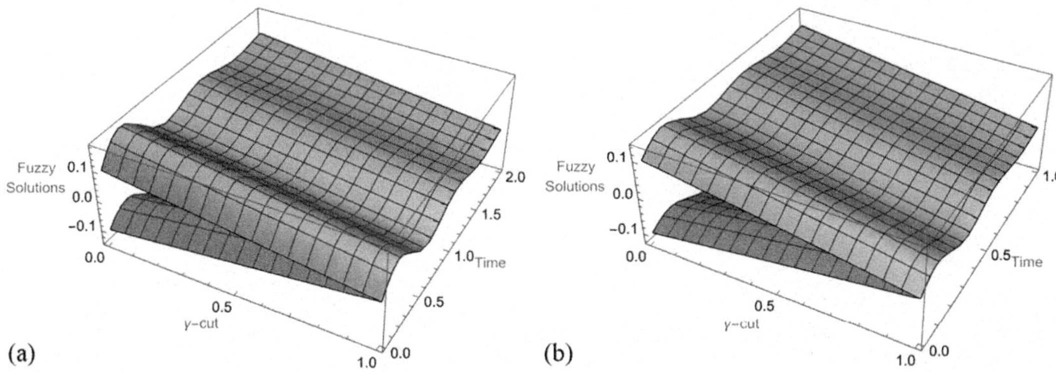


**FIGURE 5.10** Fuzzy impulse response for $\eta = 0.5$, $p = 0.4$ a) $\omega = 5$ rad/s b) $\omega = 10$ rad/s.

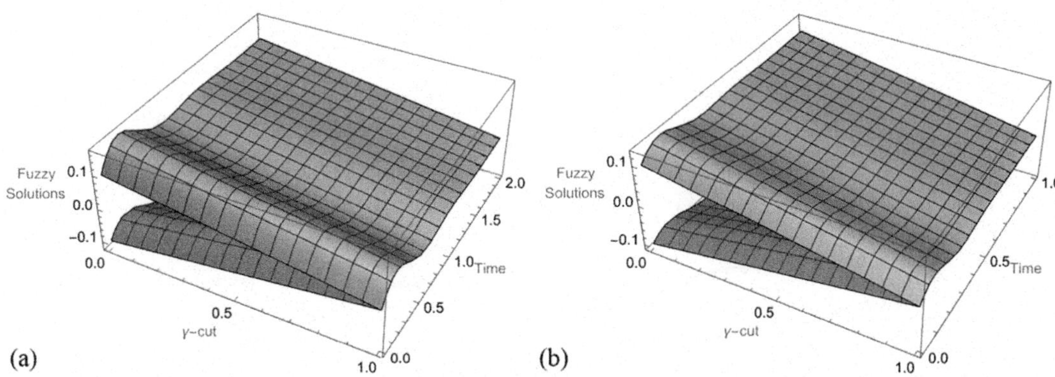


**FIGURE 5.11** Fuzzy unit impulse response for $\eta = 0.5$, $p = 0.6$ a) $\omega = 5$ rad/s b) $\omega = 10$ rad/s.

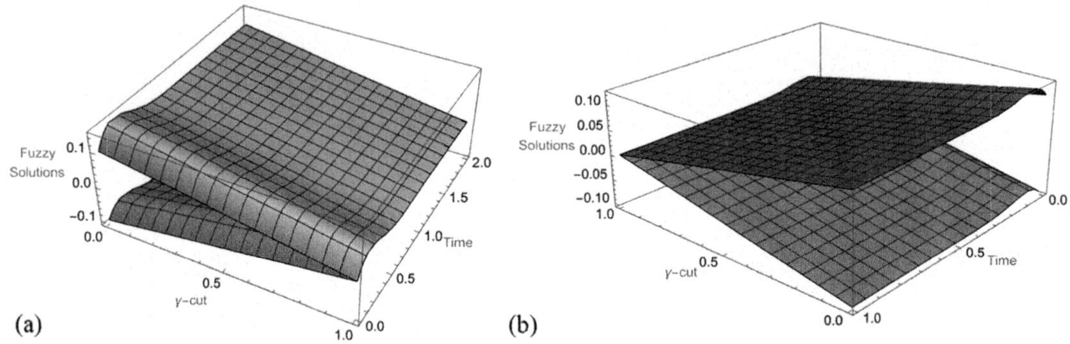


**FIGURE 5.12** Fuzzy unit impulse response for $\eta = 0.5$, $p = 0.8$ a) $\omega = 5$ rad/s b) $\omega = 10$ rad/s.

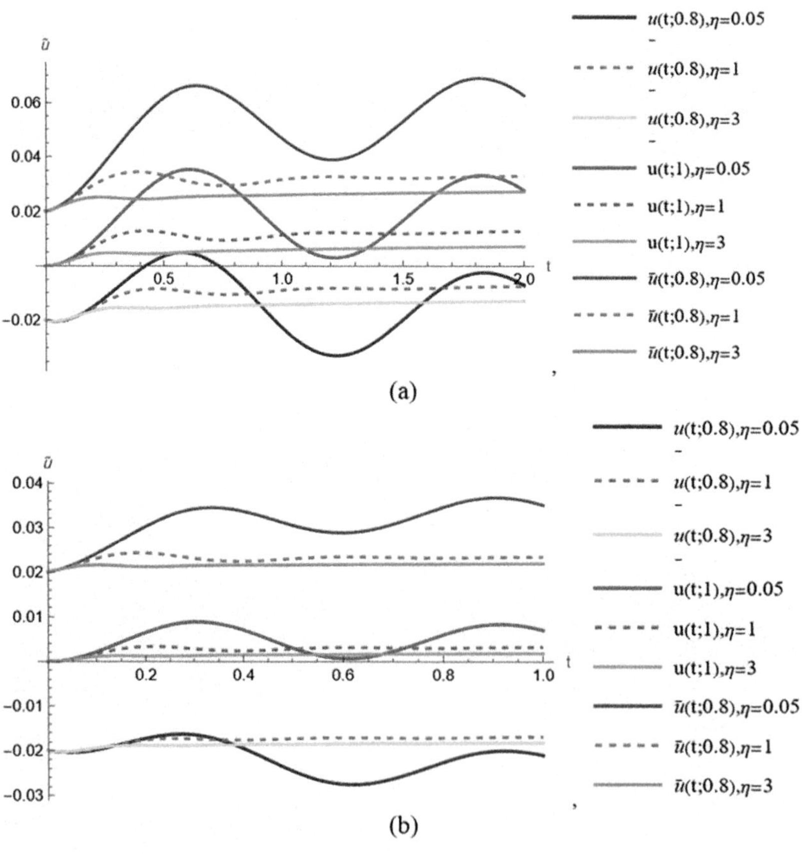


**FIGURE 5.13** Interval step response for $p = 0.4$, $\gamma = 0.8$ a) $\omega = 5$ rad/s b) $\omega = 10$ rad/s.

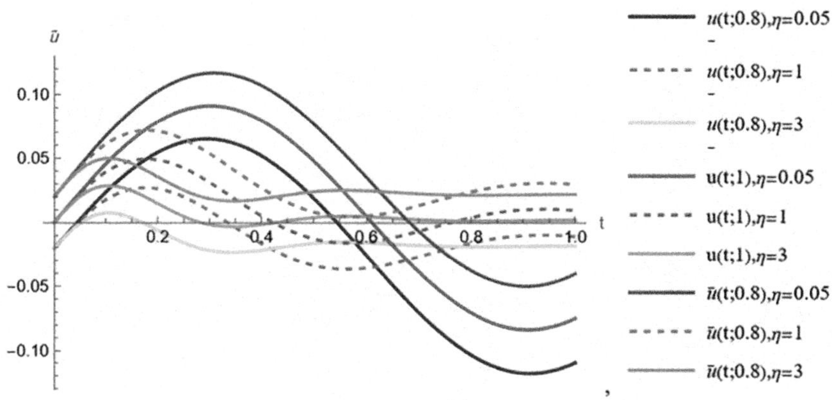


(a)

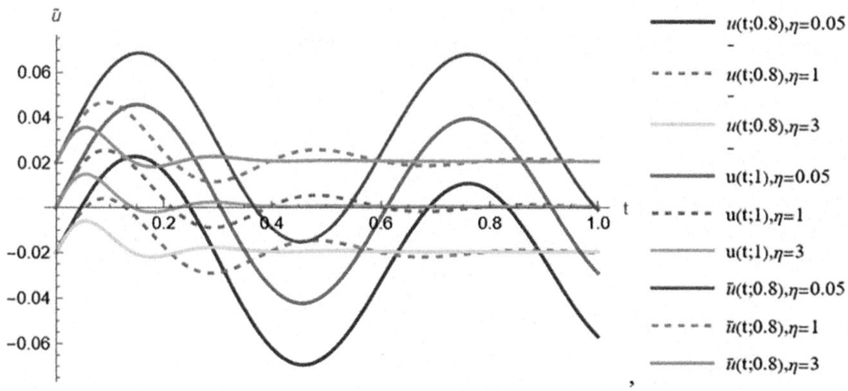


(b)

**FIGURE 5.14** Interval impulse response for $p = 0.4$, $\gamma = 0.8$ a) $\omega = 5$ rad/s b) $\omega = 10$ rad/s.

# 5.5 Numerical results and discussions

The present section considers fuzzy responses corresponding to step and impulse functions. Eqs. (5.32), (5.33), (5.40), and (5.41) offer the requisite solutions. Assuming the beam is simply supported, we have $f(x) = sin\left(\frac{\pi x}{L}\right)$. The infinite series in Eqs. (5.33), (5.34), (5.40), and (5.41) are approximated by truncating to 50 terms for numerical computation. In the simulations, let us denote $\frac{c}{m}$ and $\frac{EI}{\rho A}$, respectively, as $2\eta\omega^{3/2}$ and $\omega^2$ where, $\omega$ is the natural frequency, and $\eta$ is the damping ratio. The parameter values are chosen as $B = 1$, $\rho A = 1$, $L = \pi$, $x = \frac{1}{2}$ and $m = 1$. Notably, the lower and upper bounds of the fuzzy solutions coincide for $\gamma = 1$ in both response and our findings match the results reported by Liang and Tang (2007); Jena et al. (2019).

As shown in Fig. 5.1, Fig. 5.2, Fig. 5.3, Fig. 5.4, Fig. 5.5, and Fig. 5.6, the parameter values are chosen as $\eta = 0.05$, with different fractional order derivative $p = 0.4$, $p = 0.6$, $p = 0.8$, $\omega = \frac{5\text{ rad}}{\text{s}}$ and $\omega = \frac{10\text{ rad}}{\text{s}}$, respectively. In Fig. 5.7, Fig. 5.8, Fig. 5.9, Fig. 5.10, Fig. 5.11, and

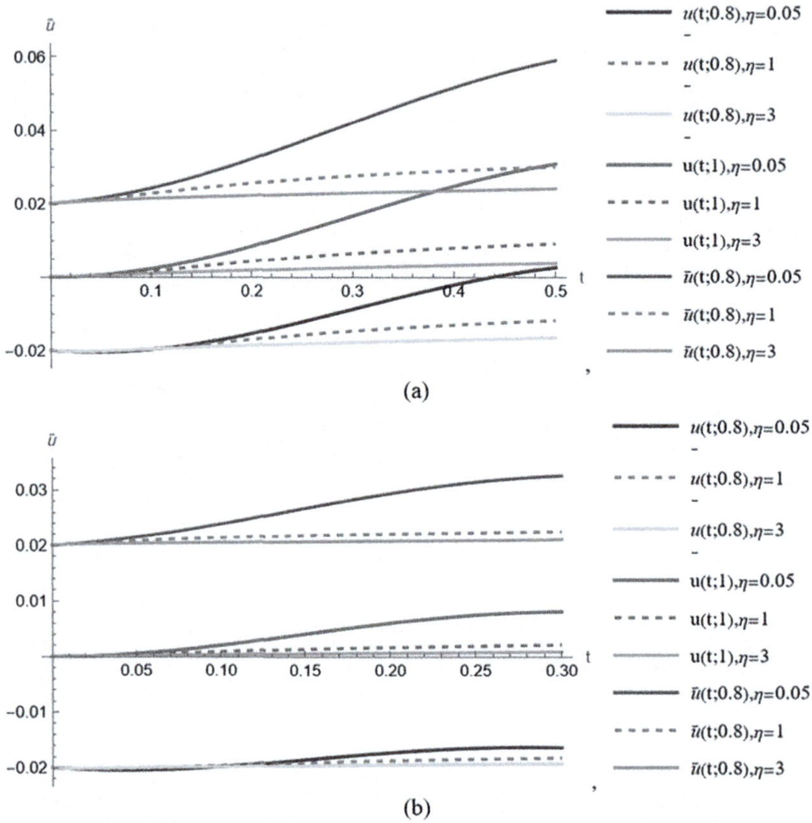


**FIGURE 5.15** Interval step response for $p = 0.8$, $\gamma = 0.8$ a) $\omega = 5$ rad/s b) $\omega = 10$ rad/s.

Fig. 5.12, the analysis focuses on $\eta = 0.5$, $p = 0.4$, $p = 0.6$, $p = 0.8$, $\omega = \frac{5 \text{ rad}}{\text{s}}$ and $\omega = 10$ rad/s. The fuzzy responses over time for all the cases are shown in Fig. 5.1 to Fig. 5.12, respectively. Following this, Fig. 5.13, Fig. 5.14, Fig. 5.15, and Fig. 5.16 display the effects of interval responses for both cases at the specified value of $\gamma = 0.8$ with fractional order $p = 0.4$, $p = 0.8$, $\eta = 0.05, 1, 3$, $\omega = 5$ rad/s, $\omega = \frac{10 \text{ rad}}{\text{s}}$, respectively. The figure shows that increasing $\eta$ leads to a reduction in oscillatory behavior. Similarly, Fig. 5.17, Fig. 5.18, Fig. 5.19, and Fig. 5.20 represent the interval responses for both the cases $\eta = 0.5$, $\gamma = 0.6$ and $\eta = 1$, $\gamma = 0.6$ and $\eta = 1$, $\gamma = 0.6$, with natural frequency $\omega = 5$ rad/s and $\omega = 10$ rad/s, varying the fractional derivative $p = 0.2$, $p = 0.6$, $p = 1$, respectively. It is observed that with increase of fractional derivative oscillation decreases. Here for $\gamma = 1$, the lower and upper bounds of the solution in both the responses are identical and assigned as $\underline{u}(t;1) = \overline{u}(t;1) = u(t;1)$.

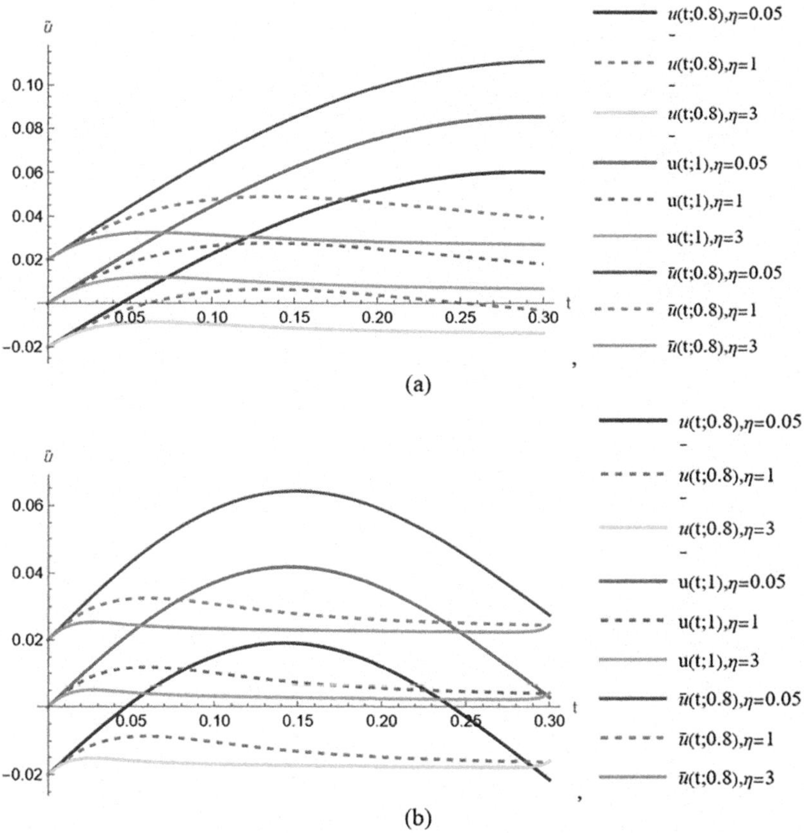


**FIGURE 5.16** Interval impulse response for $p = 0.8$, $\gamma = 0.8$ a) $\omega = 5$ rad/s b) $\omega = 10$ rad/s.

## 5.6 Conclusions

The dynamic behavior of fuzzy fractional-order damped beams under uncertainty has been investigated using the homotopy analysis method, along with double parametric fuzzy numbers. The double parametric approach facilitates the conversion of uncertain into crisp form, enabling efficient solutions. TFN are utilized in this analysis to investigate the imprecise dynamic behavior subjected to unit step and impulse forces. Notably, the results obtained using the present method show excellent agreement with those reported in Liang and Tang (2007) and Chakraverty et al. (2016b) for a special case, validating the accuracy of our approach.

## Acknowledgment

The primary and second authors sincerely acknowledge the financial support provided by OSHEC, Government of Odisha, through MRIP funding (23EM/MT/97).

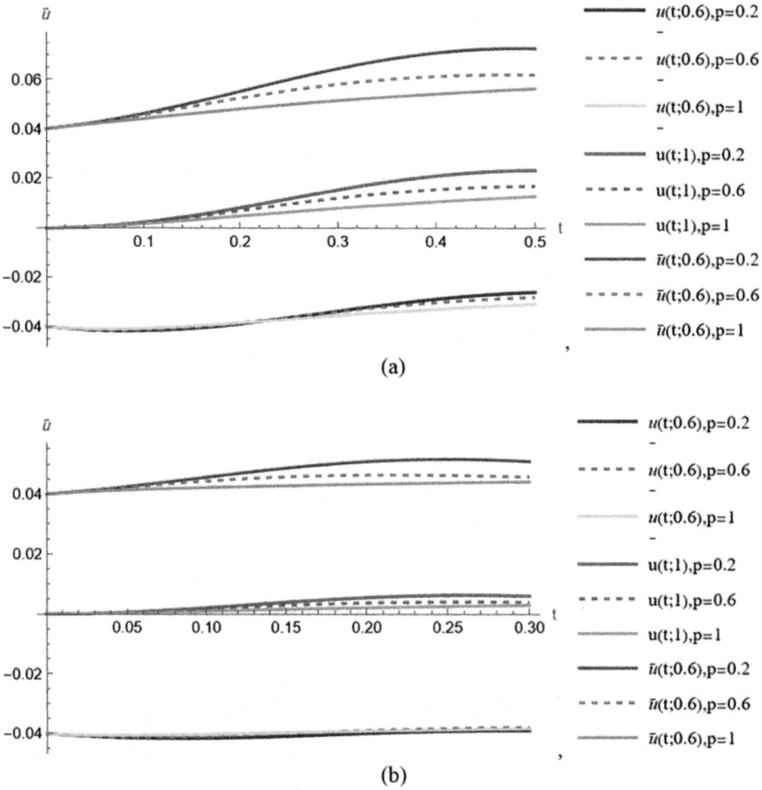


**FIGURE 5.17** Interval step response for $\eta = 0.5$, $\gamma = 0.6$ a) $\omega = 5$ rad/s b) $\omega = 10$ rad/s.

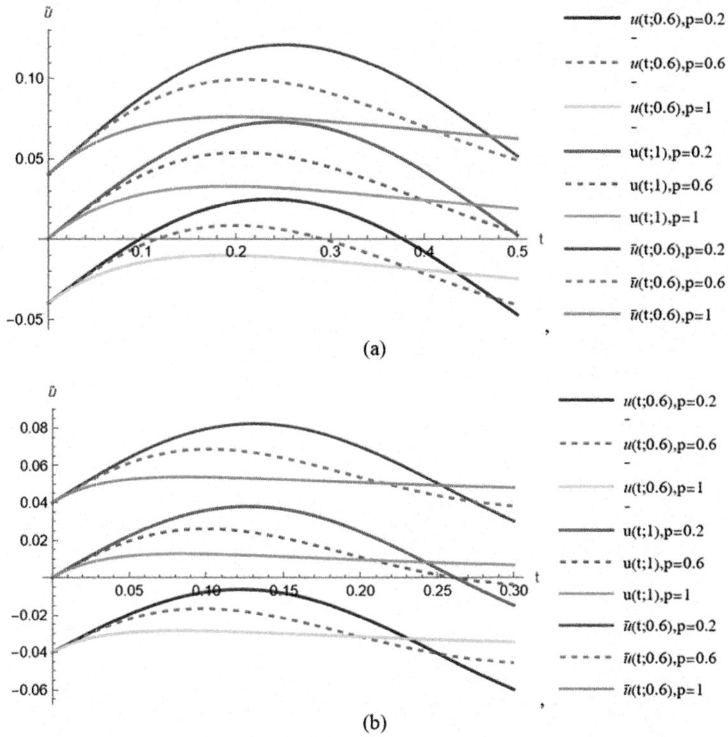


**FIGURE 5.18** Interval impulse response for $\eta = 0.5$, $\gamma = 0.6$ a) $\omega = 5$ rad/s b) $\omega = 10$ rad/s.

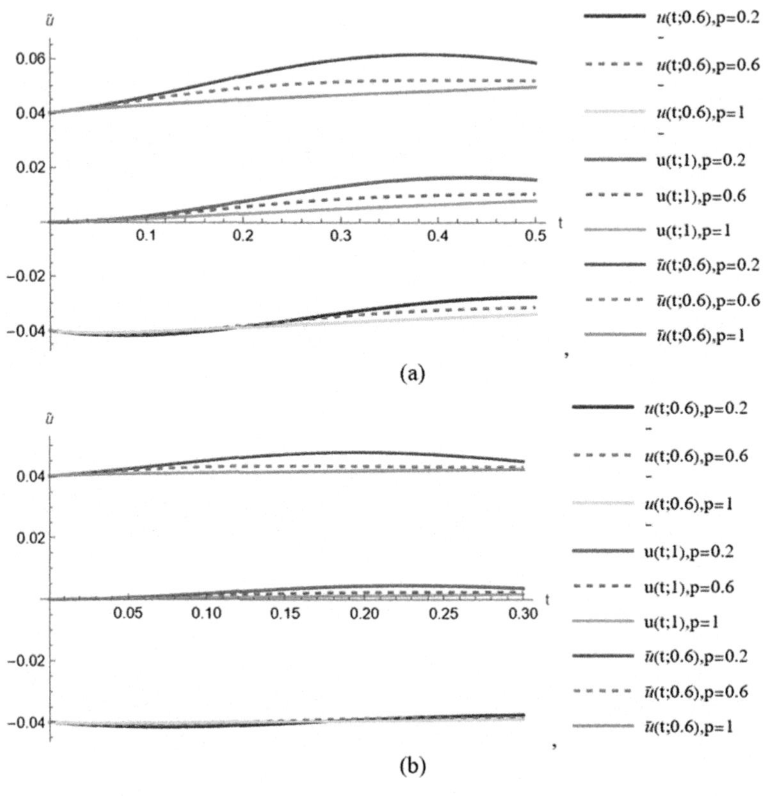


**FIGURE 5.19** Interval step response for $\eta = 1$, $\gamma = 0.6$ a) $\omega = 5$ rad/s b) $\omega = 10$ rad/s.

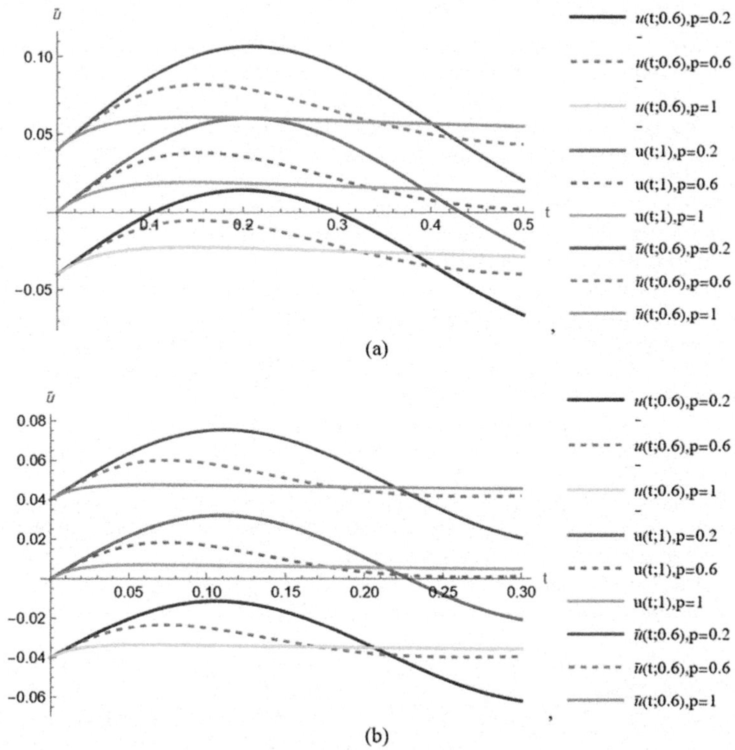


**FIGURE 5.20** Interval impulse response for $\eta = 1$, $\gamma = 0.6$ a) $\omega = 5$ rad/s b) $\omega = 10$ rad/s.

## References

Abaid Ur Rehman, Muhammad, Ahmad, Jamshad, Hassan, Ali, Awrejcewicz, Jan, Pawlowski, Witold, Karamti, Hanen, Alharbi, Fahad M., 2022. The dynamics of a fractional-order mathematical model of cancer tumor disease. Symmetry (ISSN 2073-8994) 14 (8), 1694. https://doi.org/10.3390/sym14081694.

Agrawal, R.P., Lakshmikantham, V., Nieto, J.J., 2010. On the concept of solution for fractional differential equations with uncertainty. Nonlinear Analysis 72, 2859–2862.

Ahmad, M.Z., Hasan, M.K., Abbasbandy, S., Bruzón, M., Momani, S., 2013. Solving fuzzy fractional differential equations using Zadeh's extension principle. The Scientific World Journal (ISSN 2356-6140) 2013 (1). https://doi.org/10.1155/2013/454969.

Alderremy, A.A., Khan, Hassan, Shah, Rasool, Aly, Shaban, Baleanu, Dumitru, 2020. The analytical analysis of time-fractional Fornberg–Whitham equations. Mathematics (ISSN 2227-7390) 8 (6), 987. https://doi.org/10.3390/math8060987.

Al-Qudah, Alaa, Odibat, Zaid, Shawagfeh, Nabil, 2022. A linearization-based computational algorithm of homotopy analysis method for nonlinear reaction–diffusion systems. Mathematics and Computers in Simulation (ISSN 0378-4754) 194, 505–522. https://doi.org/10.1016/j.matcom.2021.11.027.

Arshad, S., Lupulescu, V., 2011. On the fractional differential equations with uncertainty. Nonlinear Analysis: Theory, Methods & Applications (ISSN 0362-546X) 74 (11), 3685–3693. https://doi.org/10.1016/j.na.2011.02.048.

Chakraverty, S., Behera, Diptiranjan, 2013. Dynamic responses of fractionally damped mechanical system using homotopy perturbation method. Alexandria Engineering Journal (ISSN 1110-0168) 52 (3), 557–562. https://doi.org/10.1016/j.aej.2013.04.007.

Chakraverty, S., Behera, Diptiranjan, 2015. Uncertain dynamic responses of fuzzy fractionally damped spring-mass system. Journal of Intelligent & Fuzzy Systems (ISSN 1064-1246) 29 (1), 327–336. https://doi.org/10.3233/ifs-151598.

Chakraverty, Snehashish, Tapaswini, Smita, Behera, Diptiranjan, 2016a. Fuzzy Arbitrary Order System: Fuzzy Fractional Differential Equations and Applications. Wiley, Wiley.

Chakraverty, S., Smita, Tapaswini, Diptiranjan, Behera, 2016b. Fuzzy Differential Equations and Applications for Engineers and Scientists. CRC Press, Taylor and Francis Group, Florida, USA. ISBN 9781482244731-CAT#K23501.

Georgieva, A., Spasova, M., 2022. Solution of partial fuzzy integro-differential equations by double natural transform. In: AIP Conference Proceedings. 2022/04/05. American Institute of Physics Inc., Bulgaria. http://scitation.aip.org/content/aip/proceeding/aipcp2459.

Guo, Peng, 2019. The Adomian decomposition method for a type of fractional differential equations. Journal of Applied Mathematics and Physics (ISSN 2327-4352) 07 (10), 2459–2466. https://doi.org/10.4236/jamp.2019.710166.

Iyiola, Olaniyi Samuel, Zaman, F.D., 2014. A fractional diffusion equation model for cancer tumor. AIP Advances (ISSN 2158-3226) 4 (10). https://doi.org/10.1063/1.4898331.

Jamal, N., Sarwar, M., Agarwal, P., Mlaiki, N., Aloqaily, A., 2023. Solutions of fuzzy advection-diffusion and heat equations by natural Adomian decomposition method. Scientific Reports (ISSN 2045-2322) 13 (1). https://doi.org/10.1038/s41598-023-45207-y. Nature Research, Pakistan, Available from: https://www.nature.com/srep/.

Jena, R.M., Chakraverty, S., Jena, S.K., 2019. Dynamic response analysis of fractionally damped beams subjected to external loads using homotopy analysis method. Journal of Applied and Computational Mechanics (ISSN 2383-4536) 5 (2), 355–366. https://doi.org/10.22055/jacm.2019.27592.1419. Shahid Chamran University of Ahvaz, India, Available from: http://jacm.scu.ac.ir/article_14072_d4c690cf2cf2984ab3bb8e1450798d4f.pdf.

Kilbas, A.A., Srivastava, H.M., Trujillo, J.J., 2006. Theory and Applications of Fractional Differential Equations. Elsevier, New York.

Kiryakova, V.S., 1993. Generalized Fractional Calculus and Applications. Longman Scientific and Technical, Longman House, Burnt Mill, Harlow, England.

Kumar, Devendra, Singh, Jagdev, Baleanu, Dumitru, 2018. A new analysis of the Fornberg-Whitham equation pertaining to a fractional derivative with Mittag-Leffler-type kernel. The European Physical Journal Plus (ISSN 2190-5444) 133 (2). https://doi.org/10.1140/epjp/i2018-11934-y.

Liang, Zu-feng, Tang, Xiao-yan, 2007. Analytical solution of fractionally damped beam by Adomian decomposition method. Applied Mathematics and Mechanics (ISSN 0253-4827) 28 (2), 219–228. https://doi.org/10.1007/s10483-007-0210-z.

Liao, S.J., 1992. The proposed homotopy analysis technique for the solution of nonlinear problems. Ph.D. thesis. Shanghai Jiao Tong University.

Liao, S.J., 2003. Beyond Perturbation: an Introduction to the Homotopy Analysis Method. CRC Press, Chapman & Hall, Boca Raton.

Miller & Ross, 1993. An Introduction to the Fractional Calculus and Fractional Differential Equations. John Wiley and Sons, New York.

Muhammad, Ghulam, Akram, Muhammad, 2024. Fuzzy fractional generalized Bagley–Torvik equation with fuzzy Caputo gH-differentiability. Engineering Applications of Artificial Intelligence (ISSN 0952-1976) 133, 108265. https://doi.org/10.1016/j.engappai.2024.108265.

Odibat, Zaid, Baleanu, Dumitru, 2019. A linearization-based approach of homotopy analysis method for non-linear time-fractional parabolic PDEs. Mathematical Methods in the Applied Sciences (ISSN 0170-4214) 42 (18), 7222–7232. https://doi.org/10.1002/mma.5829.

Oldham, K.B., Spanier, J., 1974. The Fractional Calculus. Academic Press, NY, New York.

Podlubny, I., 1999. Fractional Differential Equations. Academic Press, NY, New York.

Qayyum, Mubashir, Ahmad, Efaza, 2024. Exploration of time-fractional cancer tumor models with variable cell killing rates via hybrid algorithm. Physica Scripta (ISSN 0031-8949) 99 (11), 115004. https://doi.org/10.1088/1402-4896/ad7f98.

Qu, Haidong, She, Zihang, Liu, Xuan, 2020. Homotopy analysis method for three types of fractional partial differential equations. Complexity (ISSN 1076-2787) 2020, 1–13. https://doi.org/10.1155/2020/7232907.

Samko, S.G., Kilbas, A.A., Marichev, O.I., 1993. Fractional Integrals and Derivatives: Theory and Application. Gordon and Breach Science Publishers PA, Langhorne.

Saratha, S.R., Bagyalakshmi, M., Sai Sundara Krishnan, G., 2020. Fractional generalised homotopy analysis method for solving nonlinear fractional differential equations. Computational & Applied Mathematics (ISSN 2238-3603) 39 (2). https://doi.org/10.1007/s40314-020-1133-9.

Sartanpara, Parthkumar P., Meher, Ramakanta, Meher, S.K., 2022. The generalized time-fractional Fornberg–Whitham equation: an analytic approach. Partial Differential Equations in Applied Mathematics (ISSN 2666-8181) 5, 100350. https://doi.org/10.1016/j.padiff.2022.100350.

Tapaswini, Smita, Mu, Chunlai, Behera, Diptiranjan, Chakraverty, Snehashish, 2017. Solving imprecisely defined vibration equation of large membranes. Engineering Computations (ISSN 0264-4401) 34 (8), 2528–2546. https://doi.org/10.1108/ec-04-2017-0118.

Tapaswini, Smita, Behera, Diptiranjan, 2020. Analysis of imprecisely defined fuzzy space-fractional telegraph equations. Pramāna (ISSN 0304-4289) 94 (1). https://doi.org/10.1007/s12043-019-1889-x.

Tapaswini, Smita, Chakraverty, S., Behera, Diptiranjan, 2015. Numerical solution of the imprecisely defined inverse heat conduction problem. Chinese Physics B (ISSN 1674-1056) 24 (5), 050203. https://doi.org/10.1088/1674-1056/24/5/050203.

Tapaswini, S., Behera, D., 2021. Imprecisely defined fractional order Fokker-Planck equation subjected to fuzzy uncertainty. Pramāna 95, 13.

Ucar, Esmehan, Özdemir, Necati, Altun, Eren, Kumar, D., Baleanu, D., Hristov, J., Nieto, J.J., Ozdemir, N., 2019. Fractional order model of immune cells influenced by cancer cells. Mathematical Modelling of Natural Phenomena (ISSN 0973-5348) 14 (3), 308. https://doi.org/10.1051/mmnp/2019002.

Ullah, Abd, Ullah, Aman, Ahmad, Shabir, Van Hoa, Ngo, 2023. Hybrid fuzzy Laplace-like transforms for solving fractional-order fuzzy differential equations. Fuzzy Sets and Systems (ISSN 0165-0114) 473, 108739. https://doi.org/10.1016/j.fss.2023.108739.

Ullah, Z., Ullah, A., Shah, K., Baleanu, D., 2020. Computation of semi-analytical solutions of fuzzy nonlinear integral equations advances in difference equations. Advances in Difference Equations, 1–11.

Zadeh, L., 1965. Fuzzy sets. Information and Control 8, 338–353.

Zhang, B.g., Li, S.y., Liu, Z.r., 2008. Homotopy perturbation method for modified Camassa-Holm and Degasperis-Procesi equations. Physics Letters, Section A: General, Atomic and Solid State Physics (ISSN 0375-9601) 372 (11), 1867–1872. https://doi.org/10.1016/j.physleta.2007.10.072. Elsevier, China, Available from: https://www.journals.elsevier.com/physics-letters-a.

Zureigat, Hamzeh, Tashtoush, Mohammad A., Al Jassar, Ali F., Az-Zo'bi, Emad A., Alomari, Mohammad W., 2023a. Dirichlet conditions using a modified Crank-Nicolson method. Advances in Mathematical Physics, 1–8.

Zureigat, H., Tashtoush, M.A., Jassar, A.F.A., Az-Zo'bi, E.A., Alomari, M.W.A., 2023b. Solution of the complex fuzzy heat equation in terms of complex. Advances in Mathematical Physics 2023, 6505227.

# 6

# A mathematical model of solute transfer through a permeable channel: impact of viscosity

M. Varunkumar and V. Pujitha
*Department of Mathematics, School of Advanced Sciences, VIT-AP University, Amaravati, Andhra Pradesh, India*

## 6.1 Introduction

The kidneys are responsible for excreting the bulk of human metabolism's end products and maintaining the concentrations of the majority of bodily fluid components, making them one of the most essential physiological organs. By producing urine, the kidney's main target is to clear the unwanted substanses from body's metabolism. The afferent arteriole transports blood to glomerulus, while efferent arteriole removes it. The Bowman's capsule and vascular space are free to exchange water and dissolved components during glomerular filtration. Glomerular filtrate is the result of Starling's effect in the glomerulus's capillaries. A net fluid flowing from the glomerular capillaries into Bowman's space occurs in these circumstances. To study the filtration mechanism in glomerular capillaries, several researchers have created mathematical models (Berman, 1958; Apelblat et al., 1974; Vander et al., 1975; Palatt et al., 1974; Guyton, 1986; James and James, 1998).

Brenner et al. (1972, 1978) examined a number of physiological parameters in the glomerulus, such as the filtration rate of nephron, solute concentration, osmotic pressure at the glomerulus's extremities, and mean and vascular pressures in Bowman's capsule. Deen et al. (1972) investigated connection between filtration and concentration of the protein brought about by filtration, and came to the conclusion that as blood volume flow increases, so does the overall filtration rate. Marshall and Trowbridge (1974) and Huss et al. (1975) created glomerular ultrafiltration models where the intraluminal pressure gradient changes with axial position by taking constant pressure gradient. Additionally, Papenfuss and Gross (1978) investigated the impact of glomerular capillary wall permeability and pressure drop on the filtration process.

Solute transport across the capillary membrane was not taken into consideration in the aforementioned studies, which were mainly concerned with fluid dynamics and pressure distribution. By examining fluid exchange and solute transfer across the capillary wall, Salathe (1988) filled this gap, assuming that the solute concentration was constant throughout each cross-section. Assuming zero osmotic pressure, Ross (1974) presented

Advanced Computational and Mathematical Approaches in Applied Differential Equations
https://doi.org/10.1016/B978-0-44-345743-2.00017-7

a mass transfer model in a porous tube with a small radial flow flux. Tyagi and Abbas (1987) developed the steady-state solution for concentration distribution during blood flow through a hollow-fiber kidney in the context of dialysis process. Chaturani and Ranganatha (1993) investigated the effects of osmotic and hydrostatic pressures on protein transport across the capillary wall and filtration velocity. Building on this, Varunkumar and Muthu (2019, 2020) created a model for the transport of solutes in a porous tube that is dependent on the local solute concentration and has spatially varying wall permeability, slip conditions, and osmotic pressure.

According to the literature review above, the concentration dependence of osmotic pressure and fluid and solute transport in the presence of variable viscosity have not been thoroughly studied. This study's main goal is to examine solute transport in a channel with a permeable boundary while taking spatially varying viscosity into account (Umavathi (2015); Awais et al. (2017)). Davis and Sherwood (1990) investigated effects of diffusivity and variable viscosity in the context of a steady-state boundary layer mass transfer model. Bowen and Williams (2001) examined the effects of viscosity and concentration-dependent diffusivity on transport behavior in the context of cross-flow ultrafiltration. Furthermore, by integrating solute viscosity as a function of concentration and investigating concentration polarization as a coupled transport phenomenon, Shamsuddin et al. (1995) provided important insights into how viscosity variations shape solute concentration profiles. The Newtonian, incompressible, and viscous laminar flow of a solute in a permeable channel with varying viscosity is investigated in this paper.

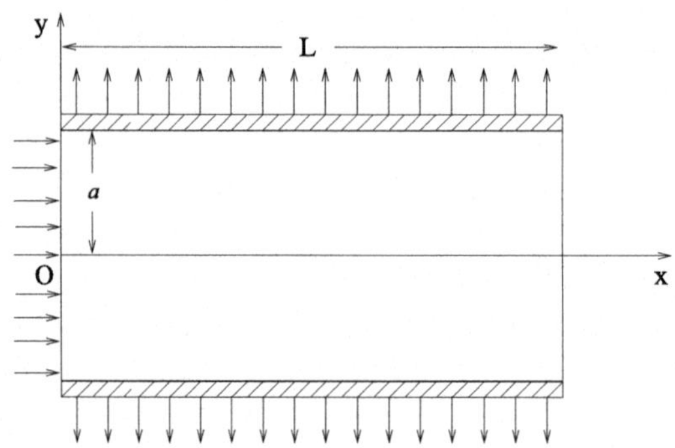


**FIGURE 6.1** Geometrical model of the problem.

## 6.2 Problem formulation

The steady, incompressible flow of a viscous Newtonian fluid and solute transport in a channel with length $L$ and half-width $a$ are examined in this study (see Fig. 6.1). Following are the equations that control fluid flow with varying viscosity $\mu'(y')$, as indicated by the fluid's momentum and continuity:

$$\rho\left(v'\frac{\partial u'}{\partial y'}+u'\frac{\partial u'}{\partial x'}\right)=-\frac{\partial p'}{\partial x'}+\frac{\partial}{\partial y'}\left(\mu'(y')\left(\frac{\partial v'}{\partial x'}+\frac{\partial u'}{\partial y'}\right)\right)+2\frac{\partial}{\partial x'}\left(\mu'(y')\frac{\partial u'}{\partial x'}\right), \tag{6.1}$$

$$\rho\left(v'\frac{\partial v'}{\partial y'}+u'\frac{\partial v'}{\partial x'}\right)=-\frac{\partial p'}{\partial y'}+\frac{\partial}{\partial x'}\left(\mu'(y')\left(\frac{\partial v'}{\partial x'}+\frac{\partial u'}{\partial y'}\right)\right)+2\frac{\partial}{\partial y'}\left(\mu'(y')\frac{\partial u'}{\partial y'}\right), \tag{6.2}$$

$$\frac{\partial u'}{\partial x'}+\frac{\partial v'}{\partial y'}=0. \tag{6.3}$$

The solute transfer in such system is

$$v'\frac{\partial c'}{\partial y'}+u'\frac{\partial c'}{\partial x'}=D\left(\frac{\partial^2 c'}{\partial x'^2}+\frac{\partial^2 c'}{\partial y'^2}\right). \tag{6.4}$$

In this case, $p'$ stands for the pressure, and $u'$ and $v'$ for the axial and transverse velocities, respectively. $c'(x', y')$ denotes solute concentration. $D$ indicates diffusion coefficient. The following are the velocity and concentration boundary conditions that apply to the current situation:

At $y'=0$ (Symmetric condition),

$$\frac{\partial u'}{\partial y'}=\frac{\partial c'}{\partial y'}=v'=0 \tag{6.5}$$

At $y'=a$,

$$u'=0,\ v'=k(\Delta P'-\sigma\Delta\pi')=V_a, \tag{6.6}$$

$$D\frac{\partial c'}{\partial y'}+h(c'-c_T)=(1-T_a)V_a\Phi, \tag{6.7}$$

where $\Phi=\begin{cases} c'; & V_a>0 \\ c_T; & V_a<0 \end{cases}$. The no-slip condition and Starling's law represented by Eq. (6.6) at $y'=a$ (Pozrikidis (2013); Pollak et al. (2014)). Eq. (6.7) (Shettinger et al. (1977)) describes the solute mass flow condition.

At $x'=0$,

$$c'=c_0, \tag{6.8}$$

$$\Delta P'=\Delta P_a, \tag{6.9}$$

$$\int_0^a u'(0,y')dy'=Q_0. \tag{6.10}$$

A uniformly distributed solute concentration is represented by condition (6.8), a constant hydrostatic pressure is represented by condition (6.9), and a constant flow at $x=0$ is represented by condition (6.10). $\Delta P'=P'$ (hydrostatic pressures inside) $-P_T$ (hydrostatic pressures outside), $\Delta P_a=P_a$ (hydrostatic pressure at $x=0$) - $P_T$, and $\Delta\pi'=\pi'$ (osmotic pressures inside) - $\pi_T$ (osmotic pressures outside). $Q_0$ represent the volume flow rate at the channel's entrance, respectively. $\pi_T$ and $P_T$ are taken to be constants. In this case, $c_0$ and $c_T$ stand for the solute concentrations outside the channel and at the channel's entrance, respectively, $h$ represents the solute wall permeability, and $T_a$ represents the transmittance coefficient. The reflection coefficient is $\sigma$, and the hydraulic wall permeability is $k$.

The osmotic pressure is assumed as (Papenfuss and Gross (1978)),

$$\pi'(x') = 0.009\, c'^3 + 0.16\, c'^2 + 2.1\, c' \tag{6.11}$$

The following expressions provide the solute mass flux ($J_s$) and total solute clearance ($J_{cl}$) across the permeable channel wall:

$$J_s'(x') = h(c'(x', a) - c_T) + T_a V_a \Phi, \tag{6.12}$$

$$J_{cl}' = a \int_{x'=0}^{x'=L} J_s'(x')\, dx'. \tag{6.13}$$

## 6.3 Nondimensionalization

We introduce the following nondimensional quantities:

$$x = x'a, \quad y = y'a, \quad u = u'U_0, \quad v = v'U_0,$$
$$V_a = V_a'U_0, \quad c = c'c_0, \quad c_T = c_T'c_0, \quad \Delta\pi = \Delta\pi'\Delta P_a,$$
$$\Delta P = \Delta P'\Delta P_a, \quad J_s = J_s'c_0 D/a, \quad J_{cl} = Q_0 c_0 J_{cl}', \quad Q = Q_0 Q', \quad \mu = \mu_0\mu'.$$

Axial diffusion is less than radial diffusion in capillaries. Because of high channel length-to-half-width ratio, end effects are minimal. Furthermore, inertial effects can be neglected, as the net radial flow is negligible compared to the axial flow, which corresponds to Reynolds number of about $10^{-3}$. Eqs. (6.3)–(6.10) are transformed (after dropping the primes) in light of these nondimensional quantities and assumptions.

$$\frac{\partial v}{\partial y} + \frac{\partial u}{\partial x} = 0, \tag{6.14}$$

$$a_p \frac{\partial(\Delta P)}{\partial x} = \frac{\partial}{\partial y}\left(\mu(y)\frac{\partial u}{\partial y}\right), \tag{6.15}$$

$$\frac{\partial(\Delta P)}{\partial y} = 0, \tag{6.16}$$

$$v\frac{\partial c}{\partial y} + u\frac{\partial c}{\partial x} = \frac{1}{Pe}\frac{\partial^2 c}{\partial y^2}. \tag{6.17}$$

The boundary conditions in dimensionless form,
At $y = 0$,

$$\frac{\partial u}{\partial y} = \frac{\partial c}{\partial y} = v = 0 \tag{6.18}$$

At $y = 1$,

$$v = \varepsilon a_p(\Delta P - \sigma \Delta \pi) = V_a, \tag{6.19}$$

$$u = 0, \tag{6.20}$$

$$\frac{\partial c}{\partial y} - Sh_w(c_T - c) = Pe(1 - T_a)V_a\Phi \tag{6.21}$$

At $x = 0$,

$$c = 1, \tag{6.22}$$

$$\Delta P = 1, \tag{6.23}$$

$$\int_0^1 u(0, y)dy = 1. \tag{6.24}$$

Eqs. (6.12) and (6.13) are presented in their nondimensional form as

$$J_s = Sh_w(c(x, 1) - c_T) + PeT_aV_a\Phi \quad \text{at} \quad y = 1, \tag{6.25}$$

$$J_{cl} = \frac{1}{Pe}\int_0^{L/a} J_s\, dx, \tag{6.26}$$

where, $\varepsilon = k\mu_0/a$, $Pe = U_0a/D$, and $Sh_w = ha/D$ are ultrafiltration parameter, Peclet number and Sherwood number, respectively.

## 6.4 Solution of the problem

The Eqs. (6.14) and (6.15) with (6.18) and (6.20) give the velocities as

$$u(x, y) = -a_p\frac{d(\Delta P)}{dx}\int_y^1 \frac{y}{\mu(y)}dy, \tag{6.27}$$

and $$v(x, y) = a_p\frac{d^2(\Delta P)}{dx^2}\left(y\int_y^1 \frac{y}{\mu(y)}dy + \int_0^y \frac{y^2}{\mu(y)}dy\right). \tag{6.28}$$

The Eq. (6.28) and the boundary condition (6.19) determine the second-order ODE for the hydrostatic pressure

$$\frac{d^2(\Delta P)}{dx^2} - \frac{\varepsilon}{I_1}(\Delta P - \sigma\Delta\pi) = 0. \tag{6.29}$$

The Eqs. (6.23) and (6.24) are written in the following way: At $x = 0$

$$\Delta P = 1, \tag{6.30}$$

$$\frac{d\Delta P}{dx} = -\frac{1}{a_pI_2}, \tag{6.31}$$

the osmotic pressure difference in nondimensional form as

$$\Delta\pi = \pi - \pi_T = 0.009[c(x,1)^3 - c_T^3] + 0.16[c(x,1)^2 - c_T^2] + 2.1[c(x,1) - c_T], \tag{6.32}$$

where $I_1 = \int_0^1 \frac{y^2}{\mu(y)} dy$ and $I_2 = \int_0^1 \int_y^1 \frac{y}{\mu(y)} dy dy$.

The viscosity affects the integrations in Eqs. (6.27)–(6.28) and (6.29) and (6.31). The exponential viscosity model is used in this investigation and is written as

$$\mu(y) = e^{-\alpha y}, \tag{6.33}$$

where $\alpha$ is the viscosity parameter. All the computations have been done with exponential form of viscosity.

The Eqs. (6.27) and (6.28) give velocities $u$ and $v$, which are in terms of $c$ and $\Delta P$. The solutions of the Eqs. (6.17) and (6.29), for solute concentration ($c$) and hydrostatic pressure ($\Delta P$) are difficult to obtain analytically, due to the coupling of equations. Therefore the Eqs. (6.17) and (6.29) are solved by using numerical techniques (RK-4 method and finite difference method) with boundary conditions (6.18), (6.21), (6.22) and (6.30), (6.31) respectively.

## 6.5 Numerical technique

The Crank–Nicolson finite difference method is used to solve solute transfer Eq. (6.17) as well as the initial and boundary conditions (6.18), (6.21), (6.22). Varunkumar and Muthu (2019) provides a detailed description of the solution method. A grid independence test was performed to verify the numerical method. For different grid points in the $y$ direction, hydrostatic pressure distributions in the axial direction were investigated. With values of 0.1, 0.2, 0.3, and 0.4, the step size was fixed in one direction and varied in the other. Grids with $51 \times 11$, $51 \times 21$, and $51 \times 31$ nodes were produced by the three test cases, respectively. For each of these grid configurations, a C++ program was compiled. After experimenting with a few different mesh sizes, the mesh size $51 \times 11$ was found to be the most accurate in predicting flow physics, and it was utilized in this study. Only the fifth decimal place separates the two sets of results. As a result, these mesh sizes are thought to be suitable for the computations at hand. In the finite-difference approximation, the truncation error is $O(\Delta y^2 + \Delta x)$ and approaches to zero as $\Delta y, \Delta x \to 0$. As a result, the system works. Varunkumar and Muthu (2019) demonstrated how the finite difference scheme is always stable. Convergence is ensured through consistency and compatibility.

## 6.6 Results and discussions

Fluid and solute transport through a permeable channel has been studied using a mathematical model of variable viscosity in the glomerular capillary. The computational findings are achieved for relevant physiological parameters (Papenfuss and Gross (1978);

Chaturani and Ranganatha (1993)). The graphs show the impacts of variable viscosity and other developing factors on pressure, velocities, solute concentration, and total solute clearance.

## 6.6.1 Hydrostatic pressure and osmotic pressure

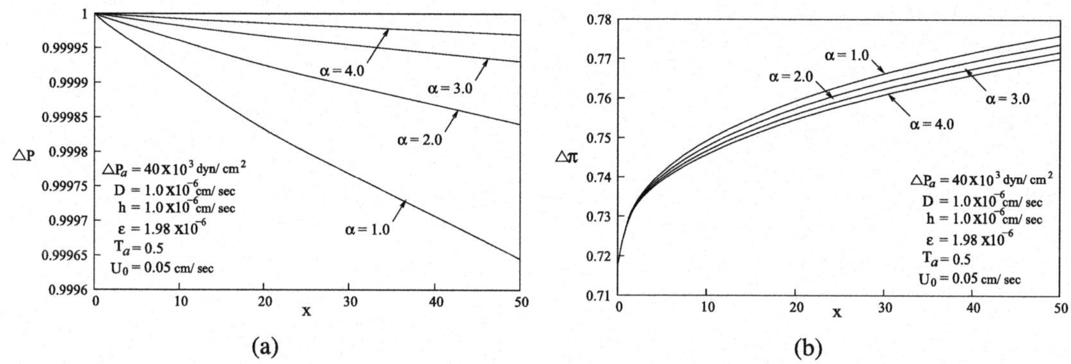


**FIGURE 6.2** The impact of $\alpha$ on the osmotic and hydrostatic pressures distribution.

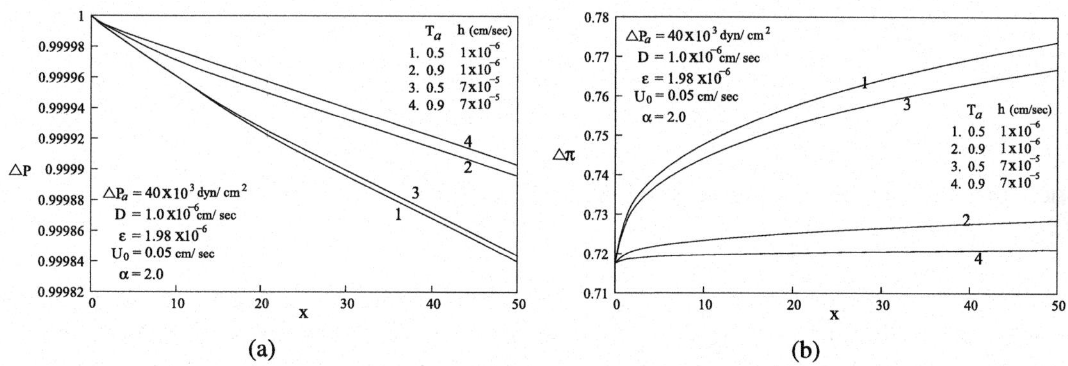


**FIGURE 6.3** Hydrostatic and osmotic pressure profiles for various $T_a$ and $h$ values.

The distributions of hydrostatic and osmotic pressure profiles with respect to axial distance $x$ for viscosity parameter ($\alpha$) and other pertinent parameters are shown in Figs. 6.2 through 6.5. The experimental data from Pollak et al. (2014) is in good agreement with the hydrostatic and osmotic pressure curves. The osmotic pressure rises nonlinearly along the capillary's axial length, whereas the hydrostatic pressure difference, $\Delta P$, falls linearly. The impact of $\alpha$ on osmotic and hydrostatic pressures is depicted in Fig. 6.2. $\Delta P$ rises along the axial length as $\alpha$ increases, indicating that higher viscosity fluids need higher pressure to be driven (see Fig. 6.2(a)). As $\alpha$ values increase along the $x$-direction, the osmotic pressure difference ($\Delta\pi$) decreases, in line with the predictions of Jacques (1921) (see Fig. 6.2(b)).

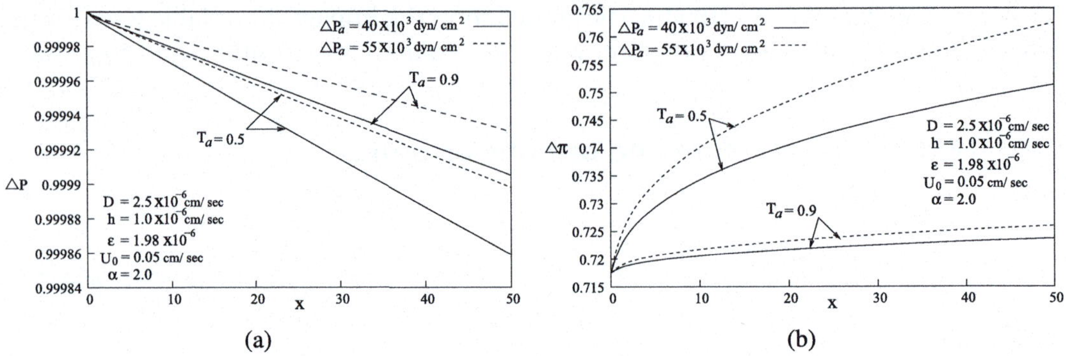


FIGURE 6.4 Distributions of hydrostatic and osmotic pressures for various $T_a$ and $\Delta P_a$.

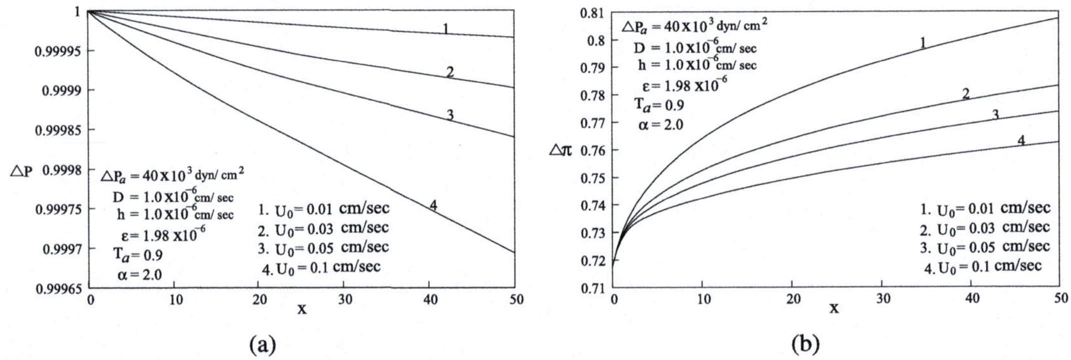


FIGURE 6.5 The impact of $U_0$ on hydrostatic and osmotic pressures.

The $\Delta P$ profiles are unaffected by $T_a$) and $h$, as seen in Fig. 6.3. The $\Delta\pi$ profiles falls along the length as rise in $h$ and $T_a$ values, suggesting that solute flows across the capillary wall. To see how the osmotic pressure ($\Delta\pi$) changes as the input pressure ($\Delta P_a$) rises, see Fig. 6.4(a). A decrease in osmotic pressure along the axial direction results from increased solute transport across the permeable wall, as indicated by the rise in $T_a$ values (see Fig. 6.4(b)). The impact of the initial average velocity ($U_0$) on osmotic and hydrostatic pressures is depicted in Fig. 6.5. In the axial direction, $\Delta P$ falls as $U_0$ rises (Fig. 6.5(a)). Likewise, as $U_0$ increases, $\Delta\pi$ decreases (see Fig. 6.5(b)).

## 6.6.2 Axial velocity ($u(x, y)$)

Fig. 6.6 shows how the axial velocity $u$ vs $y$ behaves at different axial positions for a range of viscosity parameter ($\alpha$) values. The velocity-$u$) is decrease along the length of capillary. The axial velocity close to the channel axis falls as the viscosity parameter ($\alpha$) rises.

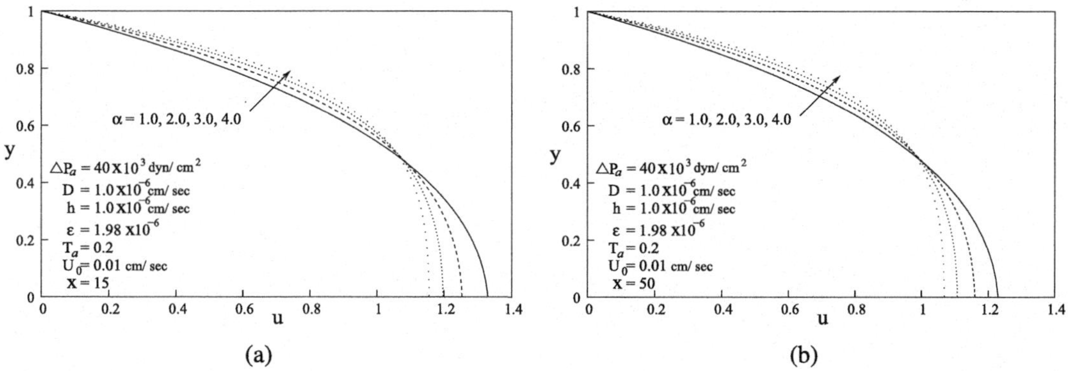


**FIGURE 6.6** Velocity-$u$ vs $y$ at (a) $x = 15$, (b) $x = 50$.

## 6.6.3 Transverse velocity $(v(x, y))$

Fig. 6.7 illustrates how the viscosity parameter $(\alpha)$ affects the velocity-$v$ at distinct points along the channel. The velocity-$v$ rises at the boundary of the capillary but decays close to channel's end as the viscosity parameter increases. The transverse velocity is greater at position $x = 15$ than it is at the channel end $(x = 50)$. Notably, flow behavior is significantly influenced by the viscosity parameter $(\alpha)$. The expected $u$- and $v$-velocity components line up nicely in the limiting case where $\mu(y) \to 1$, which corresponds to the constant viscosity case study from Chaturani and Ranganatha (1993).

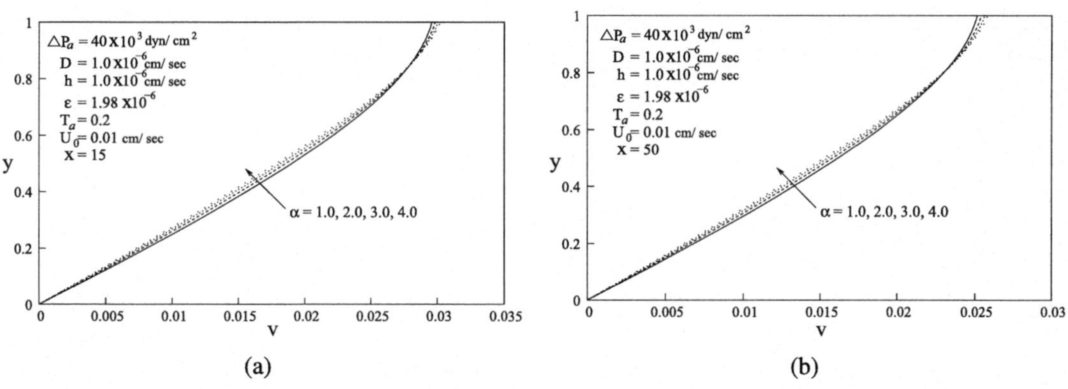


**FIGURE 6.7** Velocity-$v$ with $y$ at (a) $x = 15$, (b) $x = 50$.

## 6.6.4 Concentration profiles $(c(x, y))$

Fig. 6.8 depicts the impact of $\alpha$ on $c(x, y)$ at $x = 15$ and $x = 50$ locations of the channel. At the centerline, the concentration slightly increases as $\alpha$ rises, while the concentration close to the wall falls. As it gets closer to the wall, the solute concentration drops, reaching

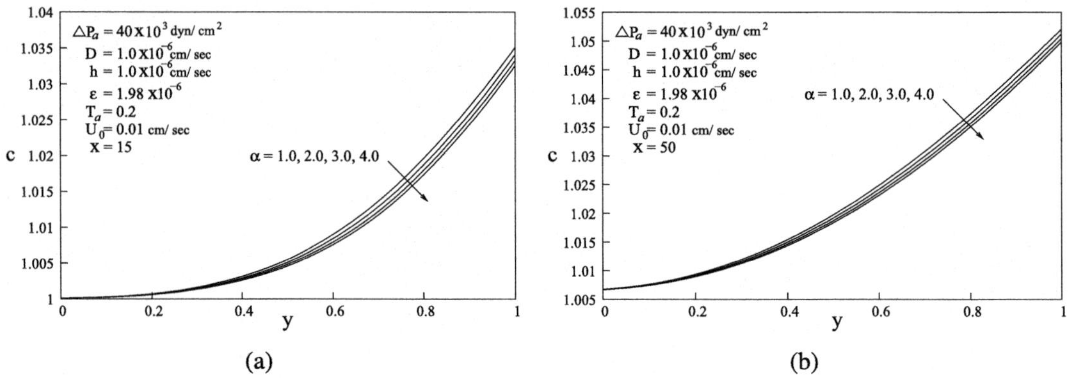


**FIGURE 6.8** $c$ profiles vs $y$ at (a) $x = 15$, (b) $x = 50$ for different $\alpha$ values.

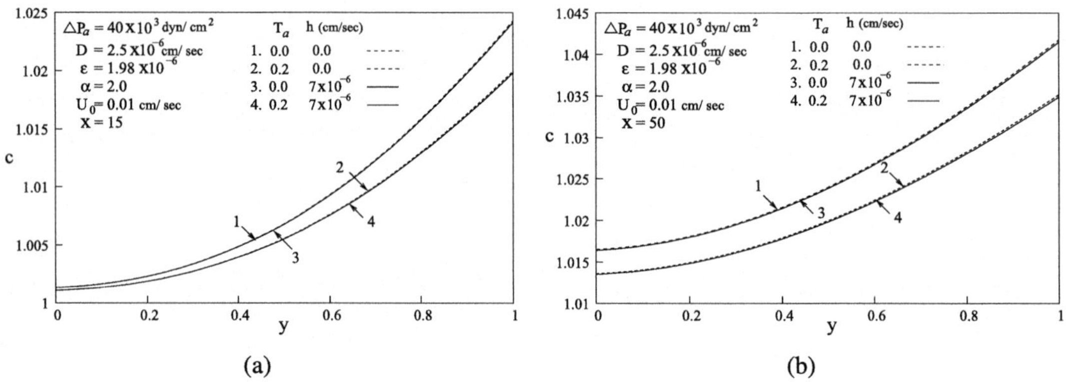


**FIGURE 6.9** Effect of $T_a$ and $h$ on $c$ vs $y$ at (a) $x = 15$, (b) $x = 50$.

its highest value there. The distribution of solute concentrations along the channel's length at two distinct locations for a range of $T_a$ and $h$ values is displayed in Fig. 6.9. Because of the balance between diffusion away from the channel wall and the convection towards it, the concentration is higher at the wall than at other points along the channel at any given axial point. At any given channel cross section, the concentration rises with transverse distance as solute particles are eliminated by ultrafiltration.

Fig. 6.10 illustrates how $\varepsilon$ and $D$ affect the concentration distribution. The solute per unit volume at channel wall has increased as $\varepsilon$ rises, as does the concentration of solute. The solute concentration falls along the axial direction as $D$ rises. When $\varepsilon = 0$ (i.e., zero ultrafiltration), there is no transport of solute at wall. This indicates that the transverse concentration profile drops dramatically from $c_0 = 1$ to zero at $x = 15$.

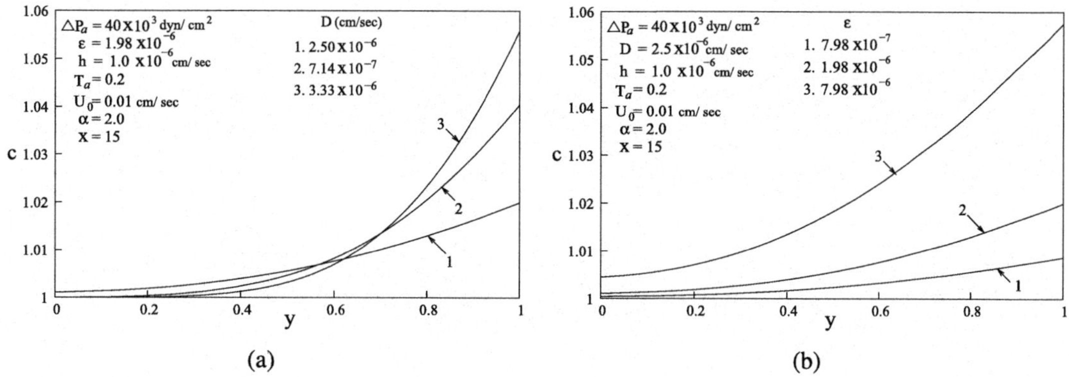


**FIGURE 6.10** Effect of $D$ and $\varepsilon$ on concentration profiles ($c$) with $y$.

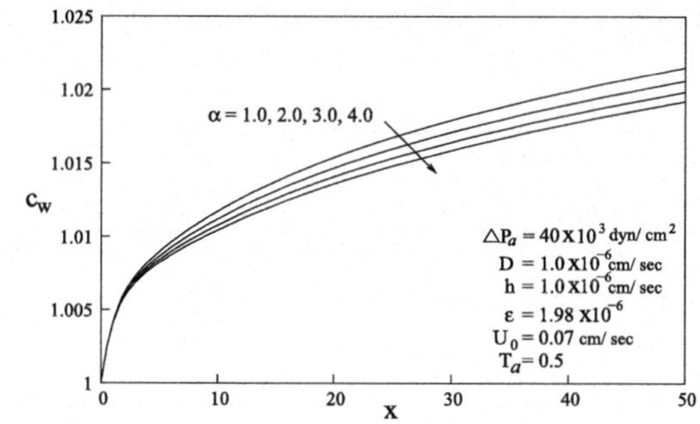


**FIGURE 6.11** Wall concentration $c_w$ distribution with effect of viscosity parameter ($\alpha$).

## 6.6.5 Wall concentration ($c_w$)

Fig. 6.11 illustrates how the viscosity parameter ($\alpha$) affects $c_w$. Note that the concentration only changes along the porous wall. Furthermore, the concentration close to the wall falls as $\alpha$ rises. The distribution of $c_w$ for different $T_a$ and $h$ values is presented in Fig. 6.12(a). The concentration of solutes on the walls is largely determined by transmittance coefficients and wall permeabilities. In particular, $c_w$ falls as $T_a$ and $h$ rise. The wall concentration as a function of $x$ for different ultrafiltration parameter ($\varepsilon$) values, is displayed in Fig. 6.12(b). The concentration at the wall ($c_w$) rises in tandem with $\varepsilon$. According to this study, hydrostatic pressure is less important than osmotic pressure and ultrafiltration.

## 6.6.6 Total solute clearance $J_{cl}$

Table 6.1 discusses the findings of total solute clearance $J_{cl}$ with varying viscosity. The current model uses a range of $\alpha$ from 1.0 to 4.0. The values of the other parameters are within the physiological data ranges. It is observed that as $\alpha$ rises, the total solute clearance increases, as do $h$, and $\varepsilon$.

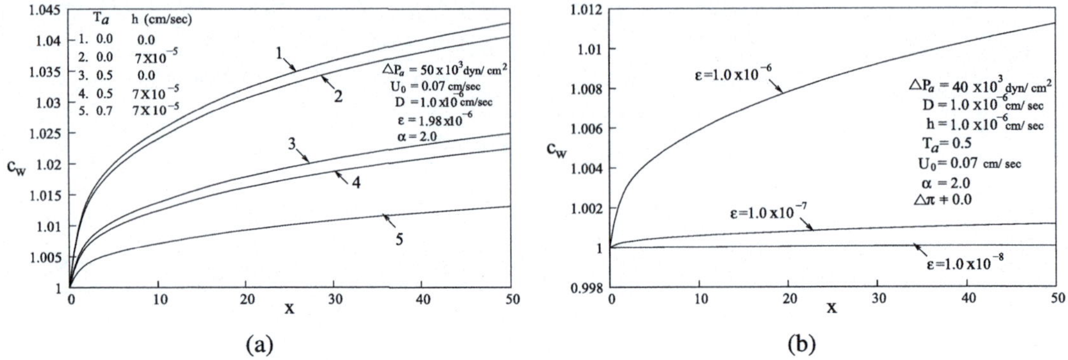


**FIGURE 6.12** Wall concentration $c_w$ distribution for various (a) $T_a$ and $h$ and (b) $\varepsilon$.

**Table 6.1** Total solute clearance ($J_{cl}$) for $\Delta P_a = 40 \times 10^3$ dyn/cm$^2$, $U_0 = 0.07$ cm/sec, $D = 1 \times 10^{-6}$ cm$^2$/sec, and $T_a = 0.5$.

| | $\varepsilon = 1.98 \times 10^{-6}$ | | $\varepsilon = 7.98 \times 10^{-7}$ | |
|---|---|---|---|---|
| $\alpha$ | $Sh = 0.035$ | $Sh = 0.005$ | $Sh = 0.035$ | $Sh = 0.005$ |
| 1.0 | 0.0674575 | 0.0535265 | 0.135206 | 0.119826 |
| 2.0 | 0.0675728 | 0.0536864 | 0.135919 | 0.120648 |
| 3.0 | 0.0676582 | 0.0538081 | 0.136481 | 0.121299 |
| 4.0 | 0.067724 | 0.0539036 | 0.136933 | 0.121822 |

## 6.7 Conclusions

A mathematical model for the transport of solutes and fluids in glomerular capillaries that takes into account variations in both permeability and viscosity is presented in this paper. The emphasis on solute transfer in a channel with variable viscosity, particularly in relation to glomerular capillary flow, is a novel feature of this work. The implicit Crank–Nicolson method was used to solve the solute transfer equation. The study investigates the effects of varying viscosity on solute concentration, fluid velocities, osmotic and hydrostatic pressures, and overall solute clearance.

- The hydrostatic pressure rises linearly along the channel's axial length as $\alpha$ is decreased. Along the channel length, the osmotic pressure curves rise nonlinearly. The osmotic pressure ($\Delta\pi$) falls as $\alpha$ rises.
- As the viscosity parameter ($\alpha$) increases, so do the axial and transverse velocities at the wall.
- The concentration rises at the axis and falls close to the channel wall as $\alpha$ increases.
- The concentration close to the channel wall rises as $\alpha$ decreases. As the parameter $\varepsilon$ rises, the disparity between the two concentrations widens.
- The total solute clearance ($J_{cl}$) rises in tandem with the values of $\alpha$, $h$, and $\varepsilon$.

It should be emphasized that the current model depicts mass transport in a channel with uniform walls. However, capillaries are not uniform in real-life physiological conditions. As a result, by expanding the above study to a nonuniform channel or tube, this model may be made more realistic.

## Funding

The author(s) received no financial support for the publication of this article.

## Declaration of competing interest

The author(s) declared no conflicts of interest.

## References

Apelblat, A., Katchasky, A.K., Silberberg, A., 1974. A mathematical analysis of capillary tissue fluid exchange. Biorheology 11, 1–49.

Awais, M., Bukhari, U., Ali, A., 2017. Convective and peristaltic viscous fluid flow with variable viscosity. Journal of Engineering Thermophysics 26, 69–78.

Berman, A.S., 1958. Laminar flow in an annulus with porous walls. Journal of Applied Physics 29, 71–75.

Bowen, W.R., Williams, P.M., 2001. Prediction of the rate of cross-flow ultrafiltration of colloids with concentration-dependent diffusion coefficient and viscosity – theory and experiment. Chemical Engineering Science 56, 3083–3099.

Brenner, B.M., Baylis, C., Deen, W.M., 1978. Transport of molecules across renal glomerular capillaries. Physiological Reviews 56, 502–534.

Brenner, B.M., Troy, J.L., Daugharty, T.M., Deen, W.M., 1972. Dynamics of glomerular ultrafiltration in the rat. II. Plasma flow dependence of gfr. The American Journal of Physiology 223, 1184–1190.

Chaturani, P., Ranganatha, T.R., 1993. Solute transfer in fluid in permeable tubes with application to flow in glomerular capillaries. Acta Mechanica 96, 139–154.

Davis, R.H., Sherwood, J.D., 1990. A similarity solution for steady-state crossflow microfiltration. Chemical Engineering Science 45, 3203–3209.

Deen, W.M., Robertson, C.R., Brenner, B.M., 1972. A model of glomerular ultrafiltration in the rat. The American Journal of Physiology 223, 1178–1183.

Guyton, A.C., 1986. Text Book of Medical Physiology, 7th edition. W.B. Saunders Company.

Huss, R.E., Marsh, D.J., Kalaba, R.E., 1975. Two models of glomerular filtration rate and renal blood flow in the rat. Annals of Biomedical Engineering 3, 72–99.

Jacques, L., 1921. The reciprocal relation between the osmotic pressure and the viscosity of gelatin solutions. The Journal of General Physiology 4 (1), 97–112.

James, K., James, S., 1998. Mathematical Physiology, 7th edition. Inter Disciplinary Applied Mathematics, vol. 8. Springer.

Marshall, E.A., Trowbridge, E.A., 1974. A mathematical model of the ultrafiltration process in a single glomerular capillary. Journal of Theoretical Biology 48, 389–412.

Palatt, J.P., Henry, S., Roger, I.T., 1974. A hydrodynamical model of a permeable tubule. Journal of Theoretical Biology 44, 287–303.

Papenfuss, H.D., Gross, J.F., 1978. Analytical study of the influence of capillary pressure drop and permeability on glomerular ultrafiltration. Microvascular Research 16, 59–72.

Pollak, M.R., Quaggin, S.E., Hoenig, M.P., Dworkin, L.D., 2014. The glomerulus: the sphere of influence. Clinical Journal of the American Society of Nephrology 9, 1461–1469.

Pozrikidis, C., 2013. Leakage through a permeable capillary tube into a poroelastic tumour interstitium. Engineering Analysis with Boundary Elements 37, 728–737.

Ross, M.S., 1974. A mathematical model of mass transport in a long permeable tube with radial convection. Journal of Fluid Mechanics 63, 157–175.

Salathe, E.P., 1988. Mathematical studies of capillary tissue exchange. Bulletin of Mathematical Biology 50 (3), 289–311.

Shamsuddin, I., Keith, A.S., Gervas, E.J.M.A., 1995. Effect of viscosity on membrane fluxes in cross-flow ultrafiltration. Separation Science and Technology 30 (7–9), 1669–1687.

Shettinger, U.R., Prabhu, H.J., Ghista, D.N., 1977. Blood ultrafiltration: a design analysis. Medical & Biological Engineering & Computing 15, 32–38.

Tyagi, V.P., Abbas, M., 1987. An exact analysis for a solute transport, due to simultaneous dialysis and ultrafiltration, in a hollow-fibre artificial kidney. Bulletin of Mathematical Biology 49 (3), 697–717.

Umavathi, J.C., 2015. Combined effect of variable viscosity and variable thermal conductivity on double-diffusive convection flow of a permeable fluid in a vertical channel. Transport in Porous Media 108, 659–678.

Vander, A.J., Sherman, J.H., Luciano, D.S., 1975. Human Physiology-the Mechanisms of Body Function, 2nd edition. TMH, New Delhi.

Varunkumar, M., Muthu, P., 2019. Fluid flow and solute transfer in a tube with variable wall permeability. Zeitschrift für Naturforschung. A, A Journal of Physical Sciences 74 (12), 1057–1067.

Varunkumar, M., Muthu, P., 2020. Fluid flow and solute transfer in a permeable tube with influence of slip velocity. An interdisciplinary. Discontinuity, Nonlinearity, and Complexity 9 (1), 153–166.

7

# Effect of viscosity variation on exponential slider bearing lubrication

V. Pujitha and M. Varunkumar
*Department of Mathematics, School of Advanced Sciences, VIT-AP University, Amaravati, Andhra Pradesh, India*

## 7.1 Introduction

The lubrication mechanism is an essential component of many mechanical systems, including biological synovial joints, engines, and machines. It is a process which is used to reduce friction and wear between interacting surfaces by introducing a liquid known as a lubricant. One of the important lubrication is slider bearing lubrication, in which surfaces move relative to each other. The relative motion between surfaces leads to formation of a wedge-shaped fluid film. This fluid film generates pressure, which supports the load and separates the surfaces, while minimizing the direct contact. Exponential slider bearing lubrication optimizes the lubricant film distribution using an exponential profile rather than a conventional uniform or linear approach. This lubrication significantly improves the load-carrying capacity and its overall performance in both mechanical and bio-mechanical applications (Shah and Bhat, 2003).

Curzon (1985) mathematically modeled the slider bearing lubrication and calculated the pressure and load carrying capacity. Singh and Gupta (2012) studied curved slider bearing lubrication using ferrofluid as a lubricant. Lin and Hung (2004) analyzed the lubrication performance in exponential slider bearing. Bujurke (1982) mathematically modeled the slider bearing lubrication in the context of synovial joints, considering second-grade fluid as a lubricant.

None of the aforementioned studies treated the bearing surface as porous. Because of its useful self-lubricating feature, the bearing's porous layer is crucial. Due to this reason, Naduvinamani and Hosmani (2018) studied the porous exponential slider bearing lubrication, treating the lubricant as a magnetohydrodynamic couple stress fluid. Bujurke and Naduvinamani (1991) further examined the lubrication mechanism in joints by modeling the bearing surfaces as permeable and deformable. The impact of velocity slip and ferrofluid on the lubrication of porous exponential slider bearings was investigated by Shah and Bhat (2003).

**Advanced Computational and Mathematical Approaches in Applied Differential Equations**
**https://doi.org/10.1016/B978-0-44-345743-2.00018-9**

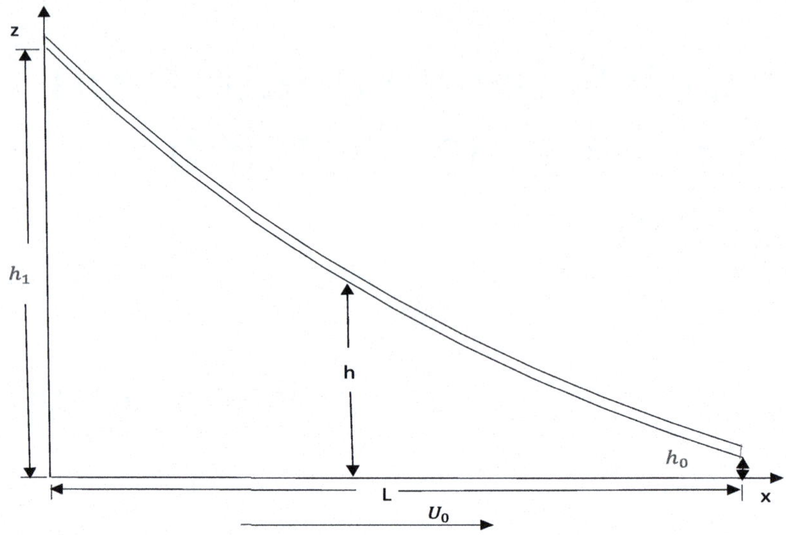


**FIGURE 7.1** Geometry of the present problem.

Physical variables, such as temperature, pressure, concentration, etc., affect the viscosity of the fluid. Several researchers have analyzed how performance of the bearing is effected by variable viscosity. Sujatha et al. (2021) developed a mathematical model to analyze the impact of viscosity changes with film thickness on the lubrication of exponential slider bearings, and Sinha et al. (1981) conducted a similar analysis for journal bearing lubrication.

Most of the earlier studies are associated with slider bearing lubrication along with permeable (or/and elastic) surface or/and variation of viscosity with film thickness. Furthermore, the variation of viscosity with solute concentration is important across several engineering applications, especially where lubrication is critical to system performance, longevity, and energy efficiency. In the context of joint lubrication, the variation in viscosity with solute concentration has critical implications for smooth functioning of the joint and joint health.

The combined effect of concentration-dependent viscosity and a poroelastic surface on the performance of slider bearing lubrication has not yet been modeled mathematically in the literature. Therefore in this paper, a mathematical model is proposed to analyze the effect of variable viscosity on lubrication behavior of an exponential slider bearing lubrication between a poroelastic surface and a rigid surface.

## 7.2 Mathematical formulation

The schematic representation of the present problem geometry is depicted in Fig. 7.1. Here coordinate system $(x, z)$ is considered. The $x$-axis is considered as the bottom plate of length $L$ and $z$-axis is perpendicular to the plates. Let $h(x)$ be the film region height which is varying with respect to $x$ and $\delta$ be the poroelastic region thickness. Bottom plate is moving with a constant velocity $U_0$ along positive $x$ direction only.

Following is the expression for $h(x)$:

$$h(x) = h_1 e^{\frac{-x}{L} \ln\left(\frac{h_1}{h_0}\right)}, \tag{7.1}$$

where $h_1$ and $h_0$ are film thickness at $x = 0$ and $x = L$, respectively. In this study, $h_1 > h_0$.

In the film region, lubricant is modeled as Newtonian viscous fluid. It is assumed that fluid viscosity varies with solute concentration. Compared with the length of the plates $L$, the film height $h(x)$ is small. Due to this reason, variation of pressure with respect to $z$ is insignificant. All derivative of the velocity components, except $\frac{\partial u}{\partial z}$ (where $u$ is fluid velocity along $x$ direction) are considered to be very small. Fluid inertia and body forces are neglected due to low Reynolds's number flow. Flow of the fluid is laminar. Following these assumptions, the equations which governs the fluid flow are given as follows:

### 7.2.1 Film region

$$\frac{\partial u}{\partial x} + \frac{\partial w}{\partial z} = 0, \tag{7.2}$$

$$\frac{\partial}{\partial z}\left[\mu \frac{\partial u}{\partial z}\right] = \frac{\partial p}{\partial x}, \tag{7.3}$$

$$\frac{\partial p}{\partial z} = 0, \tag{7.4}$$

where

$u$: fluid velocity along $x$ direction.

$w$: fluid velocity along $z$ direction.

$p$: pressure

$\mu$: coefficient of viscosity

Viscosity of the lubricant linearly depends on concentration of solute particles is presented in the works of Ferguson et al. (1968); Peeyush (1975).

$$\mu = \mu_0(1 + \lambda c), \tag{7.5}$$

where $\mu_0$ is fluid viscosity entering poroelastic region, $\lambda$ is a constant, and $c$ is solute concentration.

The convection-diffusion equation is used to represent the solute particle distribution in the film region, as shown below.

$$u\frac{\partial c}{\partial x} + w\frac{\partial c}{\partial z} = D\frac{\partial^2 c}{\partial z^2}, \tag{7.6}$$

where, $D$ is diffusion coefficient.

Newtonian fluid is considered in poroelastic region, and viscosity $\mu_0$ is constant. Coupled equations that describe fluid flow in a deformable porous surface can be written as in Bujurke and Naduvinamani (1991).

### 7.2.2 Poroelastic region

$$\textbf{Matrix:} \quad \rho_m \frac{\partial^2 \vec{U}}{\partial t^2} = \operatorname{div}(\sigma_m) - \frac{\mu_0}{k}\left(\frac{\partial \vec{U}}{\partial t} - \vec{V}\right), \tag{7.7}$$

$$\textbf{Fluid:} \quad \rho_f \frac{D\vec{V}}{Dt} = \operatorname{div}(\sigma_f) + \frac{\mu_0}{k}\left(\frac{\partial \vec{U}}{\partial t} - \vec{V}\right), \tag{7.8}$$

where
$\rho_m$: density of the poroelastic surface (cartilage);
$\vec{U}$: displacement vector of the poroelastic surface;
$\sigma_m$: stress tensor of the poroelastic surface;
$k$: poroelastic surface permeability;
$\vec{V}$: velocity of the fluid in poroelastic surface;
$\rho_f$: density of the fluid.

The stress tensors $\sigma_m$ and $\sigma_f$ are given below:

$$\sigma_m = p'I + Ne + AeI, \tag{7.9}$$

$$\sigma_f = -p'I + EeI, \tag{7.10}$$

where $p'$ is the pressure in the poroelastic region (cartilage surface), $N$, $A$, and $E$ are elastic parameters of the poroelastic surface; $e$ is dilatation of the poroelastic surface, and $I$ is the identity tensor.

After neglecting the inertial terms in the Eqs. (7.7) and (7.8) as mentioned in Bujurke and Kudenatti (2006), we have

$$\operatorname{div}(\sigma_m) = \frac{\mu_0}{k}\left(\frac{\partial \vec{U}}{\partial t} - \vec{V}\right), \tag{7.11}$$

$$\operatorname{div}(\sigma_f) = -\frac{\mu_0}{k}\left(\frac{\partial \vec{U}}{\partial t} - \vec{V}\right). \tag{7.12}$$

### 7.2.3 Boundary conditions

Fluid velocity boundary conditions are given below:

**(i)** At the bottom plate:

$$\text{at } z = 0: \quad u = U_0 \quad \text{and} \quad w = 0. \tag{7.13}$$

**(ii)** At permeable wall: No-slip boundary condition is considered for tangential velocity. The normal velocity $(w)$ is equal to normal component of the relative fluid velocity $(w_n)$ at the poroelastic surface:

$$\text{at } z = h: \quad u = 0 \quad \text{and} \quad w = -w_n. \tag{7.14}$$

The boundary conditions associated with the solute concentration are specified as follows:

**(i)** At $x = 0$, the solute concentration is constant (Collins, 1982). That is,

$$\text{at } x = 0: \quad c = c_0, \tag{7.15}$$

where $c_0$ is the uniform solute concentration distribution.

**(ii)** At the bottom solid plane, no solute particle can enter or leave the film region through this surface:

$$\text{at } z = 0: \quad \frac{\partial c}{\partial z} = 0. \tag{7.16}$$

**(iii)** At permeable surface, solute mass flux is given by Shettigar et al. (1977); Varunkumar and Muthu (2023):

$$\text{at } z = h: \quad -D\frac{\partial c}{\partial z} = (T_h - 1)w_n c + h^*(c - c_p), \tag{7.17}$$

where
$c_p$: solute concentration inside the poroelastic region;
$h^*$: permeability of the solute particles at the interface;
$D$: diffusion coefficient;
$T_h$: transmittance factor.
$T_h$ is defined as the fraction of the solute present at the interface, which actually enters the pores of the surface and which gets transferred through it by bulk flow (Shettigar et al., 1977).

## 7.3 Solution

Adding Eqs. (7.11) and (7.12) and using Eqs. (7.9) and (7.10), we get

$$\nabla e = 0. \tag{7.18}$$

The divergence of Eq. (7.18) gives

$$\nabla^2 e = 0. \tag{7.19}$$

Dilatation of the poroelastic surface (cartilage) in terms of average bulk modulus $K$ is given below as mentioned in Hori and Mockros (1976).

$$e = e_0 + \frac{p'}{K}. \tag{7.20}$$

Equation of pressure can be obtained by substituting the Eq. (7.20) in (7.19) as

$$\nabla^2 p' = 0. \tag{7.21}$$

Normal component of $w_n$ at poroelastic interface:
Substituting the Eq. (7.10) in (7.12), we have

$$\frac{\partial \vec{U}}{\partial t} - \vec{V} = (\nabla p' - E\nabla e)\frac{k}{\mu_0}. \tag{7.22}$$

Substitute the Eq. (7.20) in (7.22), we get

$$\frac{\partial \vec{U}}{\partial t} - \vec{V} = \frac{k}{\mu_0}\nabla p'\left(1 - \frac{E}{K}\right). \tag{7.23}$$

Fluid velocity $w_n$ is defined as

$$\text{normal component of}\left(\frac{\partial \vec{U}}{\partial t} - \vec{V}\right) = w_n = \left(1 - \frac{E}{K}\right)\frac{k}{\mu_0}\frac{\partial p'}{\partial z}\bigg|_{z=h}. \tag{7.24}$$

By integrating the Laplace Eq. (7.21) with respect to $z$ across the thickness of the porous region, and applying the boundary condition $\frac{\partial p'}{\partial z} = 0$ at $z = h + \delta$, the following expression is obtained:

$$\frac{\partial p'}{\partial z}\bigg|_{z=h} = \int_h^{h+\delta} \frac{\partial^2 p'}{\partial x^2}\, dz. \tag{7.25}$$

If porous region thickness ($\delta$) is assumed to be of very small, then Eq. (7.25) simplifies to

$$\frac{\partial p'}{\partial z}\bigg|_{z=h} = \delta\frac{\partial^2 p}{\partial x^2}. \tag{7.26}$$

In the limiting case where $\delta \approx 0$, Eq. (7.26) is valid (Prakash and Vij, 1973). Therefore

$$w_n = \frac{k}{\mu_0}\delta\left(1 - \frac{E}{K}\right)\frac{\partial^2 p}{\partial x^2}. \tag{7.27}$$

The following dimensionless quantities are introduced:

$$\bar{x} = \frac{x}{L}, \qquad \bar{z} = \frac{z}{h_0}, \qquad \bar{h} = \frac{h}{h_0}, \qquad \bar{u} = \frac{u}{U_0}, \qquad \bar{w} = \frac{w}{U_0},$$

$$\bar{w}_n = \frac{w_n}{U_0}, \qquad \psi_0 = \frac{k\delta}{h_0^3}, \qquad \bar{p} = \frac{p h_0^2}{\mu_0\, L\, U_0}, \qquad l = \frac{L}{h_0}, \qquad \bar{c} = \frac{c}{c_0},$$

$$\bar{c}_p = \frac{c_p}{c_0}, \qquad \bar{\mu} = \frac{\mu}{\mu_0}, \qquad h_2 = \frac{h_1}{h_0}, \qquad \bar{\lambda} = \lambda c_0.$$

Nondimensional form of governing Eqs. (7.1) to (7.27) after dropping bar are the following:

$$\frac{\partial u}{\partial x} + l\frac{\partial w}{\partial z} = 0, \tag{7.28}$$

$$\frac{\partial}{\partial z}\left[\mu \frac{\partial u}{\partial z}\right] = \frac{\partial p}{\partial x}, \tag{7.29}$$

$$\frac{\partial p}{\partial z} = 0, \tag{7.30}$$

$$\mu = (1 + \lambda c), \tag{7.31}$$

$$Pe\left[u\frac{\partial c}{\partial x} + w\frac{\partial c}{\partial z}\right] = \frac{\partial^2 c}{\partial z^2}, \tag{7.32}$$

$$h(x) = h_2 e^{-x ln(h_2)}. \tag{7.33}$$

Eq. (7.27) in nondimensional form after dropping bar is

$$w_n = \frac{\psi_0}{l}\left(1 - \frac{E}{K}\right)\frac{\partial^2 p}{\partial x^2}. \tag{7.34}$$

Boundary conditions in nondimensional form are

$$\text{at } z = 0: \quad u = 1 \quad \text{and} \quad w = 0; \tag{7.35}$$

$$\text{at } z = h: \quad u = 0 \quad \text{and} \quad w = -w_n; \tag{7.36}$$

$$\text{at } x = 0: \quad c = 1; \tag{7.37}$$

$$\text{at } z = 0: \quad \frac{\partial c}{\partial z} = 0; \tag{7.38}$$

$$\text{at } z = h: \quad \frac{\partial c}{\partial z} = Pe(1 - T_h) w_n c + Sh(c_p - c), \tag{7.39}$$

where $Sh = \dfrac{h^* h_0}{D}$ is Sherwood, $Pe = \dfrac{U_0 h_0}{D}$ is the Peclet number.

The closed-form solutions of Eqs. (7.28) to (7.32), are challenging to derive due to the coupled nature of the system. Additionally, the nonuniformity of the physical domain complicates the numerical solution of the concentration equation, as discretization in the $z$-direction results in a variable step size across different cross-sections. To address this issue, it is necessary to transform the nonuniform physical domain into a uniform computational domain, using the coordinate transformation as given below (Anderson,

1995):

$$\xi = x, \qquad \eta = \frac{z}{h(x)}. \tag{7.40}$$

The aforementioned transformation effectively straightens the upper nonuniform boundary of the physical domain. Specifically, the boundaries $\eta = 0$ and $\eta = 1$ in the computational domain correspond to $z = 0$ and $z = h(x)$ in the physical domain. Consequently, the governing equations are reformulated in terms of the transformed coordinates as the following:

$$\frac{\partial u}{\partial \xi} - \frac{\eta h'}{h}\frac{\partial u}{\partial \eta} + \frac{l}{h}\frac{\partial w}{\partial \eta} = 0, \tag{7.41}$$

$$\frac{\partial p}{\partial \xi} = \frac{1}{h^2}\frac{\partial}{\partial \eta}\left[\mu \frac{\partial u}{\partial \eta}\right], \tag{7.42}$$

$$\frac{\partial p}{\partial \eta} = 0, \tag{7.43}$$

$$Pe\left[u\left(\frac{\partial c}{\partial \xi} - \frac{\eta h'}{h}\frac{\partial c}{\partial \eta}\right) + w\frac{1}{h}\frac{\partial w}{\partial \eta}\right] = \frac{1}{h^2}\frac{\partial^2 c}{\partial \eta^2}, \tag{7.44}$$

$$h(\xi) = h_2 e^{-\xi ln(h_2)}, \tag{7.45}$$

and

$$w_n = \frac{\psi_0}{l}\left(1 - \frac{E}{K}\right)\frac{\partial^2 p}{\partial \xi^2}. \tag{7.46}$$

Similarly, the nondimensional boundary conditions are also expressed in the $(\xi, \eta)$ coordinate system as the following:

$$\text{at } \eta = 0: \quad u = 1 \quad \text{and} \quad w = 0; \tag{7.47}$$

$$\text{at } \eta = 1: \quad u = 0 \quad \text{and} \quad w = -w_n; \tag{7.48}$$

$$\text{at } \xi = 0: \quad c = 1; \tag{7.49}$$

$$\text{at } \eta = 0: \quad \frac{\partial c}{\partial \eta} = 0; \tag{7.50}$$

$$\text{at } \eta = 1: \quad \frac{\partial c}{\partial \eta} = h\left(Sh(c_p - c) + Pe(1 - T_h)w_n c\right). \tag{7.51}$$

The solution of Eqs. (7.42) and (7.41) for $u$ and $w$ are obtained as follows:

$$u = h^2\frac{\partial p}{\partial \xi}\left[\int_0^\eta \frac{\eta}{\mu}d\eta - F_1\int_0^\eta \frac{1}{\mu}d\eta\right] + 1 - \frac{1}{F_2}\int_0^\eta \frac{1}{\mu}d\eta, \tag{7.52}$$

$$w = -\frac{\partial}{\partial \xi}\left[h\int_0^\eta u d\eta\right] + h'u\eta, \tag{7.53}$$

where $F_1 = \dfrac{\int_0^1 \frac{\eta}{\mu} d\eta}{\int_0^1 \frac{1}{\mu} d\eta}$, and $F_2 = \int_0^1 \frac{1}{\mu} d\eta$.

The pressure distribution equation within the film region is derived by substituting the velocity component $u$ in Eq. (7.41), integrating with respect to $\eta$ over the interval [0, 1] and applying boundary conditions (7.47) and (7.48) associated with the velocity component $w$. The equation of pressure is called Reynolds equation:

$$\frac{\partial}{\partial \xi}\left\{h\left[h^2 \frac{\partial p}{\partial \xi}\left[\int_0^\eta \frac{\eta}{\mu} d\eta - F_1 \int_0^\eta \frac{1}{\mu} d\eta\right] + 1 - \frac{1}{F_2}\int_0^\eta \frac{1}{\mu} d\eta\right]\right\} - l w_n = 0. \tag{7.54}$$

Substituting the Eq. (7.46) in (7.54) we get,

$$\frac{\partial}{\partial \xi}\left\{h\left[h^2 \frac{\partial p}{\partial \xi}\left[\int_0^\eta \frac{\eta}{\mu} d\eta - F_1 \int_0^\eta \frac{1}{\mu} d\eta\right] + 1 - \frac{1}{F_2}\int_0^\eta \frac{1}{\mu} d\eta\right]\right\} - \psi_0\left(1 - \frac{E}{K}\right)\frac{\partial^2 p}{\partial \xi^2} = 0. \tag{7.55}$$

## 7.3.1 Boundary conditions for pressure

$$\text{at } \xi = 0: \quad p = 0 \tag{7.56}$$

$$\text{at } \xi = 1: \quad p = 0 \tag{7.57}$$

Pressure is obtained by integrating the Eq. (7.55) with respect to $\xi$ twice and using boundary conditions (7.56) and (7.57) as

$$p = \int_0^\xi \frac{h_m - G}{F} d\xi, \tag{7.58}$$

where $F = h^3 \int_0^1 \left[\int_0^\eta \frac{\eta}{\mu} d\eta - F_1 \int_0^\eta \frac{1}{\mu} d\eta\right] d\eta - \psi_0\left(1 - \frac{E}{K}\right)$,

$G = h \int_0^1 \left[1 - \frac{1}{F_2}\int_0^\eta \frac{1}{\mu} d\eta\right] d\eta$ and $h_m = \dfrac{\int_0^1 \frac{G}{F} d\xi}{\int_0^1 \frac{1}{F} d\xi}$.

Eqs. (7.58), (7.52), (7.53), and (7.31) are coupled with concentration equations. Analytical solutions for velocity components $(u, w)$, pressure $(p)$, and concentration $(c)$ are not possible in closed form. Hence these equations are solved numerically by using an iterative procedure.

Convection-diffusion Eq. (7.44) is solved by using Crank–Nicolson method along with boundary conditions (7.49)–(7.51). The Eqs. (7.52), (7.53), and (7.58) contain integrals. These integrals are solved with the help of Trapezoidal rule.

The numerical method used in this study was verified and validated in our earlier work (Muthu and Pujitha, 2020), where it was applied to a squeeze film lubrication problem involving viscosity variations. The present work extends that model by incorporating an exponential film geometry in the analysis of slider bearing lubrication, which was not considered in the previous study.

The pressure $p$ is integrated over the film region to get the load carrying capacity. It is written as

$$W = \int_0^1 p\, d\xi. \tag{7.59}$$

## 7.4 Results and discussion

The objective of the present study is to use a mathematical model to analyze how the lubrication of exponential slider bearings is affected by changes in viscosity associated with solute concentration.

In this paper, we obtained results for solute concentration ($c$) and load carrying capacity ($W$). The permeability $\psi_0$, elastic parameter $\frac{E}{K}$, and height ration $h_2$ are parameters of interest in this study.

This study focuses on analyzing the effects of the permeability parameter $\psi_0$, elastic parameter $\frac{E}{K}$, and height ration $h_2$ on the solute concentration $c$ and load carrying capacity $W$. Furthermore, the other parameters are kept fixed throughout the analysis with the following values: $Sh = 0.3$, $c_p = 0.5$, $T_a = 0.8$, $Pe = 1.0$, and $\lambda = 0.3$.

### 7.4.1 Solute concentration ($c(x, z)$)

The load-carrying capacity of a slider bearing relies on the lubricant film. An insufficient solute concentration in the lubricant can weaken the film's properties, reducing its ability to support the load effectively. Thus analyzing the impact of concentration variations on slider bearing lubrication is essential for optimizing performance and ensuring reliability.

Figs. 7.2, 7.3, and 7.4 illustrate the distribution of solute particle concentration ($c$) along normal direction $z$ at the cross sections $x = 0.25$, $x = 0.5$, and $x = 0.75$, respectively. The results clearly show that the concentration $c$ increases with an increase in the height ratio $h_2$. Moreover, this increase is more pronounced at the downstream cross-section $x = 0.75$, indicating a stronger influence of $h_2$ on solute distribution toward the exit region of the bearing.

Figs. 7.5 and 7.6 depict the influence of permeability parameter $\psi_0$ and elastic parameter $\frac{E}{K}$ on concentration distribution, respectively. It is observed that, $\psi_0$ and $\frac{E}{K}$ have not shown any significant effect on solute dispersion.

### 7.4.2 Solute concentration at interface ($c_i$)

Fig. 7.7 shows the axial variation of solute concentration ($c_i$) at the interface for different values of the height ratio $h_2$. The results indicate a decreasing trend in $c_i$ as the axial position $x$ increases. Moreover, the solute concentration at the interface increases with decreasing values of $h_2$ from $x = 0.0$ to $x = 0.5$. After the value of $x = 0.5$, $h_2$ shows reverse trending effect on $c$. That is, $c_i$ increases with increasing values of $h_2$.

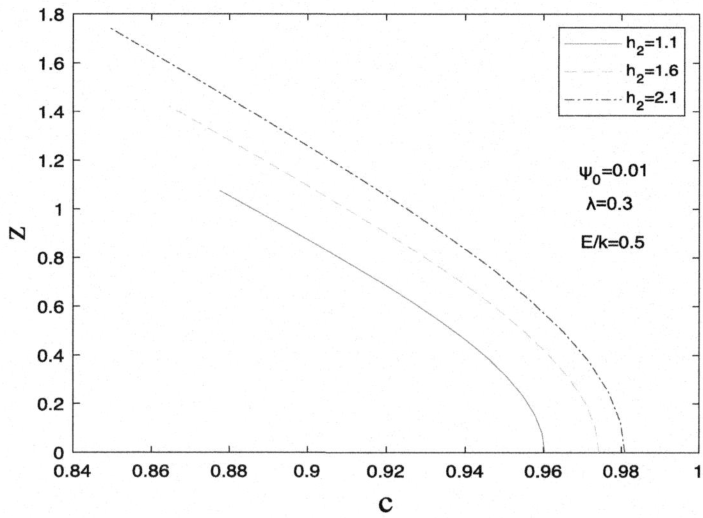


**FIGURE 7.2** Solute concentration ($c$) variation along $z$ for various $h_2$ values at the cross section $x = 0.25$.

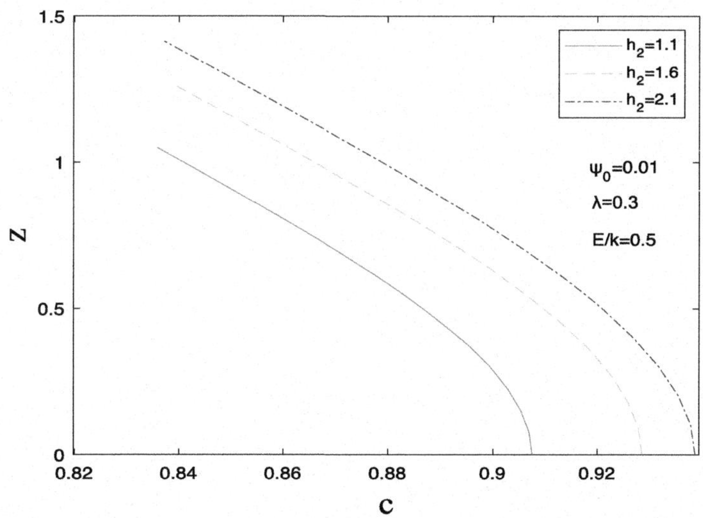


**FIGURE 7.3** Solute concentration ($c$) variation along $z$ for various $h_2$ values at the cross section $x = 0.5$.

## 7.4.3 Load carrying capacity ($W$)

Fig. 7.8 consists of graphs $W$ versus $h_2$, for different values of $\psi_0$. It is observed that, as the permeability parameter $\psi_0$ increases, $W$ decreases. With higher permeability, the surface retains less fluid within the film region, resulting in a lower load-carrying capacity. Similar results are observed in the theoretical study reported by Bujurke and Naduvinamani (1991).

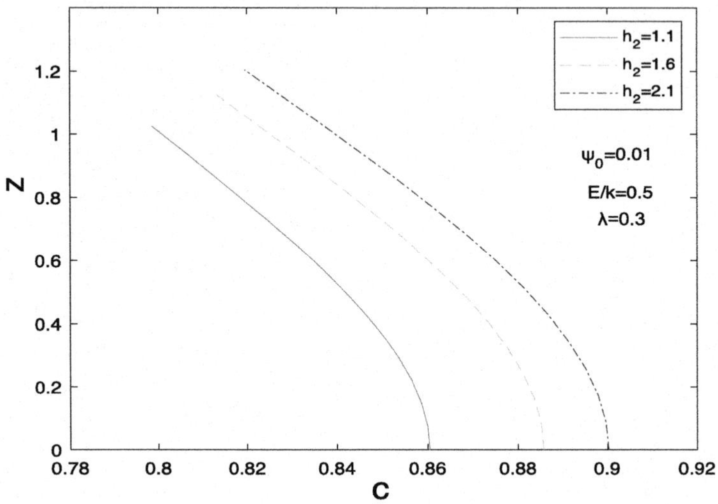


**FIGURE 7.4** Solute concentration ($c$) variation along $z$ for various $h_2$ values at the cross section $x = 0.75$.

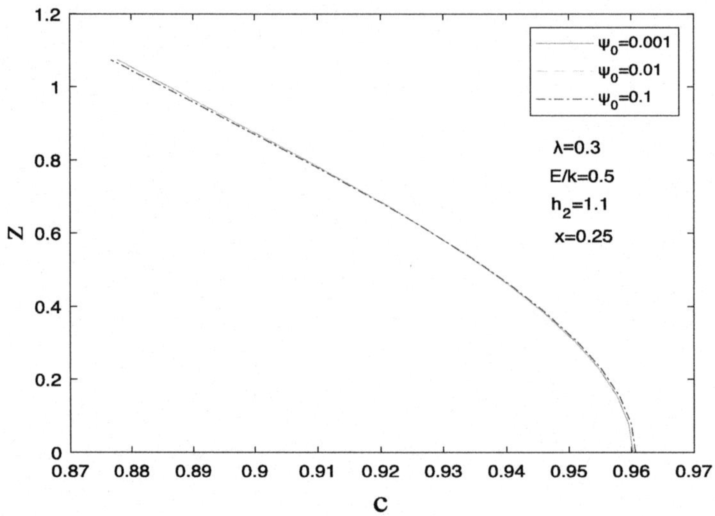


**FIGURE 7.5** Solute concentration ($c$) variation along $z$ for various $\psi_0$ values, at $x = 0.25$ and $h_2 = 1.1$.

Fig. 7.9 presents the effect of the elastic parameter $\frac{E}{K}$ on load carrying capacity $W$. $W$ increases as $\frac{E}{K}$ increases, indicating a direct correlation between the elasticity of the surface and the ability to support load. Similar results can be observed in the Table 1 and 2 of the theoretical study by Bujurke and Naduvinamani (1991).

Fig. 7.10 illustrates the impact of height ratio $h_2$ on $W$ for various values of $\lambda c_0$. As $h_2$ increases, $W$ also increases. Furthermore, the load-carrying capacity remains constant for

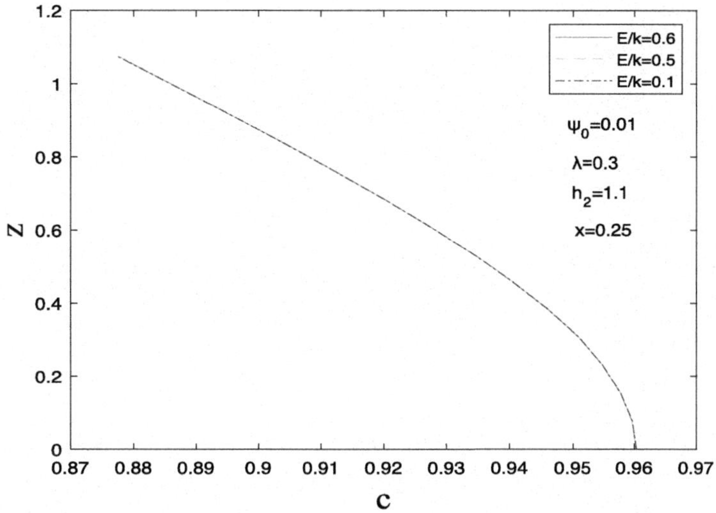


**FIGURE 7.6** Solute concentration ($c$) variation along $z$ for various $\frac{E}{K}$ values at $x = 0.25$ and $h_2 = 1.1$.

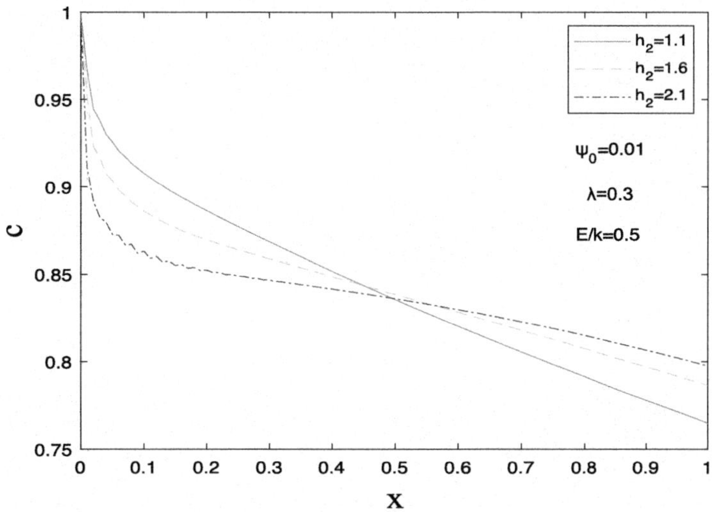


**FIGURE 7.7** Variation of solute concentration ($c_i$) at the interface for different values of $h_2$.

$h_2$ values of 2.0 and above. Also, the load carrying capacity increases with increasing values of $\lambda c_0$. This may be explained that as $\lambda c_0$ increases the concentration increases. As a result, viscosity of the lubricant increases which supports the load.

In Table 7.1, the results of the present study are compared with those of Naduvinamani and Hosmani (2018) as a limiting case. While their work analyzes porous exponential slider bearings with MHD and couple stress fluid effects, the present model introduces viscosity

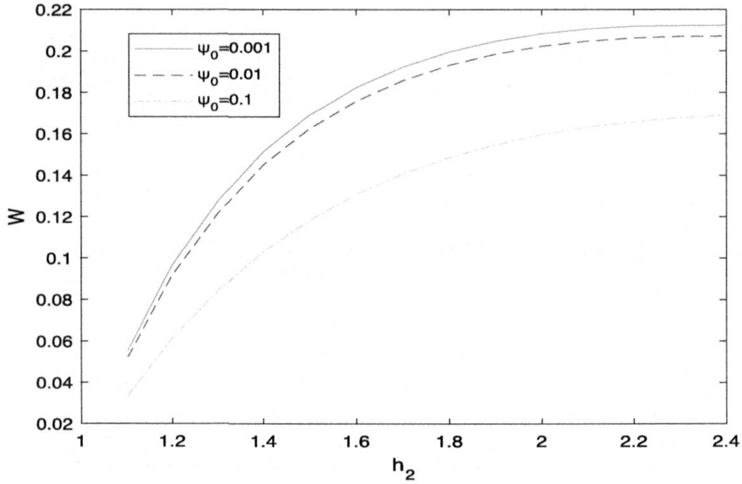


**FIGURE 7.8** Load carrying capacity ($W$) variation with $h_2$ for various $\psi_0$ values.

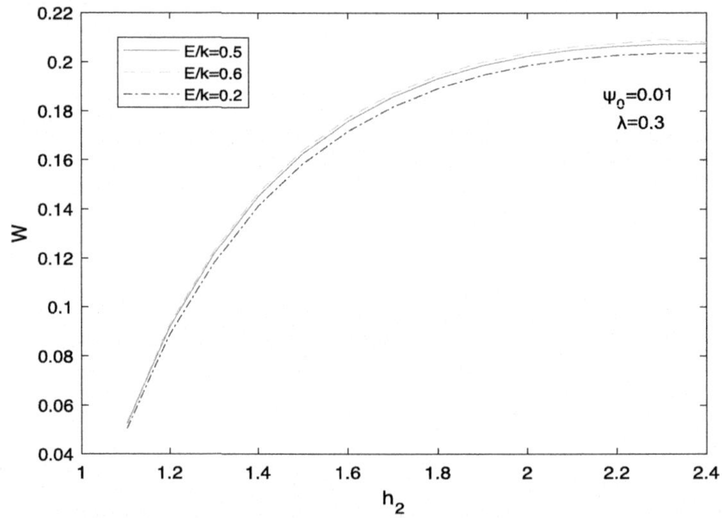


**FIGURE 7.9** Load carrying capacity ($W$) variation with $h_2$ for various $\frac{E}{K}$.

variation due to solute concentration and includes poroelastic effects, offering a novel and extended framework.

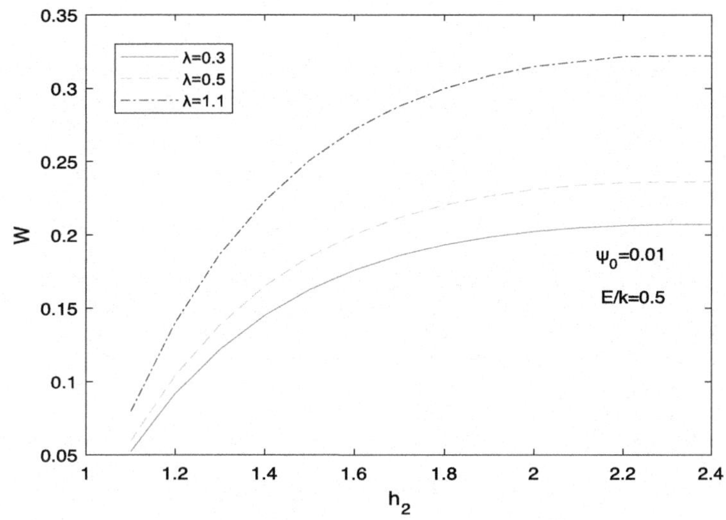


**FIGURE 7.10** Load carrying capacity ($W$) variation with $h_2$ for various $\lambda c_0$ values.

**Table 7.1** Comparison study.

| Present paper | (Naduvinamani and Hosmani, 2018) |
|---|---|
| $\frac{E}{k}=0$ (rigid) | $M=0$ (no magnetic) |
| $\lambda=0$ (no concentration) | $\bar{l}=0$ (Newtonian) |
| Aspect ratio ($h_2$) = 1.1 | Aspect ratio (a) = 1.1 |
| permeability $\psi_0=0$ | permeability $\psi=0$ |
| load carrying capacity ($W$) = 0.0434 | load carrying capacity ($W$) = 0.04 |

# 7.5 Conclusion

This paper presents a mathematical model that incorporates the effects of permeability, surface elasticity, and viscosity variations of the lubricant due to changes in solute particle concentration, aimed at analyzing the lubrication behavior of slider bearings.

The key aspect of this research is concentration dependency of viscosity. This work provides theoretical insight indicating that the load-bearing capacity decreases as the concentration of solute particles is reduced.

1. The findings for different values of parameter $h_2$ demonstrate a significant impact on both the load carrying capacity and the solute particle concentration.
2. It has been found that the solute concentration in slider bearing lubrication is unaffected by the permeability and elastic parameters.
3. It is observed that the solute particle concentration in the slider bearing lubrication influences the load carrying capability.
4. Additionally, it has been found that load carrying capacity increases with increasing elastic parameter but decreases with increasing permeability.

# References


Anderson, J.D., 1995. Computational Fluid Dynamics-the Basics with Applications. McGraw-Hill.

Bujurke, N.M., 1982. Slider bearings lubricated by a second-grade fluid with reference to synovial joints. Wear 78, 355–363.

Bujurke, N.M., Kudenatti, R.B., 2006. An analysis of rough poroelastic bearings with reference to lubrication mechanism of synovial joints. Applied Mathematics and Computation 178, 309–320.

Bujurke, N.M., Naduvinamani, N.B., 1991. Theoretical modelling of poro-elastic slider bearings lubricated by couple stress fluid with special reference to synovial joints. Applied Mathematical Modelling 15 (6), 319–324.

Collins, R., 1982. A model of lubricant gelling in synovial joints-part I: derivation and properties of mathematical model. Journal of Applied Mathematics and Physics 33, 93–123.

Curzon, A.E., 1985. A general method for calculating the pressure distribution and load capacity for finite hydrodynamic bearings with lubrication films curved in one direction. ASLE Transactions 28 (1), 54–59.

Ferguson, J., Boyle, J.A., Mcsween, R.N.M., Jasani, M.K., 1968. Observations on the flow properties of the synovial fluid from patients with rheumatoid arthritis. Biorheology 5 (2), 119–131.

Hori, R.Y., Mockros, L.L., 1976. Indentation test of human articular cartilage. Journal of Biomechanics 9, 259–268.

Lin, J.-R., Hung, C.-R., 2004. Analysis of dynamic characteristics for wide slider bearings with an exponential film profile. Journal of Marine Science and Technology 12 (3), 217–221.

Muthu, P., Pujitha, V., 2020. Effect of concentration dependence of viscosity on squeeze film lubrication. Zeitschrift für Naturforschung A 75 (6), 533–542.

Naduvinamani, N.B., Hosmani, S.S., 2018. Porous exponential slider bearings lubricated with mhd-couple stress fluid. Industrial Lubrication and Tribology 70 (5), 838–845.

Peeyush, C., 1975. Mathematical Model for Synovial Joints – a Lubrication Biomechanical Study. PhD thesis. IIT Kanpur.

Prakash, J., Vij, S.K., 1973. Load capacity and time-height relations for squeeze films between porous plates. Wear 24, 309–322.

Shah, R., Bhat, V., 2003. Analysis of a porous exponential slider bearing lubricated with a ferrofluid considering slip velocity. Journal of the Brazilian Society of Mechanical Sciences and Engineering 25, 264–267.

Shettigar, U.R., Prabhu, H.J., Ghista, D.N., 1977. Blood ultrafiltration: a design analysis. Medical & Biological Engineering & Computing 15, 32–38.

Singh, U.P., Gupta, R.S., 2012. Dynamic performance characteristics of a curved slider bearing operating with ferrofluids. Advances in Tribology, 1–6.

Sinha, P., Singh, C., Prasad, K.R., 1981. Effect of viscosity variation due to lubricant additives in journal bearings. Wear 66, 175–188.

Sujatha, E., Lavanya, G., Siva, P., 2021. Analysis of viscosity variation effect in a wide slider bearing having an exponential film profile in the presence of mhd. Advances in Mathematics: Scientific Journal 10, 1925–1936.

Varunkumar, M., Muthu, P., 2023. Mathematical model of fluid flow in a double constricted tapered tube with permeable boundary. International Journal of Nonlinear Sciences and Numerical Simulation 24 (5), 2023–2034.

# 8

# Mathematical model of velocity slip on flow in a channel with permeable walls

M. Varunkumar and Estemsetti Siva Sai
*Department of Mathematics, School of Advanced Sciences, VIT-AP University, Amaravati, Andhra Pradesh, India*

## 8.1 Introduction

Several studies on fluid flow through a tube/channel of permeability have attracted many researchers due to its importance in many experimental and theoretical studies of physiological problems (Berman, 1953, 1958; Yuan and Finkelstein, 1956; Oka and Murata, 1970; Saffman, 1971; Apelblat et al., 1974; Mariamma and Majhi, 2000).

Macey (1965, 1963) created a mathematical model that included various reabsorption mechanisms at the tubular wall for fluid flow through the proximal renal tubule. Later, Kozinski et al. (1970) expanded on Macey's work by examining flow through porous slits while assuming an exponential pressure decay along the axial direction. To account for fluid exchange, Marshall and Trowbridge (1974) used Darcy's condition at the porous wall to study the Newtonian fluid flow of through a tube. Palatt et al. (1974) presented a model where pressure gradients control fluid leakage across the wall. Acharya et al. (1981) developed a model of fluid flow in a tube with variable radius. Furthermore, Chaturani and Ranganatha (1991) studied the above model with addition of variable permeability at the wall, offering a more complete fluid flow across permeable, nonuniform tubes.

Previous studies neglected the inertial effects in the flow, which alter as the tubule radius varies along its length. To overcome this limitation, Muthu and Tesfahun (2010) proposed a mathematical model for flow in a permeable channel with nonuniformity, a small effect of Reynolds number. The reabsorption effects at the channel walls were studied by Haroon et al. (2016), who also obtained analytical solutions to the flow equations. Siddiqui et al. (2016) examined the viscous flow across a uniform slit with absorption at the walls. Haroon et al. (2017) extended the study by including the effect of periodic reabsorption, modeling the filtrate flow through a renal tubule with more physiological relevance.

In earlier research, a no-slip boundary condition, where the viscous fluid adheres entirely to the channel or tube wall, was sometimes assumed without clear explanation. Nonetheless, partial slip could occur, particularly in the presence of permeable bound-

**Advanced Computational and Mathematical Approaches in Applied Differential Equations**
**https://doi.org/10.1016/B978-0-44-345743-2.00019-0**

aries, based on both theoretical studies and experimental data. Beavers and Joshep (1967) experimentally validated the slip-flow model by displaying a great degree of agreement between the observed results and analytical predictions. Furthermore, supported by Saffman (1971) was the application of slip boundary conditions at porous media interfaces. Many other researchers (Cox and Hill, 2011; Misra and Shit, 2007; Moustafa, 2004; Shankararaman et al., 1992; Ibrahim and Abouzeid, 2022) have underlined the need of considering slip velocity at the wall. Particularly, Muthu and Tesfahun (2012) looked at how nonzero slip conditions affect flow pattern in a channel with varying width and permeability. Discourse on reabsorption mechanisms and slip at the boundary serve as the impetus for the present investigation. The combined effects of reabsorption and velocity slip on the flow in a uniform channel are investigated in detail. In this analysis, the flow is assumed to be steady, Newtonian, viscous, and incompressible.

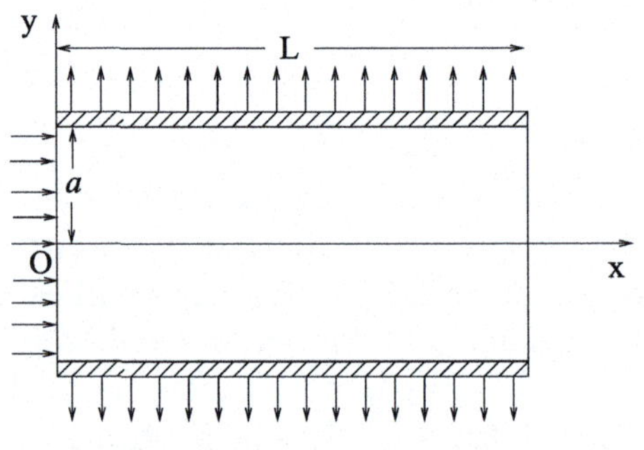


**FIGURE 8.1** Geometrical representation of the problem.

## 8.2 Problem formulation

The governing flow equations for the present situation of motion (see Fig. 8.1), describing momentum and continuity, are given by

$$U\frac{\partial U}{\partial X}+V\frac{\partial U}{\partial Y}=\nu\left(\frac{\partial^2 U}{\partial Y^2}+\frac{\partial^2 U}{\partial X^2}\right)-\frac{1}{\rho}\frac{\partial P}{\partial X}, \tag{8.1}$$

$$U\frac{\partial V}{\partial X}+V\frac{\partial V}{\partial Y}=\nu\left(\frac{\partial^2 V}{\partial Y^2}+\frac{\partial^2 V}{\partial X^2}\right)+-\frac{1}{\rho}\frac{\partial P}{\partial Y}, \tag{8.2}$$

$$\frac{\partial U}{\partial X}+\frac{\partial V}{\partial Y}=0. \tag{8.3}$$

The appropriate boundary conditions are

$$\frac{\partial U}{\partial Y}=0 \quad \text{and} \quad V=0 \quad \text{at} \quad Y=0 \quad \text{(regularity condition)}, \tag{8.4}$$

$$\frac{\sqrt{k}}{\alpha}\frac{\partial U}{\partial Y}+U=0 \quad \text{at} \quad Y=1 \quad \text{(slip velocity)}, \tag{8.5}$$

$$Q(X) = Q_0 F(\alpha X) = \int_0^1 U(X,Y)\,dY \quad \text{(flux across a cross-section)}, \tag{8.6}$$

where $U$ represents the axial velocity; $V$ represents the transverse velocity; $P$ represents the pressure; $\rho$ is the fluid density, and $\nu = \frac{\mu}{\rho}$ is the kinematic viscosity.

Nondimensional quantities are given as

$$x = X/L, \quad y = Y/a, \quad p = (Pa^2)/(\mu Q_0), \quad \psi = (\psi')(Q_0), \tag{8.7}$$

and the stream function and vorticity as

$$u = \frac{\partial \psi}{\partial y} \quad \text{and} \quad v = -\frac{\partial \psi}{\partial x}, \tag{8.8}$$

$$\omega = -\frac{\partial^2 \psi}{\partial x^2} - \frac{\partial^2 \psi}{\partial y^2}. \tag{8.9}$$

Eliminate $p$ from (8.1) and (8.2) and using Eqs. (8.7)–(8.9), the governing equations with boundary conditions become

$$\delta^2 \frac{\partial^2 \omega}{\partial x^2} + \frac{\partial^2 \omega}{\partial y^2} = \delta Re \left[ \frac{\partial \omega}{\partial x}\frac{\partial \psi}{\partial y} - \frac{\partial \omega}{\partial y}\frac{\partial \psi}{\partial x} \right], \tag{8.10}$$

$$\delta^2 \frac{\partial^2 \psi}{\partial x^2} + \frac{\partial^2 \psi}{\partial y^2} = -\omega. \tag{8.11}$$

Further the boundary conditions are

$$\frac{\partial^2 \psi}{\partial y^2} = 0 \quad \text{and} \quad \psi = 0 \quad \text{at} \quad y = 0, \tag{8.12}$$

$$\frac{\partial \psi}{\partial y} = -\phi \frac{\partial^2 \psi}{\partial y^2} \quad \text{at} \quad y = 1, \tag{8.13}$$

$$\psi = F(\alpha x) \quad \text{at} \quad y = 1, \tag{8.14}$$

where $\delta = \dfrac{a}{L}$ is the geometric parameter, $Re = \dfrac{Q_0}{\nu}$ is the Reynolds number, and $\phi = \dfrac{\sqrt{k}}{a\alpha}$ is the slip coefficient.

In this analysis, exponentially decreasing bulk flow is considered, that is (Kelman, 1962),

$$F(\alpha x) = e^{-\alpha x}. \tag{8.15}$$

Introduce the similarity variable

$$\psi = xF(y), \omega = xG(y). \tag{8.16}$$

Then the Eqs. (8.10) and (8.11) along with boundary conditions (8.12)–(8.14) in terms of $F$ and $G$ can be written as

$$\frac{d^2G}{dy^2} = \delta Re\left[G(y)\frac{dF}{dy} - F(y)\frac{dG}{dy}\right], \tag{8.17}$$

$$\frac{d^2F}{dy^2} = -G(y), \tag{8.18}$$

$$F(y) = 0 \quad \text{and} \quad \frac{d^2F}{dy^2} = 0 \quad \text{at} \quad y = 0, \tag{8.19}$$

$$\frac{dF}{dy} = -\phi\frac{d^2F}{dy^2} \quad \text{at} \quad y = 1, \tag{8.20}$$

$$F(y) = -e^{-\alpha x} \quad \text{at} \quad y = 1. \tag{8.21}$$

Now the Eqs. (8.17)–(8.21) combined to form a nonlinear boundary value $4^{th}$-order dependent ordinary differential equation.

## 8.3 Solution

To solve the nonlinear Eqs. (8.17)–(8.21), we assume a solution in a series expansion in geometric parameter $\delta$ as

$$F = F_0 + \delta F_1 + \delta^2 F_2 + \ldots\ldots, \tag{8.22}$$

$$G = G_0 + \delta G_1 + \delta^2 G_2 + \ldots\ldots \tag{8.23}$$

Using (8.22) and (8.23) for $F$ and $G$ in Eqs. (8.17)–(8.21) and collecting the coefficients of powers of $\delta$,

**(i)** Zeroth order:

$$\frac{d^2G_0}{dy^2} = 0, \tag{8.24}$$

$$\frac{d^2F_0}{dy^2} = -G_0, \tag{8.25}$$

$$F_0 = 0 \quad \text{and} \quad \frac{d^2F_0}{dy^2} = 0 \quad \text{at} \quad y = 0, \tag{8.26}$$

$$\frac{dF_0}{dy} = -\phi\frac{d^2F_0}{dy^2} \quad \text{at} \quad y = 1, \tag{8.27}$$

$$F_0 = -e^{-\alpha x} \quad \text{at} \quad y = 1. \tag{8.28}$$

**(ii)** First order:

$$\frac{d^2G_1}{dy^2} = Re\left[G_0\frac{dF_0}{dy} - F_0\frac{dG_0}{dy}\right], \tag{8.29}$$

$$\frac{d^2F_1}{dy^2}=-G_1, \tag{8.30}$$

$$F_1=0 \quad \text{and} \quad \frac{d^2F_1}{dy^2}=0 \quad \text{at} \quad y=0, \tag{8.31}$$

$$\frac{dF_1}{dy}=-\phi\frac{d^2F_1}{dy^2} \quad \text{at} \quad y=1, \tag{8.32}$$

$$F_1=0 \quad \text{at} \quad y=1. \tag{8.33}$$

**(iii)** Third order:

$$\frac{d^2G_2}{dy^2}=Re\left[G_1\frac{dF_0}{dy}+G_0\frac{dF_1}{dy}-F_0\frac{dG_1}{dy}-F_1\frac{dG_0}{dy}\right], \tag{8.34}$$

$$\frac{d^2F_2}{dy^2}=-G_2, \tag{8.35}$$

$$F_2=0 \quad \text{and} \quad \frac{d^2F_2}{dy^2}=0 \quad \text{at} \quad y=0, \tag{8.36}$$

$$\frac{dF_2}{dy}=-\phi\frac{d^2F_2}{dy^2} \quad \text{at} \quad y=1, \tag{8.37}$$

$$F_2=0 \quad \text{at} \quad y=1. \tag{8.38}$$

The Eqs. (8.24) and (8.25) are solved with boundary conditions, give the zeroth order $F_0$ and $G_0$ expressions as follows:

$$F_0=a_1(x)y^3+a_2(x)y, \tag{8.39}$$

$$G_0=-6a_1(x)y, \tag{8.40}$$

where $a_1(x)=\frac{1}{2}\frac{-e^{-\alpha x}}{(1+3\phi)}$, and $a_2(x)=\frac{3}{2}\frac{(1+2\phi)e^{-\alpha x}}{(1+3\phi)}$.

After replacing the zeroth-order components $F_0$ and $G_0$ with the expressions, the first-order components are obtained by solving Eqs. (8.29) and (8.30). This yields the following expressions for $F_1$ and $G_1$:

$$F_1=a_3(x)y^7+a_4(x)y^3+a_5(x)y, \tag{8.41}$$

$$G_1=-42a_3(x)y^5-6a_4(x), \tag{8.42}$$

where $a_3(x)=\frac{Rea_1(x)^2}{2520}$, $a_4(x)=\frac{-Rea_1(x)^2}{840}\frac{(1+7\phi)}{(1+3\phi)}$, and $a_5(x)=\frac{Rea_1(x)^2}{1260}\frac{(1+9\phi)}{(1+3\phi)}$.

Now the solution for the stream function and vorticity are given as

$$F=a_1(x)y^3+a_2(x)y+\delta\left(a_3(x)y^7+a_4(x)y^3+a_5(x)y\right), \tag{8.43}$$

$$G=-6a_1(x)y+\delta\left(-42a_3(x)y^5-6a_4(x)\right). \tag{8.44}$$

The dimensionless pressure can be obtained as

$$p(x,y)=\delta\frac{dF}{dy}+\frac{1}{\delta}\frac{d^3F}{dy^3}-Re\left[\left(\frac{dF}{dy}\right)^2-F\frac{d^2F}{dy^2}\right]. \tag{8.45}$$

The following provides the mean pressure and the associated mean pressure drop:

$$\bar{p}(x)=\int_0^1 p(x,y)dy, \tag{8.46}$$

$$\Delta\bar{p}(x_0)=\bar{p}(0)-\bar{p}(x_0). \tag{8.47}$$

The dimensionless wall shear stress is

$$\tau_w(x)=x\frac{d^2F}{dy^2}. \tag{8.48}$$

In Eq. (8.45), numerical integration is employed to calculate the integrals. This approach is chosen due to the difficulty in obtaining an analytical solution for $p(x,y)$ and deriving a closed-form expression.

## 8.4 Results and discussions

The impact of reabsorption and slip velocity at a permeable wall on the passage of a viscous incompressible Newtonian fluid through a channel is examined in this work. Through graphical analysis, the effects of the slip ($\phi$) and reabsorption ($\alpha$) are investigated, with special reference to flow in renal tubules. Important flow parameters are calculated and examined, including pressure, velocity, wall shear stress, and stream function. The reabsorption moderate values are $\alpha=1.0-2.0$, and slip coefficient values are $\phi=-0.4-0.4$ from the literature. The majority of previous research looked at low Reynolds numbers, that is, $Re=0.0$. Inertias effect and fixed $Re=1.0$ were taken into consideration in the current study. Noticeably in the limiting cases of $Re\rightarrow 0$; the current findings are in good agreement with previous published works (Acharya et al., 1981; Chaturani and Ranganatha, 1991).

The effects of the reabsorption ($\alpha$) and the slip ($\phi$) on the axial velocity at various cross-sections along the channel's length are depicted in Figs. 8.2 and 8.3. The analysis unequivocally shows that the fluid decelerates as it flows downstream, as evidenced by the axial velocity gradually decreasing along the channel's axial direction. Figs. 8.2(a), 8.2(b), and 8.2(c) show that this trend is consistently seen at three different axial positions. A range of values from $\phi=0.0$ (which represents the traditional no-slip condition) to $\phi=0.4$ was taken into consideration with regard to the slip coefficient $\phi$. It is observed that the velocity-$u(x,y)$ rises near the walls while diminishing at the center of the channel as $\phi$ values varies. The slip property that is applied at the channel walls explains this behavior. This slip permits more fluid to flow along the region of the channel walls, thereby losing the shear resistance associated with the no-slip condition.

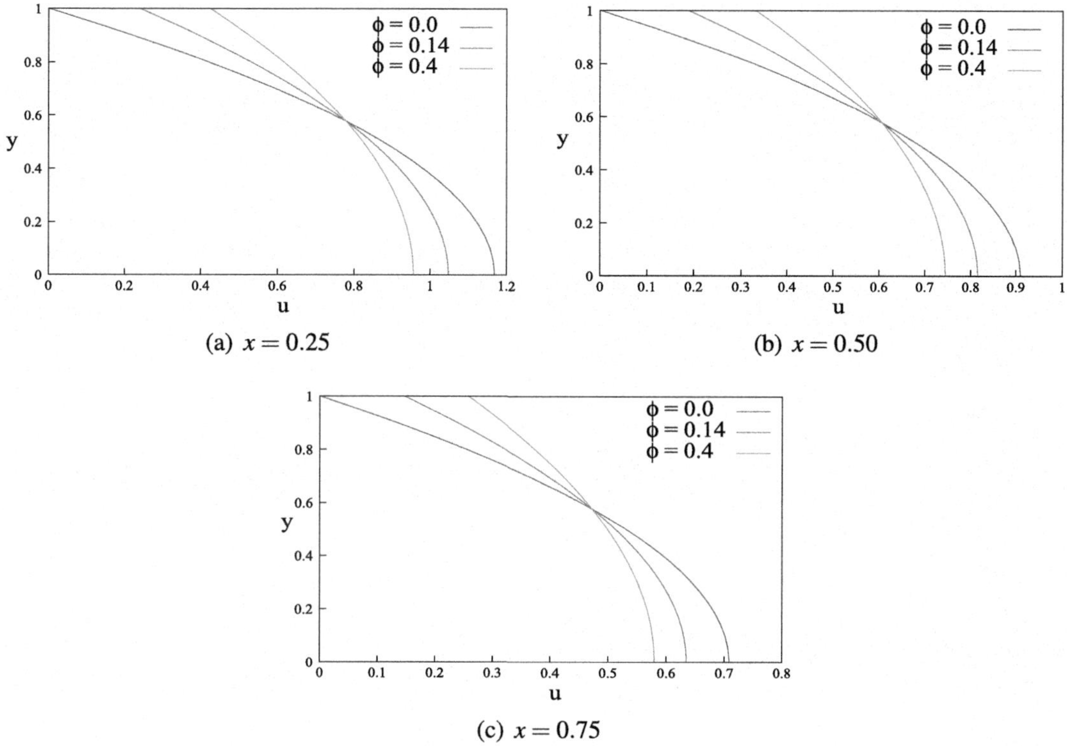


**FIGURE 8.2** $u(x, y)$ with $y$, for $\alpha = 1.0$.

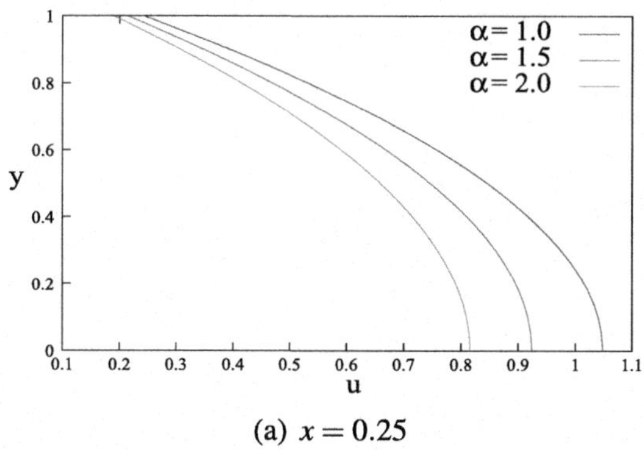


**FIGURE 8.3** $u(x, y)$ with $y$, for $\phi = 0.15$.

As the coefficient $\alpha$ rises, the velocity-$u$ decreases from its higher value, see the Fig. 8.3. The increased reabsorption at the walls, which losses the fluid volume available for fluid transfer, is noted by this velocity drop.

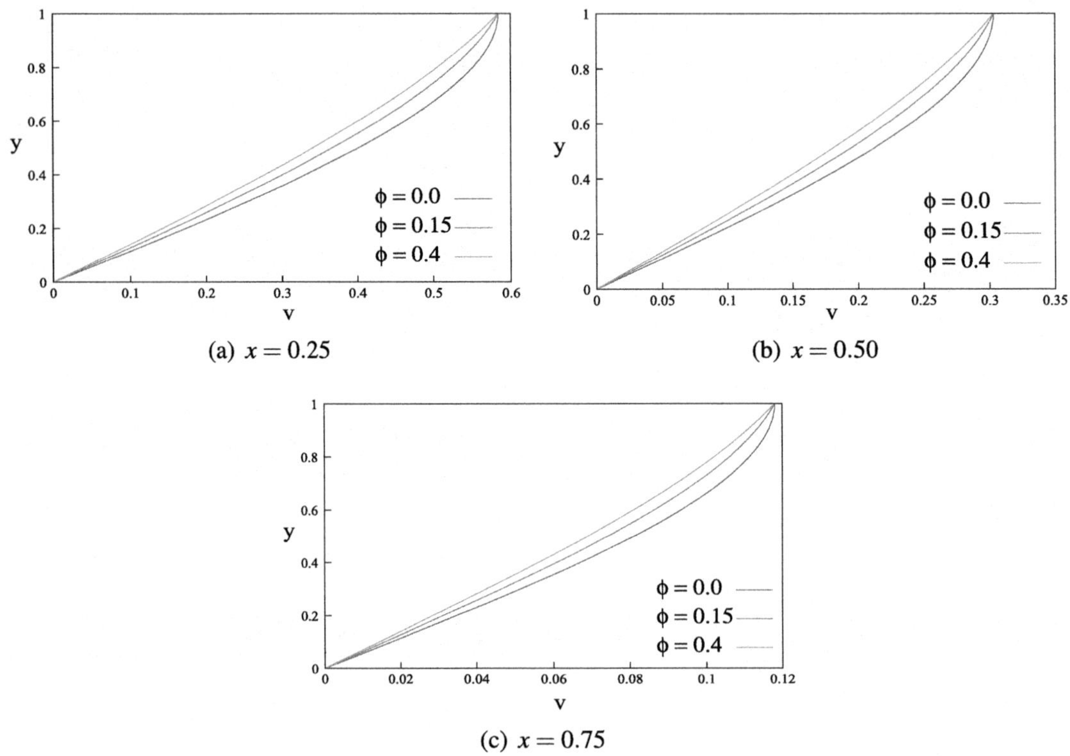


**FIGURE 8.4** $v(x, y)$ with $y$, for $\alpha = 1.0$.

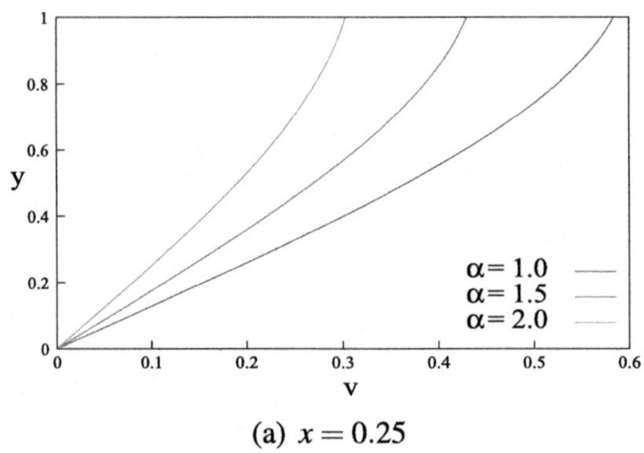


**FIGURE 8.5** $v(x, y)$ with $y$, for $\phi = 0.15$.

The coefficients $\alpha$ and $\phi$ effects on $v(x, y)$ velocity is presented in the Figs. 8.4 and 8.5. With slip condition, the fluid velocity at boundary is not necessarily zero. From the Fig. 8.4, it is seen that the effect of $\phi$ on $v(x, y)$ velocity at various positions. The rise in $\phi$ values

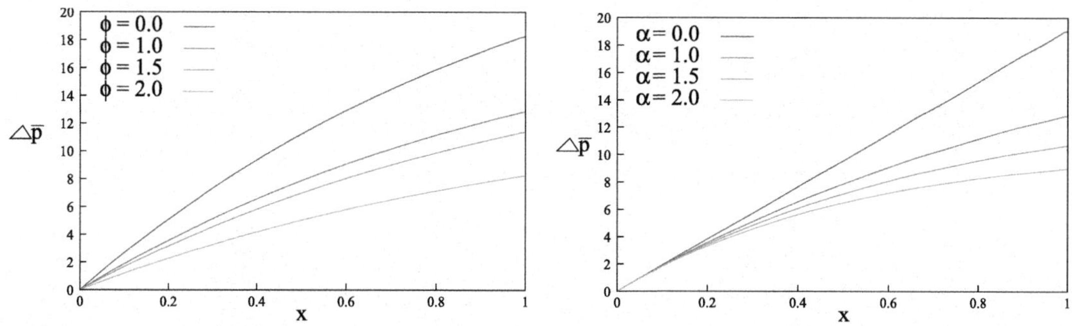


(a) $\alpha = 1.0$ (b) $\phi = 0.15$

**FIGURE 8.6** $\Delta\bar{p}$ versus $x$.

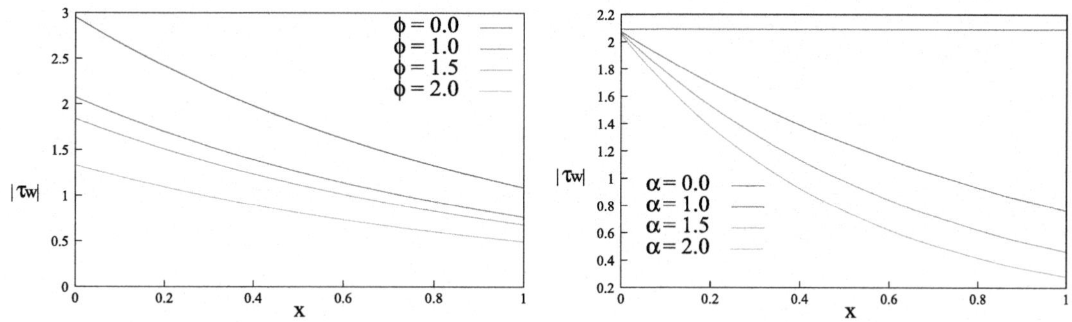


(a) $\alpha = 1.0$ (b) $\phi = 0.15$

**FIGURE 8.7** $|\tau_w|$ versus $x$.

leads to larger velocities at the wall. The slip at a permeable wall affect the relation between fluid and channel wall. It shows that with $\alpha$ from 1.0 to 2.0, the boundary $v(x, y)$ velocity increases (as given in Fig. 8.5).

The variation of the dimensionless pressure drop ($\Delta\bar{p}$) along the axial length of the permeable channel is depicted in Fig. 8.6. It is clear that the slip ($\phi$) affects the mean pressure distribution, because variations in $\phi$ alter the flow resistance close to the channel walls, which in turn affects $\Delta\bar{p}$ across the channel. Fig. 8.6(a) illustrates how the pressure drop $\Delta\bar{p}$ steadily drops as the axial coordinate $\phi$ increases. This indicates that the driving pressure gradually drops as the fluid moves downstream. Additionally, Fig. 8.6(b) shows how wall reabsorption affects the mean pressure drop. $\Delta\bar{p}$ along the channel is found to significantly decrease as the reabsorption coefficient $\alpha$ increases. This behavior implies that increased reabsorption through the channel walls lowers the pressure gradient needed to maintain the flow by reducing the net volume of fluid transported.

Another important finding is shown in Fig. 8.7(a), where it is demonstrated that the magnitude of the wall shear stress $|\tau_w|$ dramatically diminishes as the slip coefficient $\phi$

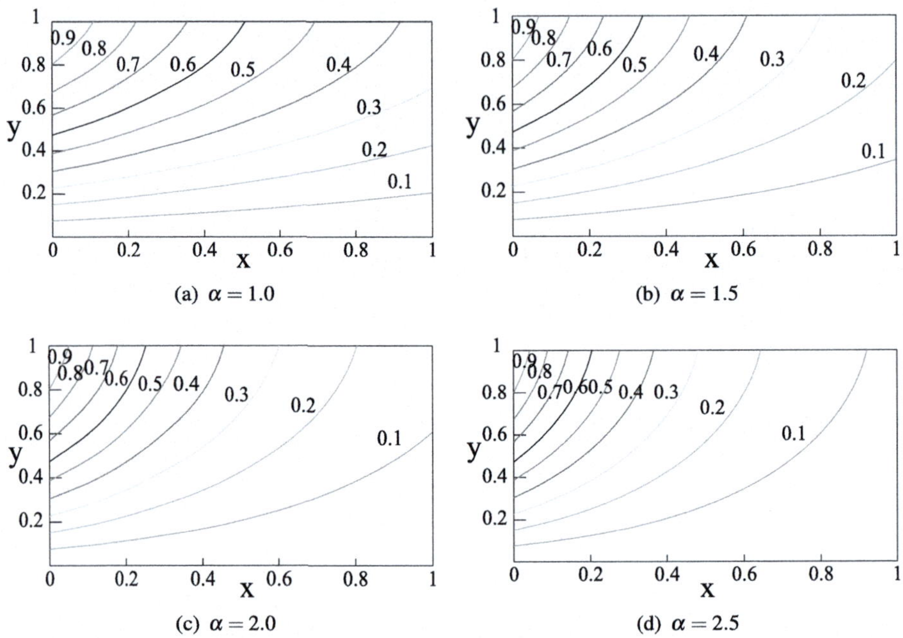


(a) $\alpha = 1.0$ (b) $\alpha = 1.5$ (c) $\alpha = 2.0$ (d) $\alpha = 2.5$

**FIGURE 8.8** Streamlines for $\phi = 0.15$ and different $\alpha$.

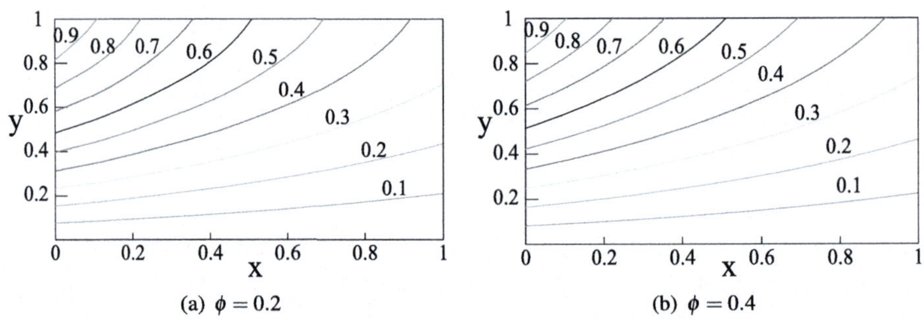


(a) $\phi = 0.2$ (b) $\phi = 0.4$

**FIGURE 8.9** Streamlines for $\alpha = 1.0$ and different $\phi$.

increases. Shear forces acting on the channel boundaries are lessened as a result of the relaxation of the no-slip at wall, which lowers the near-wall velocity gradient. The impact of the reabsorption coefficient $\alpha$ on the wall shear stress ($|\tau_w|$) along the channel's axial length is depicted in Fig. 8.7(b). Throughout the channel length, there is a noticeable drop in $|\tau_w|$ as $\alpha$ rises from 1.0 to 2.0. The improved reabsorption of fluid through the channel walls, which decreases the local flow rate and lessens the velocity gradients close to the boundary, is responsible for this decrease in shear stress.

Figs. 8.8 and 8.9 display the streamline patterns inside the channel for different reabsorption and the slip coefficient values. Improved fluid reabsorption along the boundaries is indicated by the streamlines moving closer to the channel walls as $\alpha$ increases, as shown in Figs. 8.8(a) to 8.8(d). The impact of the slip coefficient $\phi$ on the flow structure is also depicted in these figures. As $\phi$ varies, the streamlines are stretching near the walls, signifying a shift in the flow dynamics and a decrease in shear resistance. This velocity slip results in a streamlines redistribution and a variation in the overall flow behavior. Additionally, a decrease in the flow at the end ($x = 1.0$) is observed, which can be attributed to the effects of velocity slip and reabsorption.

## 8.5 Conclusions

In this work, the slip ($\phi$) and reabsorption ($\alpha$) effects on fluid flow in a permeable channel are investigated. The velocities, pressure drop, streamline patterns, and shear stress at the wall are all solved using regular perturbation method. The following are main conclusions from this analysis:

The axial and radial velocities both decrease as the reabsorption $\alpha$ increases.

The mean pressure drop decreases as both the slip coefficient ($\phi$) and the reabsorption coefficient ($\alpha$) increase.

The streamlines show how the fluid flow changes as the reabsorption $\alpha$ and slip coefficient ($\phi$) varies.

These findings provide important insight into the flow characteristics of channels with reabsorption ans slip. It should be noted that solutes are transported across the walls in addition to reabsorption, which is a important component of renal flows. In physiological systems, this solute transfer is essential to the fluid dynamics and exchange of solute, especially in the renal tubule.

## Funding

The author(s) received no financial support for the publication of this article.

## Declaration of competing interest

The author(s) declared no conflicts of interest.

## References

Acharya, G.R.K., Peeyush, C., Kaimal, M.R., 1981. A hydrodynamical study of the flow in renal tubules. Bulletin of Mathematical Biology 43, 151–163.

Apelblat, A., Katchalsky, A.K., Silberberg, A., 1974. A mathematical analysis of capillary tissue fluid exchange. Biorheology 11, 1–49.

Beavers, G.S., Joshep, D.D., 1967. Boundary conditions at a naturally permeable wall. Journal of Fluid Mechanics 30, 197–207.

Berman, A.S., 1953. Laminar flow in channels with porous walls. Journal of Applied Physics 24, 1232–1235.

Berman, A.S., 1958. Laminar flow in an annulus with porous walls. Journal of Applied Physics 29, 71–75.

Chaturani, P., Ranganatha, T.R., 1991. Flow of Newtonian fluid in non-uniform tubes with variable wall permeability with application to flow in renal tubules. Acta Mechanica 88, 11–26.
Cox, B.J., Hill, J.M., 2011. Flow through a circular tube with a permeable Navier slip boundary. Nanoscale Research Letters 6, 389.
Haroon, T., Siddiqui, A.M., Shahzad, A., 2016. Stokes flow through a slit with periodic reabsorption: an application to renal tubule. Alexandria Engineering Journal 55, 1799–1810.
Haroon, T., Siddiqui, A.M., Shahzad, A., Smeltzer, J.H., 2017. Steady creeping slip flow of viscous fluid through a permeable slit with exponential reabsorption. Applied Mathematical Sciences 11, 2477–2504.
Ibrahim, M.G., Abouzeid, M., 2022. Influence of variable velocity slip condition and activation energy on mhd peristaltic flow of Prandtl nanofluid through a non-uniform channel. Science Reports 12, 18747.
Kelman, R.B., 1962. A theoretical note on exponential flow in the proximal part of the mammalian nephron. The Bulletin of Mathematical Biophysics 24, 303–317.
Kozinski, A.A., Schmidt, F.P., Lightfoot, E.N., 1970. Velocity profiles in porous-walled ducts. Industrial & Engineering Chemistry Fundamentals 9 (3), 502–505.
Macey, R.I., 1963. Pressure flow patterns in a cylinder with reabsorbing walls. The Bulletin of Mathematical Biophysics 25, 1–9.
Macey, R.I., 1965. Hydrodynamics of renal tubule. The Bulletin of Mathematical Biophysics 27, 117–124.
Mariamma, N.K., Majhi, S.N., 2000. Flow of a Newtonian fluid in blood vessel with permeable wall – a theoretical model. Computers & Mathematics with Applications 40, 1419–1432.
Marshall, E.A., Trowbridge, E.A., 1974. Flow of a Newtonian fluid through a permeable tube: the application to the proximal renal tubule. The Bulletin of Mathematical Biophysics 36, 457–476.
Misra, J.C., Shit, G.C., 2007. Role of slip velocity in blood flow through stenosed arteries: a non-Newtonian model. Journal of Mechanics in Medicine and Biology 7, 337–353.
Moustafa, E., 2004. Blood flow in capillary under Starling hypothesis. Applied Mathematics and Computation 149, 431–439.
Muthu, P., Tesfahun, B., 2010. Mathematical model of flow in renal tubules. International Journal of Applied Mathematics and Mechanics 6, 94–107.
Muthu, P., Tesfahun, B., 2012. Flow through nonuniform channel with permeable wall and slip effect. Special Topics & Reviews in Porous Media: An International Journal 3 (4), 321–328.
Oka, S., Murata, T., 1970. A theoretical study of the flow of blood in a capillary with permeable wall. Japanese Journal of Applied Physics 9 (4), 345–352.
Palatt, J.P., Henry, S., Roger, I.T., 1974. A hydrodynamical model of a permeable tubule. Journal of Theoretical Biology 44, 287–303.
Saffman, P.G., 1971. On the boundary condition at the surface of a porous medium. Studies in Mathematics 50 (2), 93–101.
Shankararaman, C., Mark, R.W., Clint, D., 1992. Slip at uniformly porous boundary: effect on fluid flow and mass transfer. Journal of Engineering Mathematics 26, 481–492.
Siddiqui, A.M., Haroon, T., Shahzad, A., 2016. Hydrodynamics of viscous fluid through porous slit with linear absorption. Applied Mathematics and Mechanics 37 (3), 361–378.
Yuan, S.W., Finkelstein, A.B., 1956. Laminar pipe flow with injection and suction through a porous wall. Transactions of the American Society of Mechanical Engineers 78, 719–724.

9

# Stagnation point flow of trihybrid nanofluid past a shrinking sheet with induced magnetic effects

Cheong Kai Zen[a], Kohilavani Naganthran[a,b], Zailan Siri[a,b], and Ioan Pop[c]
[a]Institute of Mathematical Sciences, Faculty of Science, Universiti Malaya, Kuala Lumpur, Malaysia [b]Center for Data Analytics, Consultancy and Services, Faculty of Science, Universiti Malaya, Kuala Lumpur, Malaysia
[c]Department of Mathematics, Babes-Bolyai University, Cluj-Napoca, Romania

## Nomenclature

### Greek symbols

| | |
|---|---|
| $\alpha$ | reciprocal magnetic Prandtl number |
| $\alpha_1$ | magnetic diffusivity ($m^2s^{-1}$) |
| $\beta_f$ | thermal expansion coefficient ($K^{-1}$) |
| $\eta$ | dimensionless variable |
| $\theta$ | dimensionless temperature |
| $\lambda$ | Constant stretching/shrinking parameter |
| $\rho$ | density ($kgm^{-3}$) |
| $\rho C_p$ | thermal capacity ($Jkg^{-1}K^{-1}$) |
| $\tau_w$ | wall shear stress ($Nm^{-2}$) |
| $\mu$ | dynamic viscosity ($Nsm^{-2}$) |
| $\mu_0$ | magnetic permeability ($Hm^{-1}$) |
| $\phi_i$ | nanoparticle volume fraction |
| $\partial$ | partial differential operator |

### Roman letters

| | |
|---|---|
| $C_p$ | specific heat capacity ($Jkg^{-1}K^{-1}$) |
| $C_{fx}$ | skin friction coefficient |
| $H_0$ | upstream uniform magnetic field (T) |
| $H_e(x)$ | induced magnetic field (T) |
| $k$ | thermal conductivity ($Wm^{-1}K^{-1}$) |
| $l$ | length of stretching/shrinking sheet (m) |
| $M$ | magnetic parameter |
| $Nu_x$ | local Nusselt number |

Advanced Computational and Mathematical Approaches in Applied Differential Equations
https://doi.org/10.1016/B978-0-44-345743-2.00020-7

| | |
|---|---|
| $Pr$ | Prandtl number |
| $q_w$ | wall heat flux ($Wm^{-2}$) |
| $Re_x$ | Reynold number |
| $S$ | constant mass flux velocity/suction/injection parameter |
| $t$ | time (s) |
| $T$ | temperature (K) |
| $T_0$ | temperature at sheet surface (K) |
| $T_\infty$ | ambient temperature (K) |
| $u$ | $x$-direction velocity ($ms^{-1}$) |
| $U_w(x)$ | wall sheet velocity ($ms^{-1}$) |
| $U_\infty(x)$ | ambient velocity ($ms^{-1}$) |
| $v$ | $y$-direction velocity ($ms^{-1}$) |
| $v_w(x)$ | variable mass flux velocity ($ms^{-1}$) |

**Subscripts**

| | |
|---|---|
| $f$ | fluid |
| $w$ | surface |
| $nf$ | nanofluid |
| $hnf$ | hybrid nanofluid |
| $thnf$ | trihybrid nanofluid |

## 9.1 Introduction

Nanofluid is a substance that gained fame for its ability to enhance heat transfer with a small addition of nanoparticles. An excellent heat transfer rate is a key target in industrial processes, such as extrusion, ensuring the quality of the end product. Recognizing the interdependence of industrial process efficiency and high heat transfer rates, Choi and Eastman (Choi, 1995) introduced the concept of nanofluids, in which nanosized materials are suspended in a base fluid, which then attracted significant attention from researchers. Numerous theoretical and experimental studies have been conducted to date, as the need to enhance heat transfer rates remains a pressing demand, in parallel with technological advancements. Within the scope of laminar boundary layer flow, Khan and Pop (2010) attempted to solve the transport phenomena of nanofluid flow over an expanding sheet. Although Khan and Pop (2010) became the main reference for other extension works in boundary layer nanofluid flow, Myers et al. (2017) identified several issues with Khan and Pop (2010), including the neglect of the zero flux boundary condition and the use of unrealistic governing parameters to generate the numerical outputs. In the mathematical formulations of nanofluids, two distinct models are used: the Buongiorno model (Buongiorno, 2006) and the Tiwari–Das model (Tiwari and Das, 2007). Myers et al. (2017) highlighted the misleading aspects in both models, and it is essential for researchers to take note of the point made in Myers et al. (2017) so that the outputs from the theoretical study can serve as a reference for the experimental works.

Nanofluids then evolved into upgraded versions, where hybrid (two or more) nanoparticles are suspended in the base fluid. Devi and Devi (2017) introduced the hybrid nanofluid boundary-layer flow over an extending sheet and demonstrated greater heat transfer enhancement than the nanofluid. Then Manjunatha et al. (2022) pioneered the numerical simulation of trihybrid nanofluid boundary layer flow over an extending sheet, demonstrating that the trihybrid nanofluid exhibits a higher heat transfer rate than the hybrid nanofluid. While pioneering works focus on heat transfer rate augmentation over a stretching sheet, a series of papers by Ouyang et al. (2024, 2025a,b) show higher heat transfer performance over a shrinking sheet. Nowadays, the trend of upgrading nanofluids has progressed to tetra- and penta-hybrid nanofluids (see Singh and Nandkeolyar, 2025; Al-Arabi and Elgazery, 2025). This progress is highly welcome, as the hybrid nanofluid version exhibits better stability and can help prevent aggregation and sedimentation during extrusion. Moreover, combining different nanoparticles can synergistically improve thermal conductivity beyond what each nanoparticle could achieve alone. This enables more efficient heat transfer, which is crucial in applications such as cooling systems, heat exchangers, and electronics.

Induced magnetic field is another ideal option for enhancing heat transfer rates and improving fluid flow performance in the extrusion process (Ghosh et al., 2010). Early work on the modeling of the induced magnetic field in the magnetohydrodynamic (MHD) boundary-layer flow was conducted by Greenspan and Carrier (1959), who successfully identified a hidden, critical strait that effectively blocked transport. Glauert (1961) then created high-resolution maps for the surrounding flat areas, outlining the essential adjustments (logarithmic terms) required for precision near the sea limits. Finally, Gribben (1965) expanded on these maps to incorporate terrain variations (pressure gradients), demonstrating that these adjustments were only necessary for the initial flat regions. This streamlined the map-making process for sloped areas, making it far more efficient. Later, Kumari et al. (1990) explicitly incorporated the effect of heat flux into their analysis and demonstrated that the temperature solutions that satisfy the equations are physically unrealistic when the temperature variation index is less than $-2$ and the Prandtl number exceeds 2.5. The effect of the induced magnetic field near the stagnation region on the shrinking sheet was investigated by Ali et al. (2011), and demonstrated that it accelerates boundary-layer separation. Ali et al. (2019) revisited the work in Ali et al. (2011) to test the stability of the nonuniqueness solutions found and demonstrated that the first solution is stable while the second solution is unstable. In the context of next-generation fluids, Junoh et al. (2020) solved the problem considered by Ali et al. (2011) for a horizontally positioned stretching/shrinking sheet. The heat transfer rate, however, deteriorates as the impermeable sheet shrinks. This motivates the present study, which investigates the transport phenomena of a trihybrid nanofluid near the stagnation region over a permeable shrinking sheet, along with the impact of the induced magnetic field.

## 9.2 Mathematical model

We consider the steady, stagnation point flow of electrically conducting trihybrid nanofluid past a shrinking sheet with induced magnetic effect, where $(x, y)$ are the Cartesian coordinates with the $x$-axis measured along the sheet and $y$-axis is normal to it, the flow being at $y \geq 0$. It is assumed that the sheet moves; it can be either be constantly stretching $(c > 0)$ or shrinking $(c < 0)$. It is assumed that the sheet moves with the velocity $U_w(x) = cx$, and the ambient velocity is denoted by $U_\infty(x) = ax$, where $a$ is a positive constant (time$^{-1}$). Additionally, it is assumed that $\boldsymbol{H}$ is the induced magnetic field vector with the free stream magnetic field component $H_e(x) = H_0(x/l)$, in which $H_0$ is the upstream uniform magnetic field at infinity, and $l$ is the characteristic length of the surface. In addition, the parallel and normal components of induced magnetic field $\boldsymbol{H}$ are denoted by $H_1$ and $H_2$, respectively. At the surface, the normal component $H_2$ vanishes, whereas the parallel component $H_1$ becomes $H_0$. Hall current and ion slip effects are neglected. Furthermore, we assume $T_\infty$ is the uniform ambient temperature. Meanwhile, plate temperature is symbolized by $T_w(x) = T_\infty + T_0(x/l)$, where $T_0$ is the reference temperature. Under these assumptions, the governing conservation equations of mass, momentum, energy, and induced magnetic field can be expressed as in Cartesian coordinates $(x, y)$ (Junoh et al., 2020)

$$\frac{\partial u}{\partial x} + \frac{\partial v}{\partial y} = 0, \tag{9.1}$$

$$\frac{\partial H_1}{\partial x} + \frac{\partial H_2}{\partial y} = 0, \tag{9.2}$$

$$u\frac{\partial u}{\partial x} + v\frac{\partial u}{\partial y} = u_e\frac{\partial u_e}{\partial x} + \frac{\mu_{thnf}}{\rho_{thnf}}\frac{\partial^2 u}{\partial y^2} - \frac{\mu_0}{\rho_f}\left(H_e\frac{\partial H_e}{\partial x} - H_1\frac{\partial H_1}{\partial x} - H_2\frac{\partial H_1}{\partial y}\right), \tag{9.3}$$

$$u\frac{\partial T}{\partial x} + v\frac{\partial T}{\partial y} = \frac{k_{thnf}}{(\rho C_p)_{thnf}}\frac{\partial^2 T}{\partial y^2}, \tag{9.4}$$

$$u\frac{\partial H_1}{\partial x} + v\frac{\partial H_1}{\partial y} - H_1\frac{\partial u}{\partial x} - H_2\frac{\partial u}{\partial y} = \alpha_1\frac{\partial^2 H_1}{\partial y^2}, \tag{9.5}$$

along with the boundary conditions

$$\begin{aligned} &u = U_w(x) = cx, \quad v = v_w, \quad \frac{\partial H_1}{\partial y} = H_2 = 0, \quad T = T_w \quad \text{at} \quad y = 0, \\ &u = U_e(x) \to ax, \quad T \to T_\infty, \quad H_1 = H_e(x) \to H_0(x/l) \quad \text{as} \quad y \to \infty. \end{aligned} \tag{9.6}$$

Here $(u, v)$ are the velocity components of trihybrid nanofluid along $(x, y)$ directions, respectively; $\mu_0$ is the magnetic permeability, and $\alpha_1$ is the magnetic diffusivity. The ternary nanofluid's dynamic viscosity, density, and thermal conductivity are denoted by $\mu_{thnf}$, $\rho_{thnf}$, and $k_{thnf}$, while the ternary nanofluid's heat capacity symbolized as $(\rho C_p)_{thnf}$. These definitions are given in Table 9.1. Next, $\phi_1$, $\phi_2$, and $\phi_3$ signify the $SiO_2$'s, $TiO_2$'s, and $Al_2O_3$'s nanoparticle volume fraction, respectively. $\rho_f$ and $\rho_{thnf}$ are the densities of the base fluid

**Table 9.1** The trihybrid nanofluid correlation properties (Manjunatha et al., 2022).

| Properties | Tri-hybrid nanofluid mathematical relation |
|---|---|
| Density | $\frac{\rho_{thnf}}{\rho_f} = (1-\phi_1)\left\{(1-\phi_2)\left[(1-\phi_3)+\phi_3\frac{\rho_{s1}}{\rho_f}\right]+\phi_2\frac{\rho_{s2}}{\rho_f}\right\}+\phi_1\frac{\rho_{s1}}{\rho_f}$ |
| Dynamic viscosity | $\mu_{thnf} = \frac{\mu_f}{(1-\phi_1)^{2.5}(1-\phi_2)^{2.5}(1-\phi_3)^{2.5}}$ |
| Thermal capacity | $\frac{(\rho C_p)_{thnf}}{(\rho C_p)_f} = \phi_1\frac{(\rho C_p)_{s1}}{(\rho C_p)_f} + (1-\phi_1)\left[(1-\phi_2)\left\{(1-\phi_3)+\phi_3\frac{(\rho C_p)_{s3}}{(\rho C_p)_f}\right\} +\phi_2\frac{(\rho C_p)_{s2}}{(\rho C_p)_f}\right]$ |
| Thermal conductivity | $\frac{k_{thnf}}{k_{hnf}} = \frac{k_{s3}+2k_{hnf}-2\phi_3(k_{hnf}-k_{s3})}{k_{s3}+2k_{hnf}+\phi_3(k_{hnf}-k_{s3})}$, where $\frac{k_{hnf}}{k_f} = \frac{k_{s2}+2k_{nf}-2\phi_2(k_{nf}-k_{s2})}{k_{s2}+k_{nf}+\phi_2(k_{nf}-k_{s2})}$ and $\frac{k_{nf}}{k_f} = \frac{k_{s1}+2k_f-2\phi_1(k_f-k_{s1})}{k_{s1}+2k_f+\phi_1(k_f-k_{s1})}$ |

**Table 9.2** The thermophysical properties of selected nanoparticles and base fluid.

| Property | $SiO_2$ | $TiO_2$ | $Al_2O_3$ | $H_2O$ |
|---|---|---|---|---|
| $\rho\left(\mathrm{kgm^{-3}}\right)$ | 2270 | 4250 | 6310 | 997.1 |
| $k\left(\mathrm{Wm^{-1}K^{-1}}\right)$ | 1.4013 | 8.953 | 32.9 | 0.6071 |
| $C_p\left(\mathrm{JK^{-1}g^{-1}K^{-1}}\right)$ | 730 | 711 | 686.2 | 4179 |
| $\beta_f \times 10^5\left(\mathrm{K^{-1}}\right)$ | 1.02 | 0.9 | 0.85 | 21 |

and the hybrid nanoparticle, respectively; $k_f$, $k_{hnf}$, and $k_{thnf}$ are the thermal conductivities of the base fluid, the hybrid nanoparticles, and the ternary hybrid nanofluid, respectively; $(\rho C_p)_f$ and $(\rho C_p)_{thnf}$ are the heat capacitance of the base fluid and the ternary hybrid nanoparticle, respectively. These correlations are correct and feasible. They are based on physical assumptions and agree with the conservation of mass and energy. The physical properties of the base fluid, which is water ($H_2O$), and the utilized nanoparticles are given in Table 9.2.

Following Junoh et al. (2020), we assume that Eqs. (9.1)–(9.5), subject to the boundary conditions (9.6), enable the similarity solution

$$u = axf'(\eta), \quad v = -\sqrt{av_f}\,f(\eta), \quad \eta = y\sqrt{a/v_f}, \quad \theta(\eta) = \frac{T-T_\infty}{T_w-T_\infty},$$
$$H_1 = H_0(x/l)\,g'(\eta), \quad H_2 = -(H_0/l)\sqrt{v_f/a}\,g(\eta). \tag{9.7}$$

Thus we have

$$v_w = -\sqrt{av_f}\,S, \tag{9.8}$$

where primes denote differentiation with respect to $\eta$, and $S$ is the constant mass flux velocity with $S > 0$ for suction and $S < 0$ for injection, respectively. Substituting the similarity variables (9.7) into Eqs. (9.2) to (9.5), it is found that they are reduced to the following ordinary (similarity) differential equations:

$$\frac{\mu_{thnf}/\mu_f}{\rho_{thnf}/\rho_f} f''' + ff'' + 1 - f'^2 + M\left(g'^2 - gg'' - 1\right) = 0, \tag{9.9}$$

$$\frac{k_{thnf}/k_f}{(\rho C_p)_{thnf}/(\rho C_p)_f} \frac{1}{\Pr}\theta'' + f\theta' - f'\theta = 0, \tag{9.10}$$

$$\alpha g''' + fg'' - f''g = 0, \tag{9.11}$$

subject to the boundary conditions

$$\begin{aligned} &f(0) = S, \quad f'(0) = \lambda, \quad g(0) = 0, \quad g''(0) = 0, \quad \theta(0) = 1, \\ &f'(\eta) \to 1, \quad g'(\eta) \to 1, \quad \theta(\eta) \to 0 \quad \text{as} \quad \eta \to \infty, \end{aligned} \tag{9.12}$$

where Pr is the Prandtl number, $\alpha$ is the reciprocal magnetic Prandtl number, $M$ is the magnetic parameter and $\lambda$ is the stretching/shrinking parameter, which are defined as

$$\Pr = \frac{\mu_f (\rho C_p)_f}{k_f \rho_f}, \quad \alpha = \frac{\alpha_1}{\nu_f}, \quad M = \frac{\mu_0 H_0^2}{\rho_f l^2 a^2}, \quad \lambda = \frac{c}{a}. \tag{9.13}$$

The quantities of physical interest are the skin friction coefficient $(C_{fx})$ and the local Nusselt number $(Nu_x)$, which are defined as

$$C_{fx} = \frac{\tau_w}{\rho_f U_e^2}, \qquad Nu_x = \frac{x q_w}{k_f (T_w - T_\infty)}, \tag{9.14}$$

wherein the wall shear stress and wall heat flux are denoted by $\tau_w$ and $q_w$, respectively, which can be elaborated further as

$$\tau_w = \mu_{thnf}\left(\frac{\partial u}{\partial y}\right)_{y=0}, \qquad q_w = -k_{thnf}\left(\frac{\partial T}{\partial y}\right)\Bigg|_{y=0}. \tag{9.15}$$

Using (9.7) and (9.14), we get

$$\mathrm{Re}_x^{1/2} C_{fx} = \frac{\mu_{thnf}}{\mu_f} f''(0), \qquad \mathrm{Re}_x^{-1/2} Nu_x \frac{k_f}{k_{thnf}} = -\theta'(0). \tag{9.16}$$

Here the Reynolds number is $\mathrm{Re}_x = \frac{xU_e(x)}{\nu_f}$. It should be mentioned that for $\phi = 0$ (classical viscous fluid), Eqs. (9.9) and (9.10) reduce to the model in Ali et al. (2019) when $\lambda = 0$ (buoyancy forces are absent).

**Table 9.3** Comparison value of $f''(0)$ when $\phi_1, \phi_2 = 0$.

| λ | (Wang, 2008) | (Bachok et al., 2011) | Present study |
|---|---|---|---|
| −0.50 | 1.49567 | 1.495670 | 1.495670 |
| −1.00 | 1.32882 (0) | 1.328817 (0) | 1.328817 (0) |
| −1.15 | 1.082231 (0.116702) | 1.082231 (0.116702) | 1.082231 (0.116702) |

() indicates the second solution.

## 9.3 Results and discussion

The collocation method, which is available in MATLAB's® built-in bvp4c routine, solves the transformed mathematical model in (9.9)–(9.12). All numerical outputs have been generated by fixing the relative tolerance to $1 \times 10^{-10}$ while the Prandtl number is fixed to 7, implying pure water. By setting two different guess values, dual solutions can be obtained. The first solution explains the flow attached to the sheet, while the second solution conveys the flow far from the sheet surface (Ali et al., 2019). Before discussing the numerical solutions of the present work, it is important to validate the present model. Thus by fixing $\phi_1, \phi_2 = 0$, the values of $f''(0)$ have been generated when the sheet shrank, and then compared with Junoh et al. (2020) and Schlichting and Gersten (2016). Table 9.3 presents excellent agreement in the comparison results, and it is evident that the built-in bvp4c function is compatible with the shooting method used by Wang (2008) and Bachok et al. (2011). Next the results for varying governing parameters will be discussed physically, focusing only on the first solution's behavior, as the second solution's outputs are associated with flow separation (Wang, 2008).

Figs. 9.1 and 9.2 depict the impact of the suction parameter on the fluid flow characteristics. In Fig. 9.1, the fluid velocity increases with increasing suction intensity. The higher suction effect traps the slow moving particles at the shrinking surface, allowing the fluid velocity in the fluid flow regime to increase. An increase in fluid velocity increases wall shear stress. An increase in wall shear stress is undesirable in boundary-layer flow, as it can reduce system efficiency. The increased drag on the shrinking surface results in an increase in $Re_x^{1/2} C_{fx}$; this is shown in Fig. 9.2. Fig. 9.2 also evidences that boundary layer flow separation over the shrinking surface can be delayed by applying a higher suction intensity. Figs. 9.3 and 9.4 demonstrate the impact of suction on heat transfer characteristics during sheet shrinkage. Fig. 9.3 shows a decrease in the fluid temperature as suction intensity increases. When more particles are trapped on the sheet surface, the ambient fluid is more attracted to the flow regime, reducing the fluid temperature. Reduction in the fluid temperature heightens the temperature gradient and improves the heat transfer rate over the shrinking sheet, as shown in Fig. 9.4.

Fig. 9.5 shows that an increase in suction intensity results increase in the induced magnetic profiles $g'(\eta)$. Increasing suction intensity accelerates the fluid towards the surface, creating steeper velocity gradients. This enhances the induced magnetic field as the moving conductive fluid interacts more effectively with the magnetic field. Figs. 9.6 and 9.7 show the impact of the magnetic parameter towards the fluid flow characteristics.

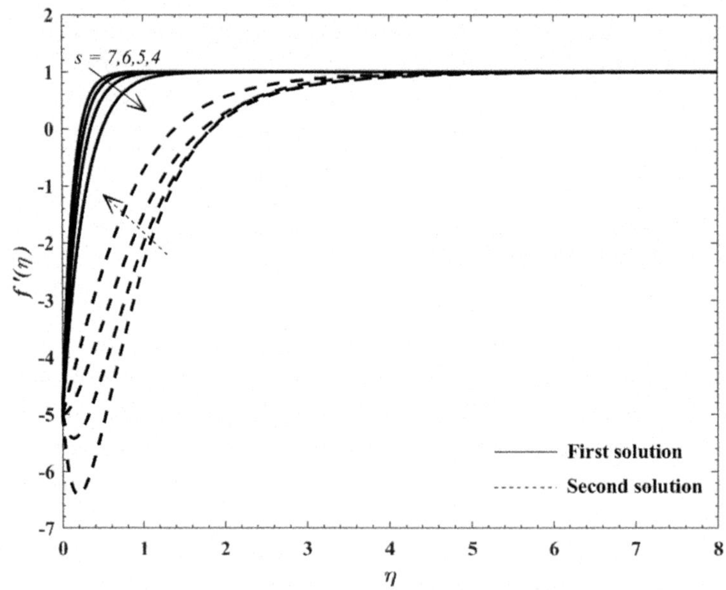


**FIGURE 9.1** Variation of $f'(\eta)$ when $\phi_1 = 0.01$, $\phi_2 = 0.02$, $\phi_3 = 0.01$, $Pr = 7$, $M = 1$, $\lambda = -5$, and $\alpha = 0.5$.

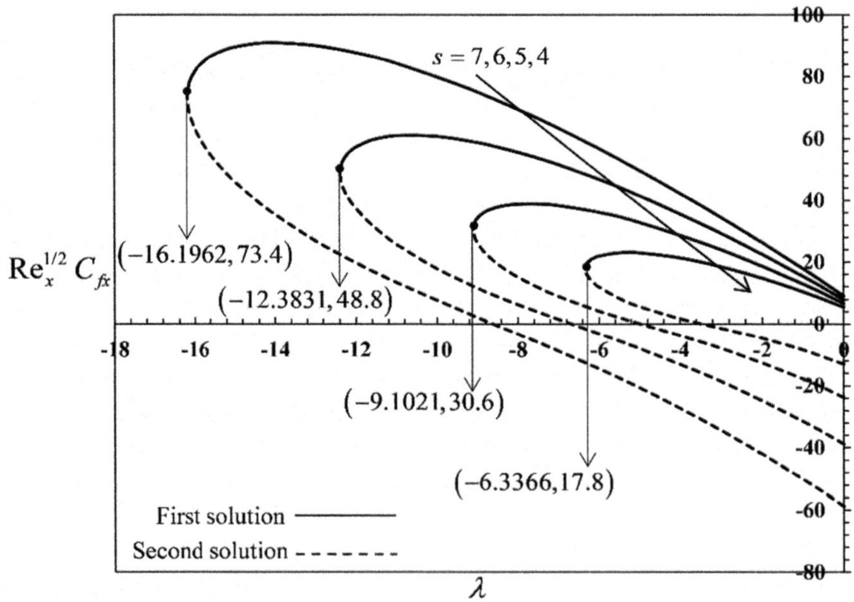


**FIGURE 9.2** Variation of $Re_x^{1/2} C_{fx}$ when $\phi_1 = 0.01$, $\phi_2 = 0.02$, $\phi_3 = 0.01$, $Pr = 7$, $M = 1$, and $\alpha = 0.5$.

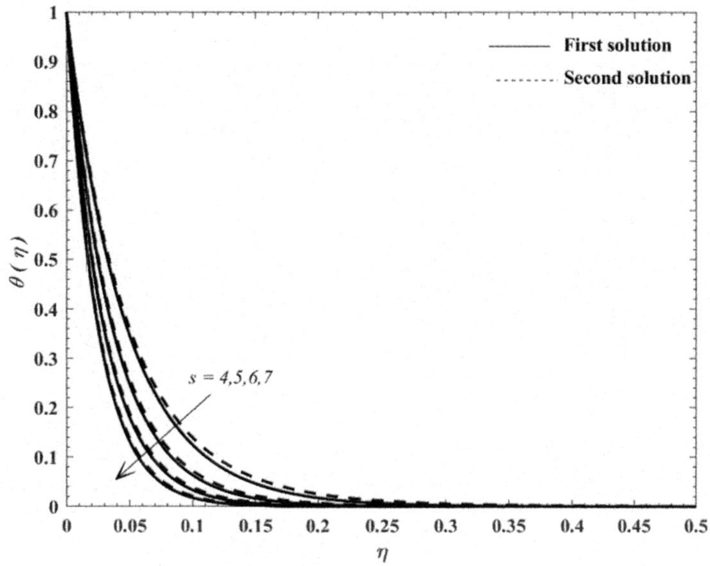


**FIGURE 9.3** Variation of $\theta(\eta)$ when $\phi_1 = 0.01$, $\phi_2 = 0.02$, $\phi_3 = 0.01$, $Pr = 7$, $M = 1$, $\lambda = -5$, and $\alpha = 0.5$.

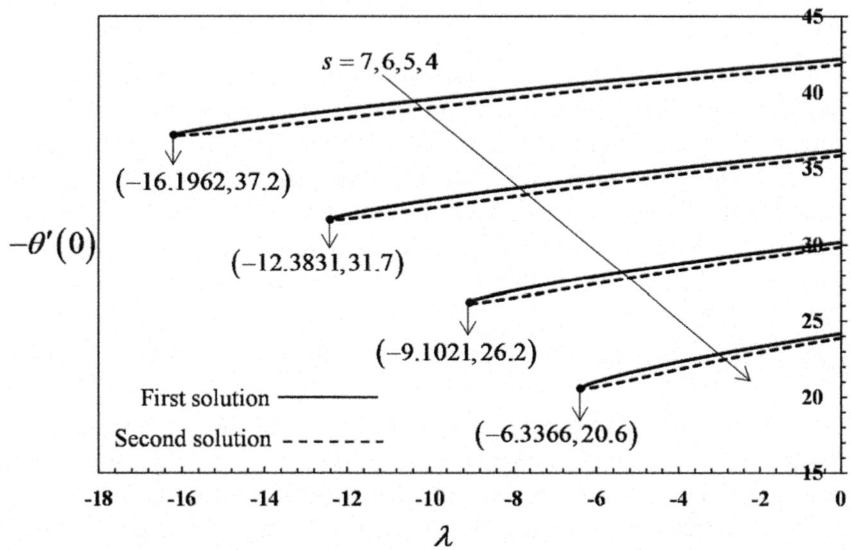


**FIGURE 9.4** Variation of $-\theta'(0)$ when $\phi_1 = 0.01$, $\phi_2 = 0.02$, $\phi_3 = 0.01$, $Pr = 7$, $M = 1$, and $\alpha = 0.5$.

Fig. 9.6 exhibits that an increment in $M$ reduces the fluid velocity. The magnetic field exerts a Lorentz force on the moving charged fluid, acting directly against the flow and reducing the fluid's velocity. A stronger magnetic field, associated with a higher magnetic

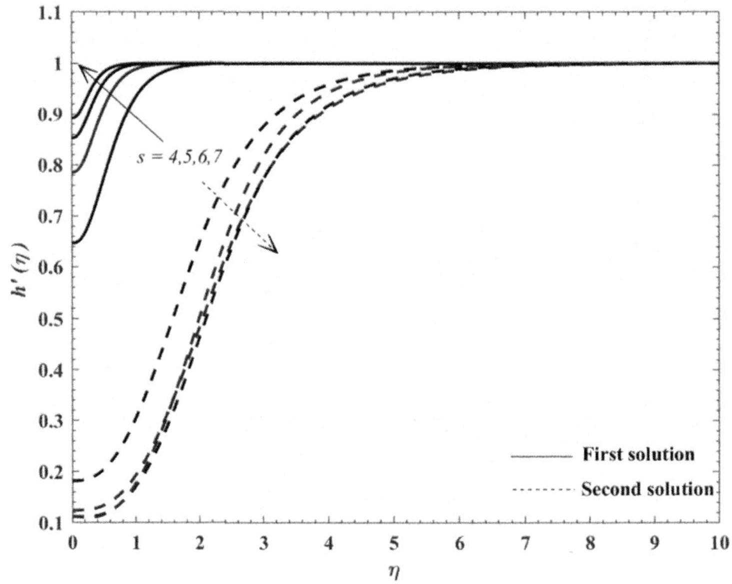


**FIGURE 9.5** Variation of $g'(\eta)$ when $\phi_1 = 0.01$, $\phi_2 = 0.02$, $\phi_3 = 0.01$, $\mathrm{Pr} = 7$, $M = 1$, $\lambda = -5$, and $\alpha = 0.5$.

parameter, significantly increases this opposing force. Eventually, this reduces the wall shear stress on the shrinking sheet and affects $Re_x^{1/2}C_{fx}$ to reduce in Fig. 9.7. Fig. 9.8 shows that the induced magnetic profiles increase when the reciprocal magnetic Prandtl number decreases. When the reciprocal magnetic Prandtl number decreases, the momentum diffusivity of the fluid becomes less effective relative to the magnetic diffusivity. This enhances the interaction between the magnetic field and fluid motion, leading to stronger induced magnetic fields, greater flow stability, and a stronger influence on flow dynamics.

## 9.4 Conclusions

The present work is devoted to examining the influence of the induced magnetic field on the trihybrid nanofluid flow near the stagnation region over a permeable shrinking surface. The numerical simulations were performed with the aid of MATLAB's bvp4c function, and dual solutions have been identified. It is found that an increment in the suction intensity enhances $Re_x^{1/2}C_{fx}$, heat transfer rate, and delays flow separation. Meanwhile, an increment in the magnetic parameter reduces $Re_x^{1/2}C_{fx}$ and fastens the flow separation.

## Acknowledgments

The Malaysian Ministry of Higher Education supported this work via the Fundamental Research Grant Scheme (FRGS) under project code FRGS/1/2023/STG06/UM/02/14.

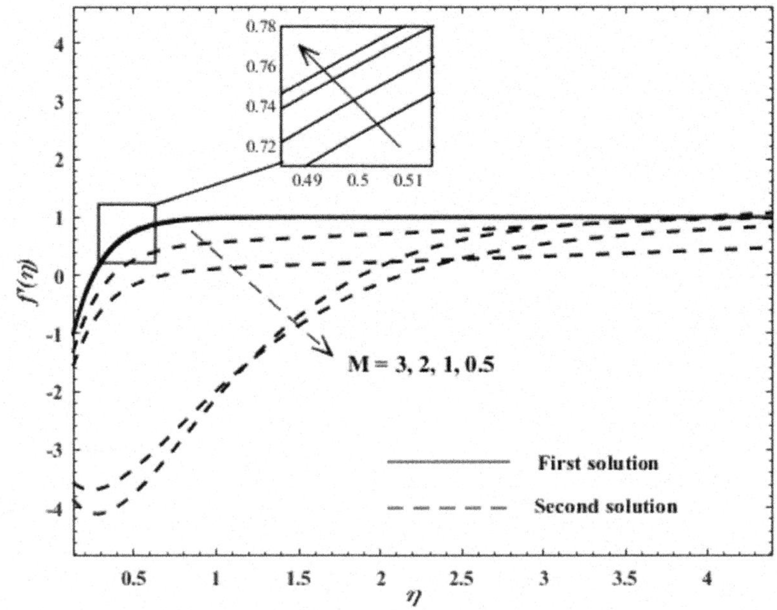


**FIGURE 9.6** Variation of $f'(\eta)$ when $\phi_1 = 0.01$, $\phi_2 = 0.02$, $\phi_3 = 0.01$, $\Pr = 7$, $s = 5$, $\lambda = -3$, and $\alpha = 0.5$.

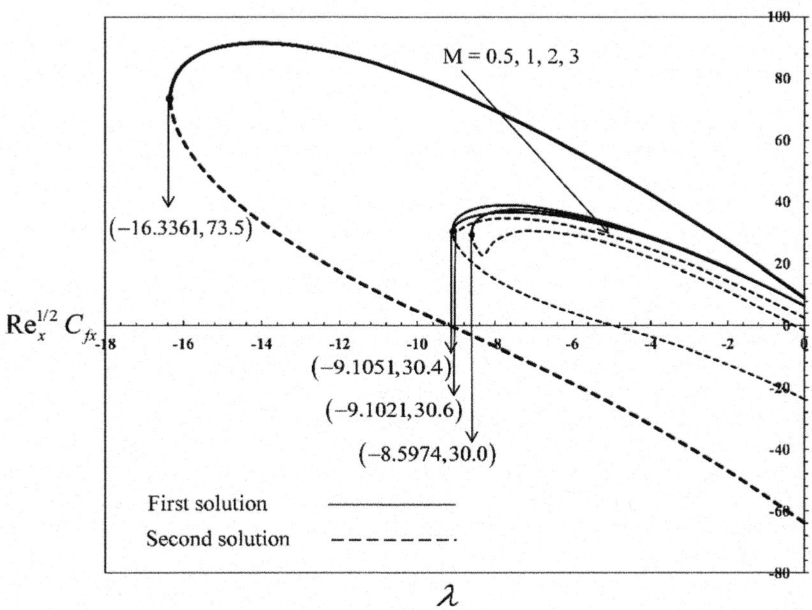


**FIGURE 9.7** Variation of $Re_x^{1/2} C_{fx}$ when $\phi_1 = 0.01$, $\phi_2 = 0.02$, $\phi_3 = 0.01$, $\Pr = 7$, $s = 5$, and $\alpha = 0.5$.

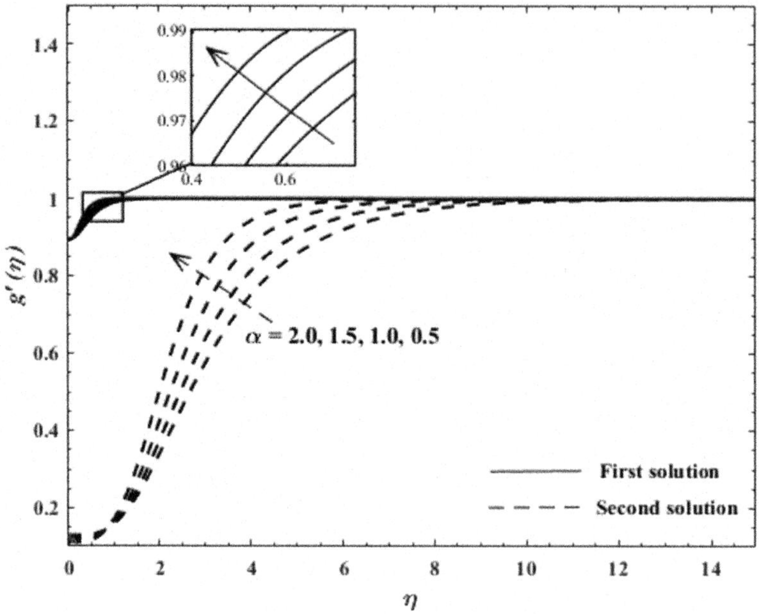


**FIGURE 9.8** Variation of $g'(\eta)$ when $\phi_1 = 0.01$, $\phi_2 = 0.02$, $\phi_3 = 0.01$, $Pr = 7$, $M = 1$, $\lambda = -5$, and $s = 7$.

# References


Al-Arabi, T.H., Elgazery, N.S., 2025. Flow of a magneto-penta-hybrid nanofluid between two disks with variable thermal conductivity/permeability: drug delivery mechanism. Journal of Thermal Analysis and Calorimetry, 1–31.

Ali, F.M., Nazar, R., Arifin, N.M., Pop, I., 2011. MHD mixed convection boundary layer flow toward a stagnation point on a vertical surface with induced magnetic field. Journal of Heat Transfer 133 (2), 022502–022502-6.

Ali, F.M., Naganthran, K., Nazar, R., Pop, I., 2019. MHD mixed convection boundary layer stagnation-point flow on a vertical surface with induced magnetic field. International Journal of Numerical Methods for Heat & Fluid Flow 10.

Bachok, N., Ishak, A., Pop, I., 2011. Stagnation-point flow over a stretching. Nanoscale Research Letters 6 (1), 1–10.

Buongiorno, J., 2006. Convective transport in nanofluids. Journal of Heat Transfer 128, 240–250.

Choi, S.U., 1995. Enhancing thermal conductivity of fluids with nanoparticles. In: ASME IMECE, vol. 17421. American Society of Mechanical Engineers, pp. 99–105.

Devi, S.U., Devi, S.A., 2017. Heat transfer enhancement of Cu-$Al_2O_3$/water hybrid nanofluid flow over a stretching sheet. Journal of the Nigerian Mathematical Society 36 (2), 419–433.

Ghosh, S.K., Anwar Bég, O., Zueco, J., 2010. Hydromagnetic free convection flow with induced magnetic field effects. Meccanica 45 (2), 175–185.

Glauert, M.B., 1961. A study of the magnetohydrodynamic boundary layer on a flat plate. Journal of Fluid Mechanics 10 (2), 276–288.

Greenspan, H.P., Carrier, G.F., 1959. The magnetohydrodynamic flow past a flat plate. Journal of Fluid Mechanics 6 (1), 77–96.

Gribben, R.J., 1965. The magnetohydrodynamic boundary layer in the presence of a pressure gradient. Proceedings of the Royal Society of London. Series A, Mathematical and Physical Sciences 287 (1408), 123–141.

Junoh, M.M., Ali, F.M., Arifin, N.M., Bachok, N., Pop, I., 2020. MHD stagnation-point flow and heat transfer past a stretching/shrinking sheet in a hybrid nanofluid with induced magnetic field. International Journal of Numerical Methods for Heat & Fluid Flow 30 (3), 1345–1364.

Khan, W.A., Pop, I., 2010. Boundary-layer flow of a nanofluid past a stretching sheet. International Journal of Heat and Mass Transfer 53 (11–12), 2477–2483.

Kumari, M., Takhar, H.S., Nath, G., 1990. MHD flow and heat transfer over a stretching surface with prescribed wall temperature or heat flux. Wärme- und Stoffübertragung 25 (6), 331–336.

Manjunatha, S., Puneeth, V., Gireesha, B.J., Chamkha, A., 2022. Theoretical study of convective heat transfer in ternary nanofluid flowing past a stretching sheet. Journal of Applied and Computational Mechanics 8 (4), 1279–1286.

Myers, T.G., Ribera, H., Cregan, V., 2017. Does mathematics contribute to the nanofluid debate? International Journal of Heat and Mass Transfer 111, 279–288.

Ouyang, Y., Basir, M.F.M., Naganthran, K., Pop, I., 2024. Triple solutions for unsteady stagnation flow of tri-hybrid nanofluid with heat generation/absorption in a porous medium. Case Studies in Thermal Engineering 61, 105027.

Ouyang, Y., Basir, M.F.M., Naganthran, K., 2025a. Slip effects and heat transfer in unsteady stagnation flow of ternary hybrid nanofluids over a rigid plate in porous media using a two-parameter scaling method. Thermal Science and Engineering Progress, 104081.

Ouyang, Y., Basir, M.F.M., Naganthran, K., Pop, I., 2025b. Stability and quadruple solution dynamics of unsteady ternary hybrid nanofluid flow over a stretching/shrinking wedge. Chinese Journal of Physics 95, 558–572.

Schlichting, H., Gersten, K., 2016. Boundary-Layer Theory. Springer.

Singh, S., Nandkeolyar, R., 2025. Investigating the MHD Casson tetra-hybrid nanofluid flow in an inclined cylinder with radiative heat transport using numerical and neural network approach. Particulate Science and Technology 43 (7), 1180–1202.

Tiwari, R.K., Das, M.K., 2007. Heat transfer augmentation in a two-sided lid-driven differentially heated square cavity utilizing nanofluids. International Journal of Heat and Mass Transfer 50, 2002–2018.

Wang, C., 2008. Stagnation flow towards a shrinking sheet. International Journal of Non-Linear Mechanics 43 (5), 377–382.

# 10

# Semianalytical approach for solving Burgers equation by Sawi transform coupled with Adomian decomposition method

**Nihar Ranjan Mallick and S. Chakraverty**
*National Institute of Technology Rourkela, Department of Mathematics, Rourkela, India*

## 10.1 Introduction

The Burgers equation is one of the most studied nonlinear partial differential equations in mathematical physics. This fundamental model may be used to explain a variety of physical phenomena, including turbulence, traffic flow, gas movement, and fluid flow (Bonkile et al., 2018; Kumar, 2023; Saad and Al-Sharif, 2017; Abd-el-Malek and El-Mansi, 2000). The equation is significant due to its connection between diffusion and nonlinear convection, which makes it a model for understanding more complex nonlinear systems (Gartia and Chakraverty, 2024; Garai and Chakraverty, 2025). Despite its relatively simple form, the nonlinearity of the equation makes it challenging to get exact analytical solutions, particularly for broad initial and boundary conditions (Liu, 2017; Cesarano et al., 2023). As a result, research continues to be conducted to develop reliable and efficient methods for approximating its solutions (Sedeeg, 2024; Anani and Folly-Gbetoula, 2024). Finding accurate analytical approximations is challenging due to the nonlinear convective factor, even with extensive investigation of the Burgers equation (Cao et al., 2017; Al-Momani and Al-Khaled, 2022). Many classic approaches include complex transformations or limited assumptions, while others are computationally inefficient or do not offer fast-converging series solutions (Harris and Garra, 2013; Dag et al., 2025). As a result, there is a growing need for straightforward, precise, and reliable analytical techniques that can effectively address the nonlinear nature of the problem (Gaillard, 2025; Kim, 2025).

In recent years, several analytical and semianalytical approaches have been applied to the Burgers equation. The Laplace, Fourier, Elzaki, Sumudu, and Shehu transforms are examples of transform-based methods that have been used to simplify the time-dependent structure of the equation (Chen et al., 2010; Yousefi and Kschischang, 2014; Elzaki and Latrach, 2025; Kapoor and Joshi, 2023; Kumar and Meher, 2024). In contrast, decomposition techniques, such as the homotopy perturbation method (HPM), variational iteration

**Advanced Computational and Mathematical Approaches in Applied Differential Equations**
**https://doi.org/10.1016/B978-0-44-345743-2.00021-9**

method (VIM), and Adomian Decomposition Method (ADM), which can manage nonlinearities without requiring linearization, have gained popularity (Chakraverty et al., 2016, 2023; Karunakar and Chakraverty, 2021, 2017; Vana and Karunakar, 2024). Integral transformations combined with decomposition approaches have been shown to improve accuracy and convergence for nonlinear PDEs (Nuruddeen et al., 2018, 2022; Wazwaz, 2010). Furthermore, several authors have investigated the error limitations, convergence criteria, and stability of series solutions obtained with these hybrid techniques (Bougoffa et al., 2013; Hosseini and Nasabzadeh, 2006).

Despite these developments, the integration of the recently announced Sawi Transform with ADM for nonlinear PDEs, such as the Burgers equation, has received little attention (Sahoo and Chakraverty, 2022; Georgieva and Pavlova, 2021). Although the Sawi Transform has demonstrated promise in minimizing time-domain problems in ordinary and partial differential equations, there is currently a lack of integration with ADM and thorough convergence evaluations for Burgers-type problems (Saadeh et al., 2025). Simultaneously, the Elzaki Transform and its combinations with HPM or ADM have been examined for Burgers-type equations, revealing successful implementation while prompting concerns about exact term-wise convergence and relative accuracy in high-order truncations (Elzaki and Kim, 2014; Hassan and Elzaki, 2020; Anac, 2022).

This gap highlights the need for an analytical approach that is straightforward to use and capable of generating rapid, highly accurate results. To ensure the success of these methods, we must examine uniqueness, convergence, and numerical accuracy in greater detail, comparing them directly with exact solutions and utilizing graphs to interpret the results (Hussain et al., 2024; Singh and Sharma, 2018).

The rest of the work is arranged as follows. The Burgers equation's mathematical formulation, as well as the Sawi Transform, are presented in Section 10.2. The approximate series solution is derived using the suggested hybrid approach, as explained in Section 10.3. Two sample examples are provided in Section 10.4 to show how the approach is applied. The convergence and uniqueness theorems are given in Section 10.5. Convergence analysis, numerical comparisons, and graphical representations of the solution are presented in Section 10.6. Lastly, a summary of the study's main conclusions is provided in Section 10.7.

## 10.2 Preliminaries

The governing Burgers equation is covered in this part, along with the fundamental Sawi transformations ideas and notations that will be utilized throughout this study.

### 10.2.1 Sawi transforms (ST) (Aggarwal and Gupta, 2019)

For $\boldsymbol{u} \in (0, \infty)$, the Sawi transform of $h(u)$ is given by (10.1)

$$S(m(u)) = M(s) = \frac{1}{s^2}\int_0^\infty m(u)\, e^{-\frac{1}{s}u} du, \quad u \geq 0$$

### 10.2.2 Linearity property of ST (Patil, 2021)

If $S(m_1(u)) = M_1(s)$ and $S(m_2(u)) = M_2(s)$, then $S[cm_1(u) + dm_2(u)] = cM_1(s) + dM_2(s)$, where c and d are constants.

### 10.2.3 ST of several important functions (Aggarwal et al., 2020)

$$S\left[u^k\right] = k!s^{k-1}, \quad k \in N,$$

$$S\left\{e^{au}\right\} = \frac{1}{s(1-as)},$$

$$S\{sin(au)\} = \frac{a}{1+a^2s^2},$$

$$S\{cos(au)\} = \frac{1}{s\left(1+a^2s^2\right)},$$

$$S\{sinh(au)\} = \frac{a}{1-a^2s^2},$$

$$S\{cosh(au)\} = \frac{1}{s\left(1-a^2s^2\right)}.$$

### 10.2.4 ST of derivatives (Singh and Aggarwal, 2019)

$$S[m(u)] = \frac{M(s)}{s} - \frac{m(0)}{s^2},$$

$$S\left[m''(u)\right] = \frac{M(s)}{s^2} - \frac{m(0)}{s^3} - \frac{m'(0)}{s^2},$$

$$S\left[m^{(k)}(u)\right] = \frac{M(s)}{s^k} - \sum_{n=0}^{k-1} \frac{m^n(0)}{s^{n-k+1}}.$$

### 10.2.5 Inverse ST (Singh and Aggarwal, 2019)

If $S[m(u)] = M(s)$, then $m(u)$ can be treated as the inverse ST of $M(s)$. Mathematically, it is defined as $m(u) = S^{-1}\{M(s)\}$, where $S^{-1}$ is called the inverse ST operator. The explicit form of inverse ST can be written as the following formula:

$$S^{-1}\{M(s)\} = m(u) = \frac{1}{2\pi i}\int_{\gamma-i\infty}^{\gamma+i\infty} e^{\frac{u}{s}} M(s)\,ds, \quad \gamma \in \Re.$$

### 10.2.6 Inverse ST of some useful functions (Singh and Aggarwal, 2019)

$$S^{-1}\left\{\frac{1}{s}\right\} = 1,$$

$$S^{-1}\{1\} = u,$$

$$S^{-1}\{s\} = \frac{u^2}{2!},$$
$$S^{-1}\left\{s^{n-1}\right\} = \frac{u^n}{n!},$$
$$S^{-1}\left\{\frac{1}{s(1-bs)}\right\} = e^{bu}.$$

### 10.2.7 Burgers equations (Hon and Mao, 1998)

The one-dimensional Burger's equation may be written as follows:

$$m_u(v,u) + m(v,u)\,m_v(v,u) = m_{vv}(v,u),$$

where, $m(v,u)$ represents the velocity of the fluid, $m_u(v,u)$ represents how the velocity field changes with time $u$ and $m_{vv}(v,u)$ represents nonlinear steepening, which can lead to the formation of shock waves.

## 10.3 Methodology (Zeidan et al., 2020)

Let us consider the Burgers equation

$$m_u(v,u) + m(v,u)\,m_v(v,u) = m_{vv}(v,u), \tag{10.1}$$

with initial conditions

$$m(v,0) = f(v). \tag{10.2}$$

By applying the Sawi transform to both sides of Eq. (10.1), we obtain

$$S\left(m_u(v,u) + m(v,u)\,m_v(v,u)\right) = S\left(m_{vv}(v,u)\right), \tag{10.3}$$
$$\Rightarrow S\left(m_u(v,u)\right) + S\left(m(v,u)\,m_v(v,u)\right) = S\left(m_{vv}(v,u)\right), \tag{10.4}$$
$$\Rightarrow \frac{H(v,s)}{s} - \frac{m(v,0)}{s^2} + S\left(m(v,u)\,m_v(v,u)\right) = S\left(m_{vv}(v,u)\right), \tag{10.5}$$
$$\Rightarrow \frac{M(v,s)}{s} - \frac{f(v)}{s^2} + S\left(m(v,u)\,m_v(v,u)\right) = S\left(m_{vv}(v,u)\right), \tag{10.6}$$
$$\Rightarrow M(v,s) = \frac{f(v)}{s} - sS\left(m(v,u)\,m_v(v,u)\right) + sS\left(m_{vv}(v,u)\right), \tag{10.7}$$
$$\Rightarrow S\left(m(v,u)\right) = \frac{f(v)}{s} - sS\left(m(v,u)\,m(v,u)_v\right) + sS\left(m(v,u)_{vv}\right). \tag{10.8}$$

By the Adomian decomposition method, we can take

$$m(v,u) = \sum_{n=0}^{\infty} m_n(v,u) \quad \text{and} \quad Nm(v,u) = \sum_{n=0}^{\infty} A_n(v,u) = m(v,u)\,m(v,u)_v, \tag{10.9}$$

where the Adomian polynomial

$$A_n(v,u) = \frac{1}{n!}\frac{\partial^n}{d\lambda^n}\left[N\left(\sum_{i=0}^{n}\lambda^i m_i(v,u)\right)\right]_{\lambda=0}. \tag{10.10}$$

Putting the values of Eq. (10.9) in Eq. (10.8), we get

$$\Rightarrow S\left(\sum_{n=0}^{\infty} m_n(v,u)\right) = \frac{f(v)}{s} - sS\left(\sum_{n=0}^{\infty} A_n(v,u)\right) + sS\left(\left(\sum_{n=0}^{\infty} m_n(v,u)\right)_{vv}\right). \tag{10.11}$$

By comparing both sides of Eq. (10.11), we obtain

$$S(m_0(v,u)) = \frac{f(v)}{s}, \tag{10.12}$$

$$S(m_n(v,u)) = -sS\{A_{n-1}(v,u)\} + sS\{m_{n-1}(v,u)\}, \quad n \geq 1. \tag{10.13}$$

Apply the inverse Sawi transform on both sides of Eqs. (10.12) and (10.13), we get

$$m_0(v,u) = f(v), \tag{10.14}$$

$$m_n(v,u) = S^{-1}\left(-sS\{A_{n-1}(v,u)\} + sS\{m_{n-1}(v,u)\}\right), \quad n \geq 1. \tag{10.15}$$

Next, the solution of Eq. (10.1) can be written as

$$m(v,u) = lim_{k\to\infty} Z_k(v,u), \tag{10.16}$$

where

$$Z_k(v,u) = \sum_{n=0}^{k} m_n(v,u). \tag{10.17}$$

## 10.4 Particular cases (Naghipour and Manafian, 2015)

**Example 1.** Consider Burger's equation, which is a nonlinear partial differential equation whose equation is given by

$$m_u(v,u) + m(v,u)\, m_v(v,u) = m_{vv}(v,u), \tag{10.18}$$

with initial condition

$$m(v,0) = \frac{1}{2} - \frac{1}{2}tanh(\frac{v}{4}). \tag{10.19}$$

By applying the Sawi transform on both sides of Eq. (10.18), we obtain

$$\frac{1}{s}M(v,s) - \frac{1}{s^2}m(v,0) + S(h(v,u)\, m_v(v,u)) = S(m_{vv}(v,u)), \tag{10.20}$$

$$\Rightarrow M(v,s) = \frac{1}{s} m(v,0) - sS(m(v,u)m_v(v,u)) + sS(m_{vv}(v,u)), \tag{10.21}$$

$$\Rightarrow M(v,s) = \frac{1}{s}\left(\frac{1}{2} - \frac{1}{2}tanh(\frac{v}{4})\right) - sS(m(v,u)m_v(v,u)) + sS(m_{vv}(v,u)), \tag{10.22}$$

$$\Rightarrow M(v,s) = \frac{1}{2s} - \frac{1}{2s}tanh(\frac{v}{4}) - sS(m(v,u)m_v(v,u)) + sS(m_{vv}(v,u)). \tag{10.23}$$

By applying the inverse Sawi transform in Eq. (10.23), we obtain

$$m(v,u) = \frac{1}{2} - \frac{1}{2}tanh(\frac{v}{4}) + S^{-1}(-sS(m(v,u)m_v(v,u)) + sS(m_{vv}(v,u))). \tag{10.24}$$

Next by the Adomian decomposition method, put $m(v,u) = \sum_{n=0}^{\infty} m_n(v,u)$ and $Nm(v,u) = m(v,u)\, m(v,u)_v = \sum_{n=0}^{\infty} A_n(v,u)$ in Eq. (10.24), we get

$$\sum_{n=0}^{\infty} m_n(v,u) = \frac{1}{2} - \frac{1}{2}tanh(\frac{v}{4}) + S^{-1}\left(-sS\left(\sum_{n=0}^{\infty} A_n(v,u)\right) + sS\left(\left(\sum_{n=0}^{\infty} m_n(v,u)\right)_{vv}\right)\right). \tag{10.25}$$

By comparing both sides of Eq. (10.25), we obtain

$$m_0(v,u) = \frac{1}{2} - \frac{1}{2}tanh(\frac{v}{4}), \tag{10.26}$$

$$m_n(v,u) = S^{-1}\left(-sS(A_{n-1}(v,u)) + sS\left((m_{n-1}(v,u))_{vv}\right)\right), \quad n \geq 1. \tag{10.27}$$

Calculating Adomian polynomial by using the formula given in Eq. (10.10), we get

$$A_0(v,u) = -(tanh(v/4)/2 - 1/2)(tanh\left(\frac{\frac{v}{4)^2}}{8} - \frac{1}{8}\right), \tag{10.28}$$

$$A_1(v,u) = -(u(tanh(v/4) - 1)^2(4tanh(v/4) + 3tanh(v/4)^2 + 1))/128, \tag{10.29}$$

$$A_2(v,u) = -(u^2(tanh(v/4) - 1)^2(2tanh(v/4) + \tag{10.30}$$

$$A_3(v,u) = (u^3(5sinh(v/2) - 13cosh(v/2) + cosh(v/2)^2 -, \quad \text{and} \tag{10.31}$$

$$\vdots$$

so on.

Next calculating $m_1(v,u)$, $m_2(v,u)$, $m_3(v,u)$, $m_4(v,u) \cdots$ so on by using the Adomian polynomial found in Eqs. (10.28)–(10.31), we get

$$m_1(v,u) = -(u(tanh(v/4)^2 - 1))/16, \tag{10.32}$$

$$m_2(v,u) = -(u^2 tanh(v/4)(tanh(v/4)^2 - 1))/128, \tag{10.33}$$

$$m_3(v,u) = -(u^3((3tanh(v/4)^4)/512 - tanh(v/4)^2/128 + 1/512))/6, \tag{10.34}$$

$$m_4(v,u) = (u^4 sinh(v/4)(sinh(v/4)^2 - 2))/(24576cosh\left(\frac{v}{4)^5}\right), \quad \text{and} \tag{10.35}$$

$$\vdots$$

so on.

The next solution of Eq. (10.18) having the initial condition in (10.19) is found by

$$m(v,u) = \sum_{n=0}^{\infty} m_n(v,u). \tag{10.36}$$

**Example 2.** Again, consider Burger's equation, having different initial conditions, is given by

$$m_u(v,u) + m(v,u)\, m_v(v,u) = m_{vv}(v,u), \tag{10.37}$$

with initial condition

$$m(v,0) = v^2. \tag{10.38}$$

By applying the Sawi transform on both sides of Eq. (10.37), we obtain

$$\frac{1}{s}M(v,s) - \frac{1}{s^2}m(v,0) + S\left(m(v,u)\, m_v(v,u)\right) = S\left(m_{vv}(v,u)\right), \tag{10.39}$$

$$\Rightarrow M(v,s) = \frac{1}{s}m(v,0) - sS\left(m(v,u)\, m_v(v,u)\right) + sS\left(m_{vv}(v,u)\right), \tag{10.40}$$

$$\Rightarrow M(v,s) = \frac{1}{s}\left(v^2\right) - sS\left(m(v,u)\, m_v(v,u)\right) + sS\left(m_{vv}(v,u)\right). \tag{10.41}$$

Applying the inverse Saw transform in Eq. (10.41), we get

$$\Rightarrow m(v,u) = v^2 + S^{-1}\left(-sS\left(m(v,u)\, m_v(v,u)\right) + sS\left(m_{vv}(v,u)\right)\right). \tag{10.42}$$

Next by the Adomian decomposition method, putting $m(v,u) = \sum_{n=0}^{\infty} m_n(v,u)$ and $Nm(v,u) = m(v,u)\, m_v(v,u) = \sum_{n=0}^{\infty} A_n(v,u)$ in Eq. (10.42), we get

$$\sum_{n=0}^{\infty} m_n(v,u) = v^2 + S^{-1}\left(-s\left(\sum_{n=0}^{\infty} A_n(v,u)\right) + sS\left(\left(\sum_{n=0}^{\infty} m_n(v,u)\right)_{xx}\right)\right). \tag{10.43}$$

By comparing both sides of Eq. (10.43), we obtain

$$m_0(v,u) = v^2, \tag{10.44}$$

$$m_n(x,t) = S^{-1}\left(-sS\left(A_{n-1}(v,u)\right) + sS\left(\left(m_{n-1}(v,u)\right)_{vv}\right)\right), \quad n \geq 1. \tag{10.45}$$

Calculating Adomian polynomial by using the formula given in Eq. (10.10), we get

$$A_0(v,u) = 2v^3, \tag{10.46}$$

$$A_1(v,u) = -2uv\left(5v^3 - 2\right), \tag{10.47}$$

$$A_2(v,u) = 6u^2v^2\left(7v^3 - 6\right), \tag{10.48}$$

$$A_3(v,u) = -8u^3\left(21v^6 - 29v^3 + 2\right), \quad \text{and} \tag{10.49}$$

$$\vdots$$

so on.

Next calculating $m_1(v,u)$, $m_2(v,u)$, $m_3(v,u)$, $m_4(v,u)\cdots$ so on by using the Adomian polynomial found in Eqs. (10.46)–(10.49), we get

$$m_1(v,u) = -2u\left(v^3 - 1\right), \tag{10.50}$$

$$m_2(v,u) = u^2v\left(5v^3 - 8\right), \tag{10.51}$$

$$m_3(v,u) = -2u^3v^2\left(7v^3 - 16\right), \tag{10.52}$$

$$m_4(v,u) = 2u^4\left(21v^6 - 64v^3 + 10\right), \quad \text{and} \tag{10.53}$$

$$\vdots$$

so on.

The next solution of Eq. (10.37) with initial condition (10.38) is found by

$$m(v,u) = \sum_{n=0}^{\infty} m_n(v,u). \tag{10.54}$$

## 10.5 Convergence analysis

In this section, we develop two theorems based on the approaches presented in Abdelrazec (2008) and Mallick et al. (2025) to establish the uniqueness and convergence of the solution.

### 10.5.1 Uniqueness and convergence of solutions for Burger's equation

Assume

$$\left|L(m(v,u)) - L\left(m^*(v,u)\right)\right| < \omega\left|m(v,u) - m^*(v,u)\right|, \quad \forall m(v,u),\ m^*(v,u) \in C[I],$$

where $C[I]$ represents a continuous function on $I$ and $L$ denote the linear operator, i.e., $L(m(v,u)) = m(v,u)_{vv} - m(v,u)\,m(v,u)_v$.

**Theorem 10.5.1.** *Uniqueness of the solution of Eq. (10.1) occurs if* $|t| < \sqrt{2}$.

*Proof.* Let us take a Banach space $(C[I], \|.\|)$ for all continuous functions on $I$ with the norm $\|.\|$. Let $F : C[I] \to C[I]$ is a mapping such that

$$F(m) = m_1(v) + S^{-1}\left(sS\left(m_{vv}(v,u) - m(v,u)\,m_v(v,u)\right)\right)$$

$$\begin{aligned}\left\|F(m)-F\left(m^*\right)\right\| &= max_{u,v\in I}\left|S^{-1}\left(sS\left(L\left(m\left(v,u\right)\right)\right)\right)-S^{-1}\left(sS\left(L\left(m^*\left(v,u\right)\right)\right)\right)\right| \\ &= max_{u,v\in I}\left|\frac{u^2}{2!}\left(L\left(m\left(v,u\right)\right)-L\left(m^*\left(v,u\right)\right)\right)\right| \\ &\le \frac{u^2}{2!}\omega\left\|m-m^*\right\| \\ &< \omega\left\|m-m^*\right\|.\end{aligned}$$

As a result, Eq. (10.1) has a unique solution. □

**Theorem 10.5.2.** *The solution of Eq. (10.1) is convergent if* $|\, t \,| < \sqrt{\frac{2}{\omega}}$.

*Proof.* Let us consider a Banach space $(C[I], \|.\|)$ for all continuous functions $I$ with the norm $\|.\|$.

For $Z_f, Z_g \in C[I]$ and $f, g \in N$, $f \ge g$,

$$\begin{aligned}&\left\|Z_f-Z_g\right\| \\ &= max_{u,v\in I}\left|Z_f\left(v,u\right)-Z_g\left(v,u\right)\right| \\ &= max_{u,v\in I}\left|\sum_{k=g+1}^{f} m_k\left(v,u\right)\right| \\ &= max_{u,v\in I}\left|\sum_{k=g+1}^{f} S^{-1}\left(s\left(S\left(L\left(m_{k-1}\left(v,u\right)\right)\right)\right)\right)\right| \\ &= max_{u,v\in I}\left|S^{-1}\left(s\left(S\left(\sum_{k=g+1}^{f} L\left(m_{k-1}\left(v,u\right)\right)\right)\right)\right)\right| \\ &= max_{u,v\in I}\left|S^{-1}\left(s\left(S\left(\sum_{k=g}^{f-1} L\left(m_{k}\left(v,u\right)\right)\right)\right)\right)\right| \\ &= max_{u,v\in I}\left|S^{-1}\left(s\left(S\left(\sum_{k=0}^{f-1} L\left(m_{k}\left(v,u\right)\right)-\sum_{k=0}^{g-1} L\left(m_{k}\left(v,u\right)\right)\right)\right)\right)\right| \\ &= max_{u,v\in I}\left|S^{-1}\left(s\left(S\left(L\left(Z_{f-1}\left(v,u\right)\right)-L\left(Z_{g-1}\left(v,u\right)\right)\right)\right)\right)\right| \\ &< max_{u,v\in I}\left|\omega\frac{u^2}{2!}\left(Z_{f-1}\left(v,u\right)-Z_{g-1}\left(v,u\right)\right)\right| \\ &< \omega\frac{u^2}{2!}\left\|Z_{f-1}-Z_{g-1}\right\|\end{aligned}$$

If $f = g + 1$,

$$\left\| Z_{g+1} - Z_g \right\| < \omega \frac{u^2}{2!} \left\| Z_g - Z_{g-1} \right\|.$$

Let $b = \omega \frac{u^2}{2!}$

$$\left\| Z_{g+1} - Z_g \right\| < b \left\| Z_g - Z_{g-1} \right\| < b^2 \left\| Z_{g-1} - Z_{g-2} \right\| < \cdots < b^g \left\| Z_1 - Z_0 \right\|$$

$$\begin{aligned} &\left\| Z_f - Z_g \right\| \\ &= \left\| (Z_f - Z_{f-1}) + (Z_{f-1} - Z_{f-2}) + \cdots\cdots + (Z_{g+1} - Z_g) \right\| \\ &\leq \left\| Z_f - Z_{f-1} \right\| + \left\| Z_{f-1} - Z_{f-2} \right\| + \cdots\cdots + \left\| Z_{g+1} - Z_g \right\| \\ &< \left( b^{f-1} + b^{f-2} + \cdots\cdots + b^g \right) \left\| Z_1 - Z_0 \right\| \\ &\leq b^g \frac{(1 - b^{f-g})}{(1 - b)} \left\| m_1 \right\| \\ &< b^g \frac{1}{(1 - b)} \left\| m_1 \right\|. \end{aligned}$$

As $g \to \infty$, $b^g \to 0$.
Therefore $\left\| Z_f - Z_g \right\| \to 0$, when $g \to \infty$.
$\therefore$ The sequence $\{Z_k\}$ is a Cauchy sequence in a Banach space $(C[I], \|.\|)$, and hence it is convergent. □

## 10.6 Numerical results and physical interpretation

This section presents the numerical results derived from the Sawi transform–Adomian decomposition method (ST–ADM) applied to the Burgers equation and examines their physical meaning. The precision of the series solution is confirmed by comparing it with the actual analytical solution, doing convergence analysis, and graphically depicting the solution's behavior.

### 10.6.1 Comparison with the exact solution

To evaluate the accuracy of the approximation solution, we calculate the series solution up to the eighth term and compare it with the precise solution at certain geographical and temporal positions. The outcomes are shown in (Table 10.1 and Table 10.2).

In both tables, the estimated values are quite close to the actual values, exhibiting little errors across the domain. This illustrates that the ST–ADM method converges very quickly and produces very precise approximations, even with a limited number of terms.

**Table 10.1** Approximate solution vs existing exact solution for u = 1 in Example 1.

| $v$ | 1st term solution | 2nd term solution | 3rd term solution | 4th term solution | 5th term solution | 6th term solution | 7th term solution | 8th term solution | Existing solution (Naghipour and Manafian, 2015) |
|---|---|---|---|---|---|---|---|---|---|
| 0.1 | 0.487503 | 0.549964 | 0.550159 | 0.549834 | 0.549832 | 0.549834 | 0.549834 | 0.549834 | 0.549834 |
| 0.2 | 0.475021 | 0.537365 | 0.537754 | 0.537432 | 0.537428 | 0.537430 | 0.537430 | 0.537430 | 0.537430 |
| 0.3 | 0.462570 | 0.524720 | 0.525301 | 0.524983 | 0.524977 | 0.524979 | 0.524979 | 0.524979 | 0.524979 |
| 0.4 | 0.450166 | 0.512045 | 0.512816 | 0.512503 | 0.512495 | 0.512497 | 0.512497 | 0.512497 | 0.512497 |
| 0.5 | 0.437823 | 0.499357 | 0.500314 | 0.499998 | 0.500000 | 0.500000 | 0.500000 | 0.500000 | 0.500000 |
| 0.6 | 0.425557 | 0.486672 | 0.487809 | 0.487501 | 0.487503 | 0.487503 | 0.487503 | 0.487503 | 0.487503 |
| 0.7 | 0.413382 | 0.474007 | 0.475320 | 0.475032 | 0.475019 | 0.475021 | 0.475021 | 0.475021 | 0.475021 |
| 0.8 | 0.401312 | 0.461378 | 0.462859 | 0.462583 | 0.462569 | 0.462570 | 0.462570 | 0.462570 | 0.462570 |

**Table 10.2** Approximate solution vs existing exact solution for u = 2 in Example 1.

| $v$ | 1st term solution | 2nd term solution | 3rd term solution | 4th term solution | 5th term solution | 6th term solution | 7th term solution | 8th term solution | Existing solution (Naghipour and Manafian, 2015) |
|---|---|---|---|---|---|---|---|---|---|
| 0.1 | 0.487503 | 0.612425 | 0.613205 | 0.610607 | 0.610575 | 0.610640 | 0.610641 | 0.610639 | 0.610639 |
| 0.2 | 0.475021 | 0.599709 | 0.601266 | 0.598688 | 0.598623 | 0.598687 | 0.598689 | 0.598688 | 0.598688 |
| 0.3 | 0.462570 | 0.586870 | 0.589196 | 0.586650 | 0.586654 | 0.586616 | 0.586619 | 0.586618 | 0.586618 |
| 0.4 | 0.450166 | 0.573924 | 0.577008 | 0.574507 | 0.574380 | 0.574440 | 0.574444 | 0.574443 | 0.574443 |
| 0.5 | 0.437823 | 0.560891 | 0.564716 | 0.562272 | 0.562116 | 0.562173 | 0.562178 | 0.562177 | 0.562177 |
| 0.6 | 0.425557 | 0.547787 | 0.552336 | 0.549959 | 0.549776 | 0.549829 | 0.549835 | 0.549834 | 0.549834 |
| 0.7 | 0.413382 | 0.534631 | 0.539882 | 0.537584 | 0.537375 | 0.537424 | 0.537431 | 0.537430 | 0.537430 |
| 0.8 | 0.401312 | 0.521443 | 0.527370 | 0.525160 | 0.524928 | 0.524973 | 0.524980 | 0.524979 | 0.524979 |

## 10.6.2 2D behavior of the solution

The profile of the solution $m(v, u)$ for fixed values of time u is displayed in the 2D plots (see Fig. 10.1). It is possible to make the following observations: The solution shows a high gradient at early stages, which is indicative of nonlinear advection events. The viscosity term diminishes sharp gradients and causes the solution profile to become smoother and more diffused with time. The numerical precision is visually confirmed by the ST-ADM findings, which closely resemble the precise solution curves. The physical characteristics of the Burgers equation, which simulates a balance between nonlinear convection and diffusive effects, are in accordance with this smoothing behavior.

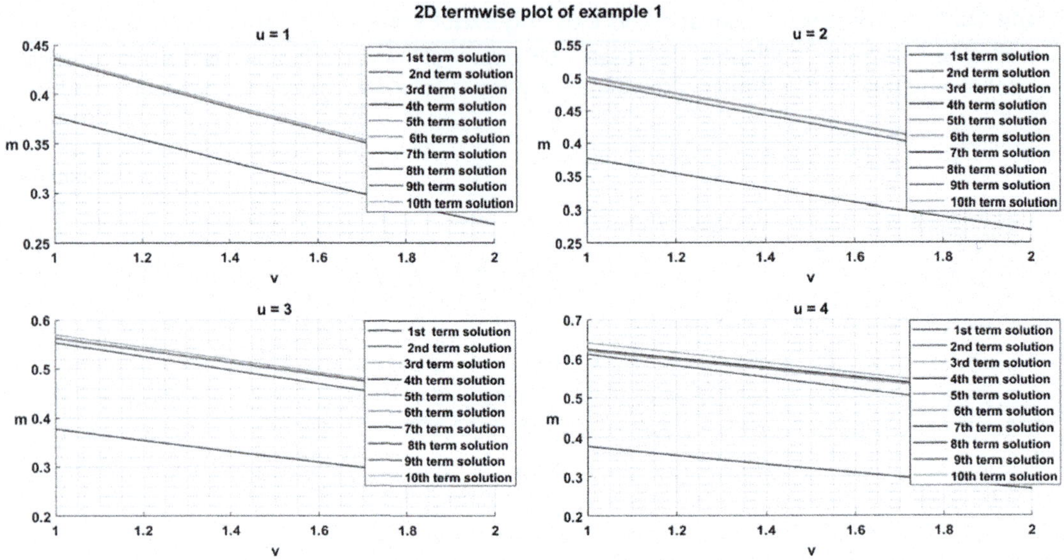


FIGURE 10.1

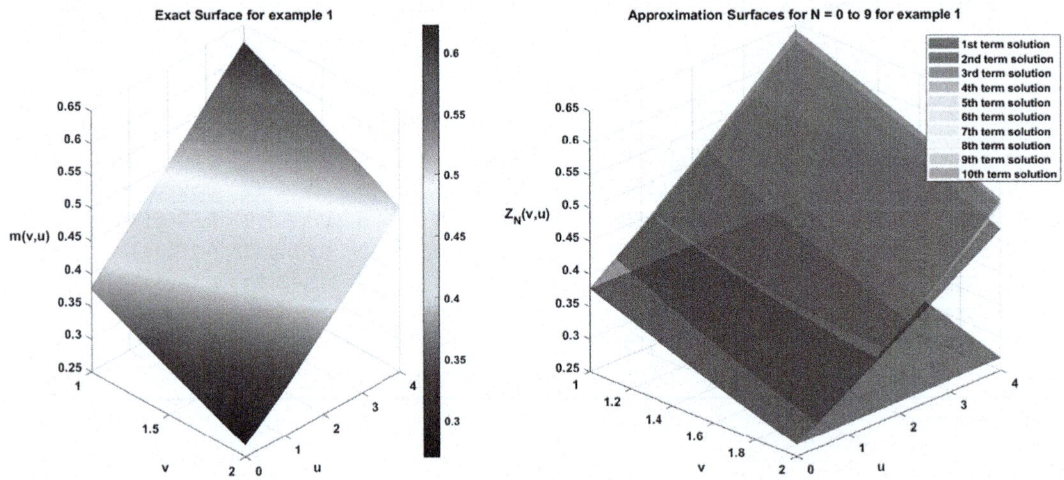


FIGURE 10.2

## 10.6.3 3D analysis of the approximate and exact solutions

The behavior in space and time may be fully viewed by comparing the approximate solution up to the tenth term with the precise solution in a 3D surface map (see Fig. 10.2). The characteristics listed below are noted. The precise solution surface and the approximate 3D surface are almost identical, with no discernible differences. Viscosity causes the nonlinear wave front to progressively flatten as it moves smoothly over time.

Both the time evolution and the spatial structure of the solution are accurately captured by the ST-ADM, as demonstrated by the consistency between the two surfaces. This shows that the entire dynamics of the Burgers equation may be well approximated, even using a finite-term truncated series.

## 10.7 Conclusion

We introduced an effective analytical-numerical approach for solving the nonlinear Burgers equation by integrating the Sawi transform with the Adomian decomposition method (ADM). The Sawi transform makes the equation's time-dependent structure easier to work with, while ADM uses Adomian polynomials to deal with the nonlinear term, which lets the solution be written as a quickly converging series. Two theorems showed that the suggested solution is unique and converges. Comparing the approximate solutions up to the 8th and 10th terms shows that they are quite close to the precise solution. The convergence study also shows that just a few terms are needed to get a high level of accuracy. The 2D and 3D graphs show even more that the approximate solution properly describes how the Burgers equation behaves in the real world, including nonlinear wave propagation and diffusive smoothing. In general, the ST–ADM method is easy to use, correct, and quick to compute. It is a dependable tool for solving nonlinear partial differential equations and may be readily adapted to more complicated or fractional versions of the Burgers equation in the future.

## References

Abd-el-Malek, M.B., El-Mansi, S.M.A, 2000. Group theoretic methods applied to Burgers' equation. Journal of Computational and Applied Mathematics 115 (1–2), 1–12. https://doi.org/10.1016/S0377-0427(99)00170-3. Elsevier Science B.V., Egypt.

Abdelrazec, A., 2008. Adomian decomposition method: convergence analysis and numerical approximations.

Aggarwal, S., Gupta, A.R., 2019. Dualities between some useful integral transforms and Sawi transform. International Journal of Recent Technology and Engineering (ISSN 2277-3878) 8 (3), 5978–5982. https://doi.org/10.35940/ijrte.C5870.098319. Blue Eyes Intelligence Engineering and Sciences Publication, India, Available from: https://www.ijrte.org/wp-content/uploads/papers/v8i3/C5870098319.pdf.

Aggarwal, S., Sharma, S.D., Vyas, A., 2020. Sawi transform of Bessel's functions with application for evaluating definite integrals. International Journal of Latest Technology in Engineering, Management & Applied Science 9, 12–18.

Al-Momani, A., Al-Khaled, K., 2022. Theory and computations for the nonlinear Burgers' equation via the use of Sinc-Galerkin method. Journal of Electrical and Computer Engineering (ISSN 2090-0155) 2022. https://doi.org/10.1155/2022/6073471. Hindawi Limited, Jordan, Available from: http://www.hindawi.com/journals/jece/.

Anac, Halil, 2022. A local fractional Elzaki transform decomposition method for the nonlinear system of local fractional partial differential equations. Fractal and Fractional (ISSN 2504-3110) 6 (3), 167. https://doi.org/10.3390/fractalfract6030167.

Anani, Kwassi, Folly-Gbetoula, Mensah, 2024. Exact solution to a class of problems for the Burgers' equation on bounded intervals. Partial Differential Equations in Applied Mathematics (ISSN 2666-8181) 12, 100977. https://doi.org/10.1016/j.padiff.2024.100977.

Bonkile, Mayur P., Awasthi, Ashish, Lakshmi, C., Mukundan, Vijitha, Aswin, V.S., 2018. A systematic literature review of Burgers' equation with recent advances. Pramāna (ISSN 0304-4289) 90 (6). https://doi.org/10.1007/s12043-018-1559-4.

Bougoffa, L., Rach, R.C., El-Manouni, S., 2013. A convergence analysis of the Adomian decomposition method for an abstract Cauchy problem of a system of first-order nonlinear differential equations. International Journal of Computer Mathematics (ISSN 1029-0265) 90 (2), 360–375. https://doi.org/10.1080/00207160.2012.718073. Taylor and Francis Ltd., Saudi Arabia, Available from: www.tandf.co.uk/journals/titles/00207160.asp.

Cao, W., Xu, Q., Zheng, Z., 2017. Solution of two-dimensional time-fractional Burgers equation with high and low Reynolds numbers. Advances in Difference Equations (ISSN 1687-1847) 2017 (1). https://doi.org/10.1186/s13662-017-1398-5. Springer International Publishing, China, Available from: http://www.springerlink.com/content/1687-1847/.

Cesarano, C., Massoun, Y., Said, A., Talbi, M.E., 2023. Analytic study of coupled Burgers' equation. Mathematics (ISSN 2227-7390) 11 (9). https://doi.org/10.3390/math11092071. MDPI, Italy, Available from: http://www.mdpi.com/journal/mathematics.

Chakraverty, S., Tapaswini, S., Behera, D., 2016. Fuzzy Arbitrary Order System: Fuzzy Fractional Differential Equations and Applications. Wiley, India, pp. 1–258. Available from: http://onlinelibrary.wiley.com/book/10.1002/9781119004233.

Chakraverty, S., Jena, R.M., Jena, S.K., 2023. Time-Fractional Order Biological Systems with Uncertain Parameters. Synthesis Lectures on Mathematics and Statistics.

Chen, S., Wu, X., Wang, Y., Kong, W., 2010. The Laplace transform method for Burgers' equation. International Journal for Numerical Methods in Fluids (ISSN 1097-0363) 63 (9), 1060–1076. https://doi.org/10.1002/fld.2116China. Available from: http://www3.interscience.wiley.com/cgi-bin/fulltext/122462714/PDFSTART.

Dag, I., Uğurluoğlu, S., Adar, N., 2025. A time-space B-spline integrator for the Burgers' equation. arXiv, Turkey arXiv. 23318422. arXiv:2510.09408. Available from: https://arxiv.org. 10.48550/arXiv.2510.09408.

Elzaki, T.M., Kim, H., 2014. The solution of Burger's equation by Elzaki homotopy perturbation method. Applied Mathematical Sciences (ISSN 1314-7552) (57–60), 2931–2940. https://doi.org/10.12988/ams.2014.44314. Hikari Ltd., Saudi Arabia, Available from: http://www.m-hikari.com/ams/ams-2014/ams-57-60-2014/kimAMS57-60-2014.pdf.

Elzaki, T.M., Latrach, M.A., 2025. The Elzaki transform method for addressing Cauchy problems in higher order nonlinear PDES. Acta Mechanica et Automatica (ISSN 2300-5319) 19 (2), 225–231. https://doi.org/10.2478/ama-2025-0028. Sciendo, Saudi Arabia, Available from: https://sciendo.com/journal/AMA.

Gaillard, Pierre, 2025. Wronskian representations of the solutions to the Burgers' equation. Journal of Applied Mathematics (ISSN 2972-4805) 3 (1). https://doi.org/10.59400/jam2285.

Garai, R., Chakraverty, S., 2025. Dynamic characteristics of non-homogeneous perforated beams with different boundary scenarios. Mechanics Based Design of Structures and Machines (ISSN 1539-7742) 53 (10), 6694–6721. https://doi.org/10.1080/15397734.2025.2488027. Taylor and Francis Ltd., India, Available from: www.tandf.co.uk/journals/titles/15397734.asp.

Gartia, A.K., Chakraverty, S., 2024. Chebyshev polynomials based Rayleigh–Ritz method for free vibration analysis of axially functionally graded cantilever nanobeam. In: Advances in Modelling and Analysis of Functionally Graded Micro- and Nanostructures. IOP Publishing, Bristol, UK, pp. 5–19.

Georgieva, Atanaska, Pavlova, Albena, 2021. Fuzzy Sawi decomposition method for solving nonlinear partial fuzzy differential equations. Symmetry (ISSN 2073-8994) 13 (9), 1580. https://doi.org/10.3390/sym13091580.

Harris, P.A., Garra, R., 2013. Analytic solution of nonlinear fractional Burgers-type equation by invariant subspace method. Nonlinear Studies (ISSN 2153-4373) 20 (4), 471–481. Italy, Available from: http://nonlinearstudies.com/index.php/nonlinear/article/download/878/585.

Hassan, Moh A., Elzaki, Tarig M., 2020. Double Elzaki transform decomposition method for solving non-linear partial differential equations. Journal of Applied Mathematics and Physics (ISSN 2327-4352) 08 (08), 1463–1471. https://doi.org/10.4236/jamp.2020.88112.

Hon, Y.C., Mao, X.Z., 1998. An efficient numerical scheme for Burgers' equation. Applied Mathematics and Computation (ISSN 0096-3003) 95 (1), 37–50. https://doi.org/10.1016/s0096-3003(97)10060-1.

Hosseini, M.M., Nasabzadeh, H., 2006. On the convergence of Adomian decomposition method. Applied Mathematics and Computation (ISSN 0096-3003) 182 (1), 536–543. https://doi.org/10.1016/j.amc.2006.04.015.

Hussain, Saddam, Arora, Gourav, Kumar, Rajesh, 2024. Semi-analytical methods for solving non-linear differential equations: a review. Journal of Mathematical Analysis and Applications (ISSN 0022-247X) 531 (1), 127821. https://doi.org/10.1016/j.jmaa.2023.127821.

Kapoor, M., Joshi, V., 2023. A comparative study of Sumudu HPM and Elzaki HPM for coupled Burgers' equation. Heliyon (ISSN 2405-8440) 9 (5). https://doi.org/10.1016/j.heliyon.2023.e15726. Elsevier Ltd, India, Available from: http://www.journals.elsevier.com/heliyon/.

Karunakar, Perumandla, Chakraverty, Snehashish, 2017. Comparison of solutions of linear and non-linear shallow water wave equations using homotopy perturbation method. International Journal of Numerical Methods for Heat & Fluid Flow (ISSN 0961-5539) 27 (9), 2015–2029. https://doi.org/10.1108/hff-09-2016-0329.

Karunakar, P., Chakraverty, S., 2021. Homotopy perturbation method for predicting tsunami wave propagation with crisp and uncertain parameters. International Journal of Numerical Methods for Heat & Fluid Flow (ISSN 0961-5539) 31 (1), 92–105. https://doi.org/10.1108/HFF-11-2019-0861. Emerald Publishing, India, Available from: http://www.emeraldinsight.com/info/journals/hff/hff.jsp.

Kim, Junseok, 2025. Nonlinear dynamic evolution of a novel normalized time-fractional Burgers equation. Partial Differential Equations in Applied Mathematics (ISSN 2666-8181) 13, 101096. https://doi.org/10.1016/j.padiff.2025.101096.

Kumar, N.K., 2023. A review on Burgers' equations and its applications. Journal of the Institute of Science Technology 28, 49–52.

Kumar, A., Meher, R., 2024. Solving time fractional partial differential equations using Shehu transform with q-homotopy analysis approach. AIP Conference Proceedings (1), 15517616. https://doi.org/10.1063/5.0234337. American Institute of Physics, India. http://scitation.aip.org/content/aip/proceeding/aipcp3242.

Liu, T.P., 2017. Hopf-Cole transformation. Bulletin of the Institute of Mathematics, Academia Sinica 12, 71–101.

Mallick, N.R., Chakraverty, S., Jena, R.M., 2025. Analysis of large membrane vibrations using fractional calculus. Fractal and Fractional 9, 219.

Naghipour, A., Manafian, J., 2015. Application of the Laplace Adomian decomposition and implicit methods for solving Burgers' equation. TWMS Journal of Pure and Applied Mathematics 6, 68–77.

Nuruddeen, R.I., Muhammad, L., Nass, A.M., Sulaiman, T.A., 2018. A review of the integral transforms-based decomposition methods and their applications in solving nonlinear PDEs. Palestine Journal of Mathematics 7, 262–280.

Nuruddeen, Rahmatullah Ibrahim, Akbar, Yasir, Kim, Hwajoon, 2022. On the application of $G_\alpha$ integral transform to nonlinear dynamical models with non-integer order derivatives. AIMS Mathematics (ISSN 2473-6988) 7 (10), 17859–17878. https://doi.org/10.3934/math.2022984.

Patil, D.P., 2021. Application of Sawi transform in Bessel functions. Aayushi International Interdisciplinary Research Journal (AIIRJ).

Saad, K., Al-Sharif, E.H., 2017. Analytical study for time and time-space fractional Burgers' equation. Advances in Difference Equations (ISSN 1687-1847) 2017 (1). https://doi.org/10.1186/s13662-017-1358-0. Springer International Publishing, Saudi Arabia, Available from: http://www.springerlink.com/content/1687--1847/.

Saadeh, Rania, Al-Wadi, Alaa, Qazza, Ahmad, 2025. Advancing solutions for fractional differential equations: integrating the Sawi transform with iterative methods. European Journal of Pure and Applied Mathematics (ISSN 1307-5543) 18 (1), 5583. https://doi.org/10.29020/nybg.ejpam.v18i1.5583.

Sahoo, Mrutyunjaya, Chakraverty, Snehashish, 2022. Sawi transform based homotopy perturbation method for solving shallow water wave equations in fuzzy environment. Mathematics (ISSN 2227-7390) 10 (16), 2900. https://doi.org/10.3390/math10162900.

Sedeeg, A.K.H., 2024. A new technique for solving regular and singular coupled Burgers' equation. Journal of Basic and Applied Sciences (ISSN 2576-5299) 31 (1), 292–309. https://doi.org/10.1080/25765299.2024.2358573. Taylor and Francis Ltd., Sudan Arab, Available from: https://www.tandfonline.com/loi/tabs20.

Singh, G.P., Aggarwal, S., 2019. Sawi transform for population growth and decay problems. International Journal of Latest Technology in Engineering, Management & Applied Science 8, 157–162.

Singh, Prince, Sharma, Dinkar, 2018. Convergence and error analysis of series solution of nonlinear partial differential equation. Nonlinear Engineering (ISSN 2192-8029) 7 (4), 303–308. https://doi.org/10.1515/nleng-2017-0113.

Vana, Rambabu, Karunakar, Perumandla, 2024. A fuzzy semi-analytical approach for modeling uncertainties in solitary wave solution of coupled nonlinear Boussinesq equations. Physica Scripta (ISSN 0031-8949) 99 (10), 105218. https://doi.org/10.1088/1402-4896/ad72aa.

Wazwaz, A.M., 2010. The combined Laplace transform-Adomian decomposition method for handling nonlinear Volterra integro-differential equations. Applied Mathematics and Computation (ISSN 0096-3003) 216 (4), 1304–1309. https://doi.org/10.1016/j.amc.2010.02.023.

Yousefi, M.I., Kschischang, F.R., 2014. Information transmission using the nonlinear Fourier transform, part I: mathematical tools. IEEE Transactions on Information Theory (ISSN 0018-9448) 60 (7), 4312–4328. https://doi.org/10.1109/TIT.2014.2321143. Institute of Electrical and Electronics Engineers Inc., Canada, Available from: http://ieeexplore.ieee.org/xpl/mostRecentIssue.jsp?punumber=18.

Zeidan, D., Chau, C.K., Lu, T.T., Zheng, W.Q., 2020. Mathematical studies of the solution of Burgers' equations by Adomian decomposition method. Mathematical Methods in the Applied Sciences (ISSN 1099-1476) 43 (5), 2171–2188. https://doi.org/10.1002/mma.5982. John Wiley and Sons Ltd, Jordan, Available from: http://onlinelibrary.wiley.com/journal/10.1002/(ISSN)1099-1476.

# 11

# Parametric mathematical programs with vanishing constraints: sensitivity analysis

Abeka Khare[a] and Triloki Nath[b]
[a]Shaheed Bhagat Singh Govt. Degree College, Ashta, India [b]Deen Dayal Upadhyay Gorakhpur University, Gorakhpur, India

## 11.1 Introduction

In this paper, we study *parametric mathematical programs with vanishing constraints*, having the following mathematical form:

$$
\begin{aligned}
(\mathrm{p-MPVC}) \qquad & \min_{x\in\mathcal{C}} f(x,p) \\
\text{subject to} \quad & g_i(x,p)\leqslant 0\,, \quad i=1,...,l, \\
& h_i(x,p)=0\,, \quad i=1,...,m, \\
& H_i(x,p)\geqslant 0\,, \quad i=1,...,n, \\
& G_i(x,p)H_i(x,p)\leqslant 0\,, \quad i=1,...,n,
\end{aligned}
\tag{11.1}
$$

where $f:\mathbb{R}^{n_1+n_2}\to\mathbb{R}$, $g_i:\mathbb{R}^{n_1+n_2}\to\mathbb{R}$, $h_i:\mathbb{R}^{n_1+n_2}\to\mathbb{R}$, $G_i,\ H_i:\mathbb{R}^{n_1+n_2}\to\mathbb{R}$, and $\mathcal{C}$ is a nonempty closed subset of $\mathbb{R}^{n_1}$. Let $\mathcal{W}_p$ be the feasible region of $p-MPVC$.

The value function and optimal solution mappings for $p-MPVC$ are defined respectively as

$$
\begin{aligned}
\mathbb{V}(p) &:= \inf\{f(x,p)\mid x\in\mathcal{W}(p)\}, \\
\mathcal{O}(p) &:= \{x\in\mathcal{W}(p)\mid f(x,p)=\mathbb{V}(p)\}.
\end{aligned}
$$

The first formal paper on MPVC was written by Achtziger and Kanzow (2008), where they elaborated the problem with some applications. Before this, we have some implicit works in engineering fields (Achtziger, 2004; Bendsøe and Sigmund, 2003; Cheng and Guo, 1997). Following the foundational work in Achtziger and Kanzow (2008), further developments in the study of mathematical programs with vanishing constraints (MPVC) has emerged in literature, notably in Hoheisel and Kanzow (2007, 2008, 2009). It can be observed that MPVC violates most of the standard constraint qualifications like LICQ, MFCQ due to its nonconvex structure. This violation and significant applications of MPVC attracted

Advanced Computational and Mathematical Approaches in Applied Differential Equations
https://doi.org/10.1016/B978-0-44-345743-2.00022-0

the researchers to further study and contribute significantly in the analysis of MPVC. For more comprehensive knowledge, we refer Achtziger et al. (2013); Hoheisel and Kanzow (2008, 2007); Izmailov and Solodov (2009); Jabr (2012); Michael et al. (2013) and references therein.

In this paper, our primary focus is on analyzing the directional derivative of the value function and computing upper estimates of the limiting and singular subdifferentials of the value function for MPVCs, which are key objects in sensitivity analysis.

In parametric optimization problems, the value function $\mathbb{V}(p)$ maps a parameter $p$ to the optimal value of the objective function. The derivative $\mathcal{D}\mathbb{V}(p)$ describes how this optimal value changes in response to variations in the parameter. Derivatives of the value function is significant in various applications, including economics, engineering design, operations research, optimal control and Markov decision processes, as they help in assessing how changes in external conditions influence the optimal outcome. Furthermore, its derivative provides essential information about decision-making and policy improvement.

There are many papers in the literature, where applications of directional derivatives of value function of MPVC is discussed. Hogan (1973) stated the necessity of directional derivative of the value function. If the value function is directionally differentiable, then these derivatives can be used to develop the algorithms for the differentiable problem, e.g., feasible direction method, which typically requires the gradient of the objective function. The directional derivatives were characterized for completely convex problems and used to construct a robust and convergent feasible direction algorithm. Subsequently, Gauvin (1979) demonstrated the significance of the directional derivative of value function, by establishing that it can become unbounded, if the set of Kuhn–Tucker multipliers is empty or unbounded at any optimal point. If this holds for all optimal points, then it violates the continuity of the value function. Hence investigating the directional derivative of the value function provides valuable insights and supports positive analysis. Rockafellar (1984) studied a parameterized nonlinear program, where all the involved functions are twice continuously differentiable and evaluated the directional derivative of optimal value function by a minimax formula. Janin (1984) approximated the directional derivative of marginal value function of a perturbed nonlinear program, under a new regularity condition, which is called constant rank constraint qualification (CRCQ). The CRCQ is trivial in presence of linear CQs or in LICQ, but it is neither weaker nor stronger than MFCQ (Janin, 1984). Later, Gauvin and Janin (1990) investigated the existence and value of the directional derivative of the value function for a finite dimensional parametric program under some certain mild assumptions. Shapiro (1995) extended this result for semiinfinite nonlinear program. He expressed the right side directional derivative of value function using a minimax formula under some regular assumptions. Minchenko and Stakhovski (2011a) introduced the relaxed constant rank (RCR) regularity conditions and studied directional derivative of value function under this new RCR framework. They also derived the local error bound property and a second-order optimality condition under this RCR regularity condition. Recently Guo et al. (2014) extended the work of Minchenko and

Stakhovski (2011a) to the parametric MPECs with the RCR-regularity and the restricted inf-compactness assumptions. They improved the result for all $C^1$ involved functions (i.e., gradient is continuous) instead of $C^{1,1}$ (gradient is locally Lipschitz). Moreover, they established a formula for the first-order directional derivative of the value function in any direction under MPEC-RCR regularity condition.

The main objective of this paper is to provide sensitivity analysis of parametric MPVC. Compared to the other classes of mathematical programs, there is a limited literature on sensitivity analysis of MPVC. Probably Izmailov and Solodov (2009) is the only paper dealing with the sensitivity analysis of the value function for MPVC. In Izmailov and Solodov (2009, Theorem 4.1–4.7), the authors established upper bounds for optimal value function and studied the stability of solutions under the strong conditions, such as MFCQ and MPVC-LICQ, for a subclass of MPVC. So it is important to investigate sensitivity of MPVC under weaker regularity assumptions. Although an MPVC can be reformulated as an MPEC, this transformation has notable drawbacks, such as an increase in the problem's dimensionality and the presence of locally nonunique solutions in the resulting MPEC. Our results are inspired by the sensitivity analysis developed for MPECs in Guo et al. (2014). However, since MPVCs are structurally different from MPECs, it is essential to investigate them independently by leveraging their distinctive structure—namely, the vanishing constraints.

Firstly, Lucet and Ye (2001, 2002) studied a sensitivity analysis of the value function for optimization problems with variational inequality constraints (OPVIC); they derived upper estimates for the limiting and singular subdifferentials of value functions in terms of normal and abnormal coderivative multipliers, and applied these results to the perturbed MPECs using the same multipliers framework. Hu and Ralph (2002) estimated both the first-order and the second-order directional derivative for the value function under inf-compactness and MPEC-LICQ assumptions. Mordukhovich and Nam (2005, 2009) studied a mathematical program with an abstract set constraint in a Banach space, which includes MPEC as a special case. He provided upper estimates for Fréchet and limiting subgradients of value function in terms of normal multipliers. Guo et al. (2013b) extended these results for a mathematical programs with geometric constraints (MPGC), for which MPEC and MPVC are particular cases, and presented upper estimates for limiting and singular subdifferentials of the value function under inf-compactness and strong quasinormality assumptions. Ye and Zhang (2013) improved the work of Mordukhovich and Nam (2009) and estimated the Fréchet, limiting, and singular subdifferentials of value function by using enhanced quasinormal multipliers, which provide a tighter characterization than normal multipliers. The present work is motivated by Guo et al. (2014), where the analysis was carried out by using enhanced MPVC-M and MPVC-C-quasinormal multipliers set, which form a smaller and more refined set than the standard Lagrange multipliers. In this paper, we adapt the framework of Guo et al. (2014) for parametric MPVCs, and for the first time, derive a formula for the directional derivative of the value function in every direction, under suitable regularity conditions. Furthermore, we establish for the first time the enhanced Fritz John stationarity condition for MPVCs in terms of the limiting subdifferential. These enhanced optimal conditions are significant and more advanced than the classical optimal conditions in the sense that they provide information about a neighborhood of extremal points in terms of converging sequence towards it, and provide new inclusion relations for the limiting and singular subdifferentials using enhanced M- and C-quasinormal multipliers.

The paper is organized as follows: Section 11.2 contains some basic results from variational analysis, which are essential to derive further results of this paper. Section 11.3 is about the analysis of directional derivative of value function. In Section 11.4, an enhanced optimality condition for parametric MPVC is derived, also some upper estimates for the limiting and singular subdifferential of the value function is established.

## 11.2 Preliminaries

In this section, some basic notations and terms of variational analysis are presented, that will be useful in subsequent sections of this paper. $\mathbb{B}_\delta(x)$ and $\bar{\mathbb{B}}_\delta(x)$ denote the open and closed ball, respectively, with center at $x$ and radius $\delta > 0$. We use $dist(x, S) = \inf\{||y - x|| \mid y \in S\}$ to denote distance of a point $x$ to a closed set $S$. Also for a given set $S \subseteq \mathbb{R}^n$ and a function $\psi : \mathbb{R}^n \to \bar{R} = \mathbb{R} \cup \{\pm\infty\}$, $x^k \to_S x^*$ means that $x^k \to x^*$ with $x^k \in S$ and $x^k \to_\psi x^*$ means $x^k \to x^*$ with $\psi(x^k) \to \psi(x^*)$.

Let $\Phi : \mathbb{R}^n \rightrightarrows \mathbb{R}^m$ be a set valued mapping. Then graph and domain of $\Phi$ will be denoted as gph$\Phi$ and dom$\Phi$, respectively, and defined same as in Guo et al. (2014).

We borrow some other definitions from Guo et al. (2014), which are essential to derive the results of this article, but we avoid to write the full mathematical representations of them.

$\limsup_{x \to_S x^*} \Phi(x)$ and $\liminf_{x \to_S x^*} \Phi(x)$ are the Painlevé–Kuratowski outer and inner limit of $\Phi$ relative to a set $S$ at any point $x^*$, respectively. The *tangent* and *inner tangent* cones for a set $S$ at $x^* \in S$, denoted as $\mathcal{T}_S(x^*)$ and $\mathcal{T}_S^i(x^*)$, respectively.

Furthermore, for a point $x^* \in \mathbb{R}^n$, let $\psi(x^*)$ be finite. Then we have various subdifferentials of $\psi$ at $x^*$, defined as follows:

**1.** *Regular* or *Fréchet subdifferential* of $\psi$ at $x^*$

$$\partial^F \psi(x^*) := \{u \mid \psi(x) \geqslant \psi(x^*) + u^T(x - x^*) + o(||x - x^*||)\}.$$

**2.** *Limiting* or *Mordukhovich subdifferential* of $\psi$ at $x^*$

$$\partial\psi(x^*) := \limsup_{x \to_\psi x^*} \partial^F \psi(x) = \{u \mid \exists x^k \to_\psi x^*, \exists u^k \in \partial^F \psi(x^k) \text{ s.t. } u^k \to u\}.$$

**3.** *Horizon* or *singular Mordukhovich subdifferential* of $\psi$ at $x^*$

$$\partial^\infty \psi(x^*) := \{u \mid \exists x^k \to_\psi x^*, \exists u^k \in \partial^F \psi(x^k), \exists t_k \downarrow 0 \text{ s.t. } t_k u^k \to u\}.$$

Here are the concepts of some known standard cones, defined in variational analysis. Let $S \subset \mathbb{R}^n$ be a nonempty closed set and $x^* \in S$.

1. The *Fréchet normal cone* of $S$ at $x^*$ is defined as

$$N_S^F(x^*) := T_S(x^*)^\circ.$$

2. The convex cone

$$N_S^\pi(x^*) := \{d \in \mathbb{R}^n \mid \exists\, \sigma > 0 \; s.t. \; \langle d, x - x^* \rangle \leqslant \sigma ||x - x^*||^2 \,;\; \forall\, x \in S\}$$

is called the *proximal normal cone* to $S$ at $x^*$.

3. The *limiting normal cone* of $S$ at $x^*$ is defined as

$$\begin{aligned} N_S(x^*) &:= \{d \in \mathbb{R}^n \mid \exists \{x^k\} \to_S x^*, d^k \in N_S^F(x^k) \text{ s.t. } d^k \to d\} \\ &= \{d \in \mathbb{R}^n \mid \exists \{x^k\} \to_S x^*, d^k \in N_S^\pi(x^k) \text{ s.t. } d^k \to d\}. \end{aligned}$$

Next we have concept of directional derivative, which is the most discussing part of this paper.

**Definition 11.2.1.** For a direction $d_p$, we denote the upper and lower Dini directional derivative of $\mathbb{V}$ at $p^*$, respectively as

$$\mathcal{D}^+\mathbb{V}(p^*; d_p) := \limsup_{t \downarrow 0} \frac{\mathbb{V}(p^* + t d_p) - \mathbb{V}(p^*)}{t},$$

$$\mathcal{D}_+\mathbb{V}(p^*; d_p) := \liminf_{t \downarrow 0} \frac{\mathbb{V}(p^* + t d_p) - \mathbb{V}(p^*)}{t}.$$

We define the directional derivative of $\mathbb{V}$ at $p^*$ in direction $d_p$ as

$$\mathcal{D}\mathbb{V}(p^*; d_p) = \lim_{t \downarrow 0} \frac{\mathbb{V}(p^* + t d_p) - \mathbb{V}(p^*)}{t}.$$

## 11.3 Directional differentiability of the value function for $p - MPVC$

This section is devoted to the directional derivative of the value function for the parametric $MPVC$. Suppose all the involved functions $f$, $g$, $h$, $G$, $H$ are smooth, i.e., continuously differentiable and $C = \mathbb{R}^{n_1}$.

We borrow the following index set notations from the literature Hoheisel and Kanzow (2008); Hu et al. (2017), for any given feasible point $x^* \in \mathcal{W}(p^*)$:

$$\begin{aligned} I_g^* &:= \{i \mid g_i(x^*, p^*) = 0\}, \\ I_+^* &:= \{i \mid H_i(x^*, p^*) > 0\}, \\ I_{+0}^* &:= \{i \mid H_i(x^*, p^*) > 0, \; G_i(x^*, p^*) = 0\}, \\ I_{+-}^* &:= \{i \mid H_i(x^*, p^*) > 0, \; G_i(x^*, p^*) < 0\}, \end{aligned}$$

$$I_{00}^* := \{i \mid H_i(x^*, p^*) = 0,\ G_i(x^*, p^*) = 0\},$$
$$I_{0+}^* := \{i \mid H_i(x^*, p^*) = 0,\ G_i(x^*, p^*) > 0\},$$
$$I_{0-}^* := \{i \mid H_i(x^*, p^*) = 0,\ G_i(x^*, p^*) < 0\}.$$

Furthermore, we define the MPVC-generalized Lagrangian function for $p - MPVC$,

$$\mathscr{L}^r(x, p; \lambda, \mu, \eta^G, \eta^H) = rf(x, p) + g(x, p)^T\lambda + h(x, p)^T\mu + G(x, p)^T\eta^G - H(x, p)^T\eta^H, \quad r \geqslant 0$$

and the linearized cone is structured as follows; it is borrowed from Achtziger and Kanzow (2008, Lemma 4), and defined in the same context:

$$\mathcal{L}(x^*, p^*) = \left\{ d \,\middle|\, \begin{array}{lll} \nabla_x g_i(x^*, p^*)^T d & \leqslant 0, & \forall i \in I_g^* \\ \nabla_x h_i(x^*, p^*)^T d & = 0, & \forall i = 1, \ldots, m \\ \nabla_x H_i(x^*, p^*)^T d & = 0, & \forall i \in I_{0+}^* \\ \nabla_x H_i(x^*, p^*)^T d & \geqslant 0, & \forall i \in I_{00}^* \cup I_{0-}^* \\ \nabla_x G_i(x^*, p^*)^T d & \leqslant 0, & \forall i \in I_{+0}^* \end{array} \right\}.$$

Using index sets stated above, we can define some stationary concepts for $p - MPVC$, which are parametric version of Khare and Nath (2019, Theorem 3.2):

1. **Fritz John type** $p - MPVC$ **generalized S-stationarity:** For any feasible point $x^* \in \mathcal{W}(p^*)$, generalized strong stationarity (generalized S-stationarity) condition for $p - MPVC$ holds, if there is a multiplier $(\lambda, \mu, \eta^G, \eta^H) \in \mathbb{R}^l \times \mathbb{R}^m \times \mathbb{R}^n \times \mathbb{R}^n$ and $r \geqslant 0$ such that $(r, \lambda, \mu, \eta^G, \eta^H) \neq 0$ and

$$\begin{aligned} &\nabla_x \mathscr{L}^r(x^*, p^*; \lambda, \mu, \eta^G, \eta^H) = 0, \\ &\lambda \geqslant 0, \quad g(x^*, p^*)^T\lambda = 0, \\ &\eta_i^G = 0, i \in I_{+-}^* \cup I_{0-}^* \cup I_{0+}^*, \\ &\eta_i^H = 0, i \in I_{+}^*, \\ &\eta_i^G \geqslant 0, i \in I_{+0}^* \cup I_{00}^*, \\ &\eta_i^H \geqslant 0, i \in I_{0-}^*, \\ &\eta_i^H \text{ is free}, i \in I_{0+}^*, \\ &\eta_i^H \geqslant 0, \quad \eta_i^G = 0, i \in I_{00}^*. \end{aligned} \tag{11.2}$$

We define the set of generalized S-multipliers at $x^* \in \mathcal{W}(p^*)$ as

$$\mathcal{M}_S^r(x^*, p^*) = \{(\lambda, \mu, \eta^G, \eta^H) \mid 0 \neq (r, \lambda, \mu, \eta^G, \eta^H) \text{ satisfies system (11.2)}\}.$$

2. $p - MPVC$ **B-stationarity (Izmailov and Solodov, 2009):** Any feasible point $x^* \in \mathcal{W}(p^*)$ is said to satisfy $B - \textit{stationarity}$ or *piecewise stationarity* condition for $p - MPVC$, if for all $I_{00} \subseteq I_{00}^*$, there exist multipliers $(\lambda, \mu, \eta^G, \eta^H) \in \mathbb{R}^l \times \mathbb{R}^m \times \mathbb{R}^n \times \mathbb{R}^n$ such that

$$\nabla_x \mathscr{L}^r(x^*, p^*; \lambda, \mu, \eta^G, \eta^H) = 0,$$

$$
\begin{aligned}
&\lambda \geqslant 0, \quad g(x^*, p^*)^T \lambda = 0,\\
&\eta_i^G = 0, \quad i \in I_{+-}^* \cup I_{0-}^* \cup I_{0+}^* \cup I_{00},\\
&\eta_i^H = 0, \quad i \in I_{+}^*,\\
&\eta_i^G \geqslant 0, \quad i \in I_{+0}^* \cup I_{00}^c,\\
&\eta_i^H \geqslant 0, \quad i \in I_{0-}^* \cup I_{0+}^* \cup I_{00}^c,
\end{aligned}
\tag{11.3}
$$

where $I_{00}^c = I_{00}^* \backslash I_{00}$ is the complement of $I_{00}$ in $I_{00}^*$. The set of $B-stationarity$ multipliers can be defined as

$$
\mathcal{B}_{I_{00}}(x^*, p^*) := \{(\lambda, \mu, \eta^G, \eta^H) \mid (\lambda, \mu, \eta^G, \eta^H) \text{ satisfies system (11.3) for each } I_{00} \subseteq I_{00}^* \text{ at } x^* \in \mathcal{W}(p^*)\}.
$$

It can be observed that $p - MPVC$-S-stationarity implies the $p - MPVC$-B-stationarity, but the converse need not be true.

Now we define some constraint qualifications (CQs) for parametric MPVC, which are useful in subsequent sections for further analysis. These CQs can be considered as counter part to MPEC's constraint qualifications, defined in Guo et al. (2013a).

**Definition 11.3.1.** A point $x^* \in \mathcal{W}(p^*)$ is said to satisfy $p - MPVC - LICQ$ (or linearly independent constraint qualification), if the set of gradients

$$
\{\nabla_x g_i(x^*, p^*), \nabla_x h_j(x^*, p^*), \nabla_x G_k(x^*, p^*), \nabla_x H_l(x^*, p^*) | i \in I_g^*, j = 1, ..., m_2, \\ k \in I_{+0}^* \cup I_{00}^*, l \in I_{00}^* \cup I_{0+}^* \cup I_{0-}^*\}
$$

is linearly independent.

**Definition 11.3.2.** At a feasible point $x^* \in \mathcal{W}(p^*)$, $p - MPVC$-CRCQ (or constant rank constraint qualification) is said to hold if there exists $\delta > 0$ such that for any $I_1 \subseteq I_g^*$, $I_2 \subseteq \{1, 2, ..., m\}$, $I_3 \subseteq I_{+0}^* \cup I_{00}^*$, $I_4 \subseteq I_{0+}^* \cup I_{0-}^* \cup I_{00}^*$, the family of gradients

$$
\{\nabla_x g_i(x, p), \nabla_x h_j(x, p), \nabla_x G_k(x, p), \nabla_x H_l(x, p) \mid i \in I_1,\ j \in I_2,\ k \in I_3,\ l \in I_4\}
$$

has the same rank for all $x \in \mathbb{B}_\delta(x^*)$ and for all $p \in \mathbb{B}_\delta(p^*)$.

**Definition 11.3.3.** The set valued map $\mathcal{W}$ is said to be $p - MPVC$-RCR regular (or relaxed constant rank regular) at $x^* \in \mathcal{W}(p^*)$ if there exists $\delta > 0$ such that for any $I_1 \subseteq I_g^*$, $I_3, I_4 \subseteq I_{00}^*$, the family of gradients

$$
\{\nabla_x g_i(x, p), \nabla_x h_j(x, p), \nabla_x G_k(x, p), \nabla_x H_l(x, p) \mid i \in I_1, j = 1, ...m, \\ k \in I_3 \cup I_{+0}^*,\ l \in I_4 \cup I_{0+}^* \cup I_{0-}^*\}
$$

has the same rank for all $x \in \mathbb{B}_\delta(x^*)$ and for all $p \in \mathbb{B}_\delta(p^*)$.

Next we provide the $x$-projection of $p - MPVC$-linearized cone of $gph\mathcal{W}$ at $(x^*, p^*)$, which is denoted as $\mathbb{L}(x^*, p^*, d_p)$ and defined as

$$\mathbb{L}(x^*, p^*; d_p) = \left\{ d_x \left| \begin{array}{ccc} \nabla g_i(x^*, p^*)^T d \leqslant 0 & , & \forall i \in I_g^* \\ \nabla h_i(x^*, p^*)^T d = 0 & , & \forall i = 1, ..., m \\ \nabla H_i(x^*, p^*)^T d = 0 & , & \forall i \in I_{0+}^* \\ \nabla H_i(x^*, p^*)^T d \geqslant 0 & , & \forall i \in I_{00}^* \cup I_{0-}^* \\ \nabla G_i(x^*, p^*)^T d \leqslant 0 & , & \forall i \in I_{+0}^* \end{array} \right. \right\},$$

and the constraint functions involved in $p - MPVC$ can also be represented in vector form as

$$F(x, p) = \begin{pmatrix} g(x, p) \\ h(x, p) \\ \psi(x, p) \end{pmatrix}, \quad \Delta := \mathbb{R}_-^l \times \{0\}^m \times \Omega^{2n}, \tag{11.4}$$

where $\mathbb{R}_- = \{a \in \mathbb{R} | a \leqslant 0\}$

$$\psi(x, p) = \begin{pmatrix} G_1(x, p) \\ H_1(x, p) \\ \vdots \\ \vdots \\ G_n(x, p) \\ H_n(x, p) \end{pmatrix}, \tag{11.5}$$

and

$$\Omega = \{(a, b) \in \mathbb{R}^2 \mid b \geqslant 0,\ ab \leqslant 0\}. \tag{11.6}$$

Hence the feasible region of parametric-MPVC can be written as

$$\mathcal{W}(p) = \{x \mid F(x, p) \in \Delta\}. \tag{11.7}$$

Based on this reformulation of parametric MPVC, we have one more regularity concept.

**Definition 11.3.4.** The set valued map $\mathcal{W}$ is said to be $p - MPVC - R$-regular at $x^* \in \mathcal{W}(p^*)$ in direction $d_p$ if there exist $k > 0$ and $\delta > 0$ such that

$$dist(x, \mathcal{W}(p^* + td_p)) \leqslant k\, dist(F(x, p^* + td_p), \Delta), \quad \forall t \in (0, \delta), \text{ and } \forall x \in \mathbb{B}_\delta(x^*),$$

that is,

$$\begin{aligned} dist(x, \mathcal{W}(p^* + td_p)) \leqslant k(&\|\max(g(x, p^* + td_p), 0)\| + \|h(x, p^* + td_p)\| \\ &+ \max\{0, -H_i(x, p^* + td_p), \min\{G_i(x, p^* + td_p), H_i(x, p^* + td_p)\}\}), \\ &; \forall t \in (0, \delta), \text{ and } \forall x \in \mathbb{B}_\delta(x^*). \end{aligned}$$

Now motivated by Guo et al. (2013a, Proposition 3.2), we can conclude an important result for $p-MPVC$ analogous to parametric MPEC (Guo et al., 2014). For this we have to write tangent cone of $\Omega$ as

$$\mathcal{T}_{\Omega}(a,b) = \left\{ (d_1,d_2) \in \mathbb{R}^2 \middle| \begin{array}{ccc} d_1 \in \mathbb{R},\ d_2 = 0 & , & a>0,\ b=0 \\ d_1 \in \mathbb{R},\ d_2 \geqslant 0 & , & a<0,\ b=0 \\ (d_1,d_2) \in \Omega & , & a=0=b \\ d_1 \leqslant 0,\ d_2 \in \mathbb{R} & , & a=0,\ b>0 \end{array} \right\}.$$

Hence

$$\mathcal{T}_{\Delta}(a,b) = \left\{ d_x \middle| \begin{array}{ccc} \nabla g_i(x^*,p^*)^T d \leqslant 0 & , & \forall i \in I_g^* \\ \nabla h_i(x^*,p^*)^T d = 0 & , & \forall i = 1,\dots,m \\ \nabla H_i(x^*,p^*)^T d = 0 & , & \forall i \in I_{0+}^* \\ \nabla H_i(x^*,p^*)^T d \geqslant 0 & , & \forall i \in I_{00}^* \cup I_{0-}^* \\ \nabla G_i(x^*,p^*)^T d \leqslant 0 & , & \forall i \in I_{+0}^* \end{array} \right\}.$$

Also $x$-projection of $p-MPVC$-linearized cone can be given as in Guo et al. (2013a).

$$\mathbb{L}(x^*,p^*;d_p) = \{d_x \mid \nabla F(x^*,p^*)^T d \in \mathcal{T}_{\Delta}(F(x^*,p^*))\} \tag{11.8}$$

**Lemma 11.3.1.** *Let $x^* \in \mathcal{W}(p^*)$, then*

$$\mathcal{D}_+\mathcal{W}(x^*,p^*,d_p) \subseteq \mathbb{L}(x^*,p^*,d_p).$$

*If the set valued mapping $\mathcal{W}$ is MPVC-R-regular at $x^* \in \mathcal{W}(p^*)$ in direction $d_p$, then*

$$\mathcal{D}_+\mathcal{W}(x^*,p^*,d_p) = \mathbb{L}(x^*,p^*,d_p).$$

*Proof.* To prove the first inclusion, let $d_x \in \mathcal{D}_+\mathcal{W}(x^*,p^*,d_p)$, then by the definition of lower Dini directional derivative of set valued map, there exists $o(t)$ such that

$$\begin{aligned} & x^* + td_x + o(t) \in \mathcal{W}(p^* + td_p)\,, \quad \forall\, t \geqslant 0 \\ \Longrightarrow\ & F(x^* + td_x + o(t),\ p^* + td_p) \ \in\ \Delta. \end{aligned}$$

Now using Taylor's expansion gives

$$F(x^* + td_x + o(t),\ p^* + td_p) = F(x^*,p^*) + t\nabla F(x^*,p^*)^T \begin{pmatrix} d_x \\ d_p \end{pmatrix} + o(t) \ \in\ \Delta,$$

$$\Longrightarrow \nabla F(x^*,p^*)^T \begin{pmatrix} d_x \\ d_p \end{pmatrix} \ \in\ \mathcal{T}_{\Delta}(F(x^*,p^*)).$$

Hence we have $d_x \in \mathbb{L}(x^*,p^*,d_p)$ by (11.8). Therefore

$$\mathcal{D}_+\mathcal{W}(x^*,p^*,d_p) \subseteq \mathbb{L}(x^*,p^*,d_p).$$

To show the second result, it is sufficient to show $\mathbb{L}(x^*, p^*, d_p) \subseteq \mathcal{D}_+\mathcal{W}(x^*, p^*, d_p)$ under MPVC-R-regularity. For this, first we have to show

$$\mathcal{T}_\Delta(F(x^*, p^*)) = \mathcal{T}^i_\Delta(F(x^*, p^*)).$$

Now we have from Nath and Khare (2019, Lemma 4.2)

$$T_\Delta(F(x^*, p^*)) = \prod_{i=1}^{l} T_{\mathbb{R}_-}(g_i(x^*, p^*)) \times \prod_{i=1}^{m} T_{\{0\}}(h_i(x^*, p^*)) \times \prod_{i=1}^{n} T_\Omega(G_i(x^*, p^*), H_i(x^*, p^*)).$$

Since $\mathbb{R}_-$ and $\{0\}$ are convex and closed, therefore Bonnas and Shapiro (2000, Proposition 2.55) gives

$$\mathcal{T}_{\mathbb{R}_-}(g_j(x^*, p^*)) = \mathcal{T}^i_{\mathbb{R}_-}(g_j(x^*, p^*)),$$
$$\mathcal{T}_{\{0\}}(h_j(x^*, p^*)) = \mathcal{T}^i_{\{0\}}(h_j(x^*, p^*)),$$

and by the definition of inner tangent cone, we have

$$\mathcal{T}^i_\Omega(G_j(x^*, p^*), H_j(x^*, p^*)) = \mathcal{T}_\Omega(G_j(x^*, p^*), H_j(x^*, p^*)).$$

Now we have to show that

$$T^i_\Delta(F(x^*, p^*)) = \prod_{i=1}^{l} T^i_{\mathbb{R}_-}(g_j(x^*, p^*)) \times \prod_{i=1}^{m} T^i_{\{0\}}(h_j(x^*, p^*)) \times \prod_{i=1}^{n} T^i_\Omega(G_j(x^*, p^*), H_j(x^*, p^*)).$$

Let $d \in \mathcal{T}^i_\Delta(F(x^*, p^*))$, then

$$dist(F(x^*, p^* + td), \Delta) = 0(t), \quad \forall t \geqslant 0; \tag{11.9}$$

according to the structure of $\Delta$ and $F$, we have

$$d = (d_{g_i}, d_{h_j}, d_{G_l}, d_{H_l}), \quad i = 1, ..., l, \;\; j = 1, ..., m, \;\; l = 1, ...n.$$

Then (11.9) is true for all components. Therefore we have

$$d \in \prod_{i=1}^{l} T^i_{\mathbb{R}_-}(g_j(x^*, p^*)) \times \prod_{i=1}^{m} T^i_{\{0\}}(h_j(x^*, p^*)) \times \prod_{i=1}^{n} T^i_\Omega(G_j(x^*, p^*), H_j(x^*, p^*)).$$

Similarly, we can show that the converse is also true. Therefore we have

$$\begin{aligned} T^i_\Delta(F(x^*, p^*)) &= \prod_{i=1}^{l} T^i_{\mathbb{R}_-}(g_j(x^*, p^*)) \times \prod_{i=1}^{m} T^i_{\{0\}}(h_j(x^*, p^*)) \times \prod_{i=1}^{n} T^i_\Omega(G_j(x^*, p^*), H_j(x^*, p^*)) \\ &= \prod_{i=1}^{l} T_{\mathbb{R}_-}(g_j(x^*, p^*)) \times \prod_{i=1}^{m} T_{\{0\}}(h_j(x^*, p^*)) \times \prod_{i=1}^{n} T_\Omega(G_j(x^*, p^*), H_j(x^*, p^*)), \end{aligned}$$

i.e., $T^i_\Delta(F(x^*, p^*)) = T_\Delta(F(x^*, p^*))$.

Now, let $d_x \in \mathbb{L}(x^*, p^*; d_p)$, then

$$\nabla F(x^*, p^*)^T \begin{pmatrix} d_x \\ d_p \end{pmatrix} \in T_\Delta(F(x^*, p^*)) = T^i_\Delta(F(x^*, p^*));$$

then by the definition of inner tangent cone

$$dist\left(F(x^*, p^*) + t\nabla F(x^*, p^*)^T \begin{pmatrix} d_x \\ d_p \end{pmatrix}, \Delta\right) = o(t)\,, \quad \forall\, t \geqslant 0.$$

Now by the MPVC-R-regularity in direction $d_p$, we have

$$\begin{aligned} dist(x^* + td_x, \mathcal{W}(p^* + td_p)) &\leqslant k\, dist(F(x^* + td_x, p^* + td_p),\ \Delta) \\ &= k\, dist\left(F(x^*, p^*) + t\ \nabla F(x^*, p^*)^T \begin{pmatrix} d_x \\ d_p \end{pmatrix} + 0(t),\ \Delta\right) \\ &\qquad \text{(by the Taylor's expansion)} \\ &= k\, dist\left(F(x^*, p^*) + t\ \nabla F(x^*, p^*)^T \begin{pmatrix} d_x \\ d_p \end{pmatrix},\ \Delta\right) + 0(t) \\ &= o(t), \\ &\Longrightarrow\ d_x \in \mathcal{D}_+\mathcal{W}(x^*, p^*; d_p). \end{aligned}$$

Therefore $\mathbb{L}(x^*, p^*; d_p) \subseteq \mathcal{D}_+\mathcal{W}(x^*, p^*; d_p)$.
Hence $\mathbb{L}(x^*, p^*; d_p) = \mathcal{D}_+\mathcal{W}(x^*, p^*; d_p)$. □

Now to derive some further results, which are sufficient for $x^* \in \mathcal{W}(p^*)$ to be $p-MPVC$-R-Regular, we define a tightened $p-MPVC$, whose constraint region is given by

$$\mathcal{W}_T(p) = \left\{ x \,\middle|\, \begin{array}{ccc} g(x, p) \leqslant 0 & , & h(x, p) = 0, \\ G_{I^*_{+0} \cup I^*_{00}}(x, p) = 0 & , & H_{I^*_{00} \cup I^*_{0+} \cup I^*_{0-}}(x, p) = 0 \end{array} \right\}.$$

Clearly, $\mathcal{W}_T(p) \subseteq \mathcal{W}(p)$ near $(x^*, p^*)$, hence $\mathcal{D}_+\mathcal{W}_T(x^*, p^*; d_p) \subseteq \mathcal{D}_+\mathcal{W}(x^*, p^*; d_p)$ for all $d_p$.

Now here is the main result, we need the following key lemma:

**Lemma 11.3.2.** *For any* $(a, b) \in \mathbb{R}^2$, *let*

$$\Phi(a, b) := \min(|b| + \max(-a, 0),\ \max(a, 0) + \max(-b, 0)).$$

*Then we have*

$$\Phi(a, b) = dist_\Omega(a, b),$$

*where* $\Omega = \{(a, b) \in \mathbb{R}^2 | b \geqslant 0, ab \leqslant 0\}$.

*Proof.* We consider the following cases:

- $a \geqslant b \geqslant 0$: $dist_\Omega(a,b) = b$ and $\Phi(a,b) = \min(|b|, a) = b$.
- $b \geqslant a \geqslant 0$: $dist_\Omega(a,b) = a$ and $\Phi(a,b) = \min(|b|, a) = a$.
- $b \geqslant 0, a \leqslant 0$: $dist_\Omega(a,b) = 0$ and $\Phi(a,b) = \min(-a + |b|, 0 + 0) = 0$.
- $a \leqslant b \leqslant 0$: $dist_\Omega(a,b) = |b|$ and $\Phi(a,b) = \min(-a + |b|, 0 + (-b)) = -b$.
- $b \leqslant a \leqslant 0$: $dist_\Omega(a,b) = |b|$ and $\Phi(a,b) = -b = |b|$.
- $a \geqslant 0,\ b = 0$: $dist_\Omega(a,b) = 0$ and $\Phi(a,b) = \min(0 + 0, a + 0) = 0$.
- $a = 0,\ b \geqslant 0$: $dist_\Omega(a,b) = 0$ and $\Phi(a,b) = \min(0 + |b|, 0 + 0) = 0$.
- $a \geqslant 0, b \leqslant 0$: $dist_\Omega(a,b) = |b|$ and $\Phi(a,b) = \min(0 + |b|, a + (-b)) = |b|$.

Hence the result is true for all cases. □

Now here is the main result regarding the sufficiency of $p - MPVC - R$-regularity, which is derived using the above key lemma:

**Lemma 11.3.3.** *Suppose the set valued map $\mathcal{W}$ is MPVC-RCR regular at $x^* \in \mathcal{W}(p^*)$. If $\mathcal{D}_+\mathcal{W}_T(x^*, p^*; d_p) \neq \emptyset$, then $\mathcal{W}$ is MPVC-R-regular at $x^* \in \mathcal{W}(p^*)$ in direction $d_p$ and*

$$\mathcal{D}_+\mathcal{W}(x^*, p^*; d_p) = \mathbb{L}(x^*, p^*; d_p) \neq \emptyset.$$

*Proof.* This result is derived using the concept of mathematical induction. First, we have to show that MPVC-R-regularity holds when the number of constraints is zero.
**Step 1.** If $n = 0$, then this MPVC becomes a standard nonlinear program, and we can show

MPVC-R-regularity for this case similar as in Guo et al. (2014, Lemma 3.5).
**Step 2.** Now to show the same result for $n = 1$, consider the following two constraint

regions:

$$\mathcal{W}_1(p) = \left\{ x \;\middle|\; \begin{array}{ccc} g(x,p) \leqslant 0 & , & h(x,p) = 0 \\ H_1(x,p) = 0 & , & G_1(x,p) \geqslant 0 \end{array} \right\}$$

$$\mathcal{W}_2(p) = \left\{ x \;\middle|\; \begin{array}{ccc} g(x,p) \leqslant 0 & , & h(x,p) = 0 \\ H_1(x,p) \geqslant 0 & , & G_1(x,p) \leqslant 0 \end{array} \right\},$$

then $\mathcal{W}_1(p) \cup \mathcal{W}_2(p) = \mathcal{W}(p)$ and $\mathcal{W}_T(p) \subseteq \mathcal{W}_i(p)$ around $(x^*, p^*)$ for $i = 1, 2$, and by the definition of lower Dini directional derivative.

$$0 \neq \mathcal{D}_+\mathcal{W}_T(x^*, p^*; d_p) \subseteq \mathcal{D}_+\mathcal{W}_i(x^*, p^*; d_p)\ , \qquad i = 1, 2. \tag{11.10}$$

Now to complete the proof, we consider the following cases:
**Case I.** $x^* \in \mathcal{W}_1(p^*) \backslash \mathcal{W}_2(p^*)$; in this case,

$$dist(x, \mathcal{W}(p)) = dist(x, \mathcal{W}_1(p))$$

for any $(x, p)$ sufficiently close to $(x^*, p^*)$.

We also have $H_1(x^*, p^*) = 0$, $G_1(x^*, p^*) > 0$, therefore

$$0 = H_1(x^*, p^*) < G_1(x^*, p^*). \tag{11.11}$$

Moreover, the MPVC-RCR regularity assumption shows that $\mathcal{W}_1$ is RCR regular at $x^* \in \mathcal{W}_1(p^*)$, and then the set $\mathcal{W}_1(p) - \mathcal{W}_2(p)$ represents set of equality and strict inequality as

$$\mathcal{W}_1(p)\backslash\mathcal{W}_2(p) = \left\{ x \;\middle|\; \begin{array}{ccc} g(x,p) \leqslant 0 & , & h(x,p) = 0 \\ H_1(x,p) = 0 & , & -G_1(x,p) < 0 \end{array} \right\}.$$

This fact along with the first part of this proof indicates that MPVC-R-regularity holds at $x^* \in \mathcal{W}_1(p)$ in direction $d_p$; then from (11.11), there exist $k > 0$ and $\delta > 0$ such that for all $x \in \mathcal{B}_\delta(x^*)$ and $t \in (0, \delta)$, we have

$$\begin{aligned} dist(x, \mathcal{W}_1(p^* + td_p)) &\leqslant k(||\max(g(x, p^* + td_p), 0)|| + ||h(x, p^* + td_p)|| \\ &\quad + \max(-G_1(x, p^* + td_p), 0) + |H_1(x, p^* + td_p)|) \\ &= k(||\max(g(x, p^* + td_p), 0)|| + ||h(x, p^* + td_p)|| + |H_1(x, p^* + td_p)|) \\ &= k(||\max(g(x, p^* + td_p), 0)|| + ||h(x, p^* + td_p)|| \\ &\quad + dist((G_1(x, p^* + td_p), H_1(x, p^* + td_p)), \Omega). \end{aligned}$$

**Case II.** $x \in \mathcal{W}_2(p^*)\backslash\mathcal{W}_1(p^*)$; in this case, the constraint region is

$$\mathcal{W}_2(p)\backslash\mathcal{W}_1(p) = \left\{ x \;\middle|\; \begin{array}{ccc} g(x,p) \leqslant 0 & , & h(x,p) = 0 \\ \text{either (a). } H_1(x,p) = 0 & , & G_1(x,p) < 0 \\ \text{or (b). } H_1(x,p) > 0 & , & G_1(x,p) = 0 \\ \text{or (c). } H_1(x,p) > 0 & , & G_1(x,p) < 0 \end{array} \right\}.$$

In all three cases, we have $G_1(x, p) < H_1(x, p) \geqslant 0$, and in all cases MPVC-RCR regularity assumption implies that $\mathcal{W}_2$ is RCR-regular at $x^* \in \mathcal{W}_2(p^*)$. Similarly as in Case I, MPVC-R-regularity holds at $x^* \in \mathcal{W}_2$ in direction $d_p$ and $\exists\, k > 0$ and $\delta > 0$ such that for each $x \in \mathcal{B}_\delta(x^*)$ and $t \in (0, \delta)$, we have

**(a).**

$$\begin{aligned} dist(x, \mathcal{W}_2(p^* + td_p)) &\leqslant k(||\max(g(x, p^* + td_p), 0)|| + ||h(x, p^* + td_p)|| \\ &\quad + \max(G_1(x, p^* + td_p), 0) + |H_1(x, p^* + td_p)|) \\ &= k(||\max(g(x, p^* + td_p), 0)|| + ||h(x, p^* + td_p)|| + |H_1(x, p^* + td_p)|) \\ &= k(||\max(g(x, p^* + td_p), 0)|| + ||h(x, p^* + td_p)|| \\ &\quad + dist((G_1(x, p^* + td_p), H_1(x, p^* + td_p)), \Omega); \end{aligned}$$

**(b).**

$$dist(x, \mathcal{W}_2(p^* + td_p)) \leqslant k(||\max(g(x, p^* + td_p), 0)|| + ||h(x, p^* + td_p)||$$

$$
\begin{aligned}
&+\max(-H_1(x,p^*+td_p),0)+|G_1(x,p^*+td_p)|)\\
&=k(\|\max(g(x,p^*+td_p),0)\|+\|h(x,p^*+td_p)\|+|G_1(x,p^*+td_p)|)\\
&=k(\|\max(g(x,p^*+td_p),0)\|+\|h(x,p^*+td_p)\|\\
&\quad+dist((G_1(x,p^*+td_p),H_1(x,p^*+td_p)),\Omega);
\end{aligned}
$$

**(c).**

$$
\begin{aligned}
dist(x,\mathcal{W}_1(p^*+td_p)) &\leqslant k(\|\max(g(x,p^*+td_p),0)\|+\|h(x,p^*+td_p)\|\\
&\quad+\max(G_1(x,p^*+td_p),0)+\max(-H_1(x,p^*+td_p),0))\\
&=k(\|\max(g(x,p^*+td_p),0)\|+\|h(x,p^*+td_p)\|+0\\
&=k(\|\max(g(x,p^*+td_p),0)\|+\|h(x,p^*+td_p)\|\\
&\quad+dist((G_1(x,p^*+td_p),H_1(x,p^*+td_p)),\Omega).
\end{aligned}
$$

**Case III.** $x\in\mathcal{W}_1(p^*)\cap\mathcal{W}_2(p^*)$; in this case, the constraint region is

$$
\mathcal{W}_1(p)\cap\mathcal{W}_2(p)=\left\{x\left|\begin{array}{ccc} g(x,p)\leqslant 0 & , & h(x,p)=0\\ H_1(x,p)=0 & , & G_1(x,p)=0\end{array}\right.\right\}.
$$

Here also, MPVC-RCR regularity assumption implies that both $\mathcal{W}_1$ and $\mathcal{W}_2$ are RCR-regular at $x^*$. Hence MPVC-R-regularity holds at $x^*\in\mathcal{W}_1(p^*)\cap\mathcal{W}_2(p^*)$, and therefore there exist $k>0$ and $\delta>0$ such that for each $x\in\mathbb{B}_\delta(x^*)$ and $t\in(0,\delta)$, we have

$$
\begin{aligned}
dist(x,\mathcal{W}_1(p^*+td_p)) &\leqslant k(\|\max(g(x,p^*+td_p),0)\|+\|h(x,p^*+td_p)\|\\
&\quad+\|\max(-G_1(x,p^*+td_p),0)\|+|H_1(x,p^*+td_p)|)
\end{aligned}
$$

and

$$
\begin{aligned}
dist(x,\mathcal{W}_2(p^*+td_p)) &\leqslant k(\|\max(g(x,p^*+td_p),0)\|+\|h(x,p^*+td_p)\|\\
&\quad+\|\max(G_1(x,p^*+td_p),0)\|+\|\max(-H_1(x,p^*+td_p),0)\|).
\end{aligned}
$$

Therefore from above inequalities and Lemma 11.3.2, we have for any $x\in\mathcal{B}_\delta(x^*)$ and $t\in(0,\delta)$,

$$
\begin{aligned}
dist(x,\mathcal{W}(p^*+td_p)) &=\min\{dist(x,\mathcal{W}_1(p^*+td_p)),\ dist(x,\mathcal{W}_2(p^*+td_p))\}\\
&\leqslant k(\|\max(g(x,p^*+td_p),0)\|+\|h(x,p^*+td_p)\|\\
&\quad+dist((G_1(x,p^*+td_p),H_1(x,p^*+td_p)),\Omega)).
\end{aligned}
$$

Hence MPVC-R-regularity holds at $x^*\in\mathcal{W}(p^*)$ in direction $d_p$, when $n=1$.

In a similar way, we can show that if the MPVC-R-regularity holds for $n\leqslant k$, then the MPVC-R-regularity holds also for $n=k+1$.

Hence by using mathematical induction, it is proved that $\mathcal{W}$ is MPVC-R-regular at $x^*\in\mathcal{X}(p^*)$ in direction $d_p$, and then from Lemma 11.3.1, we have $\mathcal{D}_+\mathcal{X}(x^*,p^*;d_p)=\mathbb{L}(x^*,p^*;d_p)$. □

**Remark 11.3.1.** If $I_{00}^* \cup I_{+-}^* = \emptyset$, then for tightened MPVC program, the constraint region $\mathcal{W}_T(p)$ coincides with $\mathcal{W}(p)$, and for this case, the Lemma 11.3.3 will remain same for $\mathcal{W}(p)$ (We have analogous result for MPEC in Guo et al. (2014, Corollary 3.6)). But in this case, a significant portion of feasible region will remain ignored, so we can not leave aside these two important index sets in our study.

Now, we have an MPVC version of Minchenko and Stakhovski (2011a, Lemma 5).

**Lemma 11.3.4.** *Let $x^* \in \mathcal{W}(p^*)$, and $\mathcal{W}$ is MPVC-RCR regular at $x^* \in \mathcal{W}(p^*)$, then $\exists\ \delta > 0$ and $k > 0$ such that for any $p \in \mathbb{B}_\delta(p^*)$ and $x \in \mathbb{B}_\delta(x^*) \cap \mathcal{W}(p)$, $\exists$ a point $\bar{x} \in \mathcal{W}(p^*)$ such that*

1. $||x - \bar{x}|| \leqslant k||p - p^*||$ *and* $g_i(x, p) - g(\bar{x}, p^*) \leqslant 0, i \in I_g^*$;
2. $(G_i(x, p) - G_i(\bar{x}, p^*), H_i(x, p) - H_i(\bar{x}, p^*)) \in \Omega, i \in I_{00}^*$.

*Proof.* For each $I_{00} \subseteq I_{00}^*$, we can have following piecewise constraint region:

$$\mathcal{W}_{I_{00}}(p) = \left\{ x \in \mathcal{W}(p) \left| \begin{array}{lcl} g(x,p) \leqslant 0 & , & h(x,p) = 0 \\ H_i(x,p) \geqslant 0 & , & i \in I_{00}^c \\ H_i(x,p) = 0 & , & i \in I_{00} \cup I_{0+}^* \cup I_{0-}^* \\ G_i(x,p) \leqslant 0 & , & i \in I_{00}^c \cup I_+^* \\ G_i(x,p) \in \mathbb{R} & , & i \in I_{00} \end{array} \right. \right\}, \tag{11.12}$$

then $x^* \in \mathcal{W}_{I_{00}}(p^*)$, and the RCR-regularity holds at $x^* \in \mathcal{W}_{I_{00}}(p^*)$, $\forall\ I_{00} \subseteq I_{00}^*$.

Then for each $I_{00} \subseteq I_{00}^*$, by Minchenko and Stakhovski (2011a, Lemma 5), $\exists\ \delta_{I_{00}}$ and $k_{I_{00}}$ such that for any $p \in \mathbb{B}_{\delta_{I_{00}}}$ and $x \in \mathbb{B}_{\delta_{I_{00}}} \cap \mathcal{W}_{I_{00}}(p)$, $\exists$ a point $\bar{x} \in \mathcal{W}_{I_{00}}$ such that

1. $||x - \bar{x}|| \leqslant k_{I_{00}}||p - p^*||$ and $g_i(x, p) - g(\bar{x}, p^*) \leqslant 0, i \in I_g^*$,
2. 
   - $H_i(x,p) \geqslant 0$, $H_i(\bar{x}, p^*) \geqslant 0, \forall\, i \in I_{00}^c$
     $-H_i(x,p) \leqslant 0$, $-H_i(\bar{x}, p^*) \leqslant 0, \forall\, i \in I_{00}^c$
     $-H_i(x,p) \leqslant -H_i(\bar{x}, p^*) \leqslant 0$, $\forall\ i \in I_{00}^c$ by Minchenko and Stakhovski (2011a, Lemma 5)
     $0 \leqslant H_i(\bar{x}, p^*) \leqslant H_i(x,p), \forall\, i \in I_{00}^c$
     $\Longrightarrow H_i(x,p) - H_i(\bar{x}, p^*) \geqslant 0, \forall\, i \in I_{00}^c$
   - $H_i(x,p) = H_i(\bar{x}, p^*) = 0, \forall\, i \in I_{00} \cup I_{0+}^* \cup I_{0-}^*$
     $H_i(x,p) - H_i(\bar{x}, p^*) = 0, \forall\, i \in I_{00} \cup I_{0+}^* \cup I_{0-}^*$
   - $G_i(x,p) \leqslant G_i(\bar{x}, p^*) \leqslant 0, \forall\, i \in I_{00}^c \cup I_+^*$
     $G_i(x,p) - G_i(\bar{x}, p^*) \leqslant 0, \forall\, i \in I_{00}^c \cup I_+^*$
   - $G_i(x,p) \in \mathbb{R}$, $G_i(\bar{x}, p^*) \in \mathbb{R}, \forall\, i \in I_{00}$
     $G_i(x,p) - G_i(\bar{x}, p^*) \in \mathbb{R}, \forall\, i \in I_{00}$

   Since

$$I_{00}^* = I_{00} \cup I_{00}^c,$$

therefore

$$(G_i(x,p) - G_i(\bar{x}, p^*))(H_i(x,p) - H_i(\bar{x}, p^*)) \leqslant 0, \quad \forall\, i \in I_{00}^*,$$

$$\text{that is } (G_i(x,p) - G_i(\bar{x},p^*), H_i(x,p) - H_i(\bar{x},p^*)) \in \Omega, \quad \forall i \in I_{00}^*.$$

Since we can see that $\mathcal{W}(p) = \bigcup_{I_{00}\subseteq I_{00}^*} \mathcal{W}_{I_{00}}(p)$ around $(x^*, p^*)$ and if we set

$$k = \max_{I_{00}\subseteq I_{00}^*} k_{I_{00}}$$

$$\text{and } \delta = \min_{I_{00}\subseteq I_{00}^*} \delta_{I_{00}},$$

hence required result is true for above $k$ and $\delta$. □

Now our objective here is to show that the value function $\mathbb{V}$ is directionally differentiable in every direction. This result has been established earlier for nonlinear programs in Minchenko and Stakhovski (2011a, Theorem 5). Later, Guo et al. (2014) converted this result for parametric MPEC under a restricted inf-compactness assumption. Here we are transforming the result of Guo et al. (2014, Theorem 3.9) for parametric MPVC. For this, we need following MPVC version of restricted inf-compactness:

**Definition 11.3.5. Restricted inf-compactness:** For any $p - MPVC$, restricted inf-compactness holds around $p^*$ if $\mathbb{V}(p^*)$ is finite and there exist a compact set $S$ and a positive number $\epsilon$ such that for all $p \in \mathbb{B}_\epsilon(p^*)$, for which $\mathbb{V}(p) < \mathbb{V}(p^*) + \epsilon$, the problem $p - MPVC$ has a solution in $S$.

**Theorem 11.3.1.** *Suppose restricted inf-compactness holds around $p^*$. Let $\mathcal{W}$ be MPVC-RCR regular at each $x^* \in \mathcal{O}(p^*)$, then the value function $\mathbb{V}$ is directionally differentiable at $p = p^*$ in every direction $d_p \in \bigcap_{x^*\in\mathcal{O}(p^*)} \mathrm{dom}\mathcal{D}_+\mathcal{W}_T(x^*, p^*; .)$ and*

$$\begin{aligned}
\mathcal{D}\mathbb{V}(p^*; d_p) &= \min_{x^*\in\mathcal{O}(p^*)} \min_{d_x\in\mathbb{L}(x^*,p^*;d_p)} \nabla f(x^*, p^*)^T \begin{pmatrix} d_x \\ d_p \end{pmatrix} \qquad (11.13)\\
&= \min_{x^*\in\mathcal{O}(p^*)} \min_{I_{00}\subseteq I_{00}^*} \max_{(\lambda,\mu,\eta^G,\eta^H)\in\mathcal{B}_{I_{00}}(x^*,p^*)} \nabla_p \mathscr{L}^1(x^*, p^*; \lambda, \mu, \eta^G, \eta^H)^T d_p.
\end{aligned}$$

*Proof.* The proof can be established using the result derived above, and techniques as used in the proof of Guo et al. (2014, Theorem 3.9). □

The next result extends the range $d_p \in \bigcap_{x^*\in\mathcal{O}(p^*)} \mathrm{dom}\mathcal{D}_+\mathcal{W}_T(x^*, p^*, .)$ for every direction $d_p \in \mathbb{R}^{n_2}$. But before stating this result, we need one more definition of MPVC-NNAMCQ. We define it by using Khare and Nath (2019, Remark 2.1) and Nath and Khare (2019, Definition 4.2, Lemma 4.1) as

**Definition 11.3.6.** Let $x^*$ be feasible for p-MPVC, then the no nonzero abnormal multiplier constraint qualification (MPVC-NNAMCQ) holds at $x^*$ if the following holds:

$$\left.\begin{aligned} \nabla_x F(x^*, p^*)^T \lambda = 0 \\ \lambda \in N_\Delta(F(x^*, p^*)) \end{aligned}\right\} \Rightarrow \lambda = 0. \qquad (11.14)$$

**Theorem 11.3.2.** *Suppose the restricted inf-compactness holds around $p^*$; also assume that the MPVC-RCR regularity and the MPVC-NNAMCQ hold at each $x^* \in \mathcal{O}(p^*)$. Then the value function $\mathbb{V}$ is directionally differentiable at $p = p^*$ in every direction $d_p \in \mathbb{R}^{n_2}$ and the formula (11.13) established in Theorem 11.3.1 also holds.*

*Proof.* To prove this result, it is enough to show that

$$\mathcal{D}_+\mathcal{W}(x^*, p^*; d_p) = \mathbb{L}(x^*, p^*; d_p), \quad \forall\, d_p \in \mathbb{R}^{n_2},$$

and any one of these two is nonempty for each $x^* \in \mathcal{O}(p^*)$.

Since MPVC-NNAMCQ holds, so from Lemma 11.3.1 and Guo et al. (2012, Lemma 3.1), we have

$$\mathcal{D}_+\mathcal{W}(x^*, p^*; d_p) = \mathbb{L}(x^*, p^*; d_p), \quad \forall\, d_p \in \mathbb{R}^{n_2}.$$

Now we have the following result from Rockafellar and Wets (1998, Exercise 6.39):

$$-\nabla_x F(x^*, p^*)^T \mathbb{R}^{n_1} + \mathcal{T}_\Delta(F(x^*, p^*)) = \mathbb{R}^{l+m+2n};$$

also since

$$\nabla_p F(x^*, p^*)^T d_p \in \mathbb{R}^{l+m+2n},$$

therefore for any given $d_p$, $\exists\, d_x$ such that

$$\begin{aligned} \nabla_p F(x^*, p^*)^T d_p &\in -\nabla_x F(x^*, p^*)^T d_x + \mathcal{T}_\Delta(F(x^*, p^*)) \\ &\Longrightarrow \quad \nabla F(x^*, p^*)^T d \in \mathcal{T}_\Delta(F(x^*, p^*)). \end{aligned}$$

Now by Eq. (11.8), $d \in \mathbb{L}(x^*, p^*; d_p)$, that is, this set is nonempty. Hence $\mathcal{D}_+\mathcal{W}(x^*, p^*; d_p) \neq \emptyset$, and then by using proof of previous Theorem 11.3.1, we obtained the desired result. □

**Remark 11.3.2.** For MPVC, we have $\mathcal{M}^1_S(x^*, p^*) \subseteq \mathcal{B}_{I_{00}}(x^*, p^*)$ for each $I_{00} \subseteq I^*_{00}$. Therefore we have the following lower estimate for the directional derivative of the value function:

$$\min_{x^* \in \mathcal{O}(p^*)} \max_{y \in \mathcal{M}^1_S(x^*, p^*)} \nabla_p \mathcal{L}^1(x^*, p^*; y)^T d_p \leqslant \mathcal{D}\mathbb{V}(p^*; d_p).$$

Under MPVC-LICQ, the above result will becomes relaxed from combinatorial construction and can be written as follows:

**Corollary 11.3.1.** *Let $p - MPVC$-LICQ holds at $x^* \in \mathcal{O}(p^*)$, and restricted inf-compactness also holds around $p^*$. Then the value function $\mathbb{V}$ is directionally differentiable at $p = p^*$ in every direction $d_p \in \mathbb{R}^{n_2}$ and*

$$\mathcal{D}\mathbb{V}(p^*; d_p) = \min_{x^* \in \mathcal{O}(p^*)} \nabla_p \mathcal{L}^1(x^*, p^*, \lambda, \mu, \eta^G, \eta^H)^T d_p,$$

*where $(\lambda, \mu, \eta^G, \eta^H)$ is the unique S-multiplier vector at $x^* \in \mathcal{W}(p^*)$.*

*Proof.* It is easily observed here that MPVC-LICQ implies MPVC-RCR-regularity and MPVC-NNAMCQ, then similar to Theorem 11.3.1, Eq. (11.13) holds. Also MPVC-LICQ implies MPVC-B-stationarity at $x^* \in \mathcal{O}(p^*)$ (see Izmailov and Solodov, 2009, Theorem 2.1), that is,

$$\mathcal{B}_{I_{00}} = \{(\lambda, \mu, \eta^G, \eta^H)\},$$

for each $I_{00} \subseteq I^*_{00}$, where $(\lambda, \mu, \eta^G, \eta^H)$ is the unique S-multiplier, hence the required result holds. □

Along with the MPVC-LICQ, the concept of inf-compactness for parametric MPVC ensures the nonemptiness and uniform compactness of $\mathcal{O}(p)$ around $p^*$. It is a property that provides information about the continuity, and the stability of solution under perturbations, which are obvious to change when analyzing the value function, its directional derivatives and solutions behavior with changing parameters.

**Definition 11.3.7. Inf-compactness for** $p - MPVC$: For any parametric MPVC, inf-compactness holds around $p^*$ if there exist two positive number $\{\alpha, \delta\}$ and a bounded set S such that $\alpha > \mathbb{V}(p^*)$ and

$$\{x \in \mathcal{W}(p) \mid f(x, p) \leqslant \alpha, \ p \in \mathbb{B}_\delta(p^*)\} \subseteq S.$$

It can be observed that the inf-compactness implies the restricted inf-compactness, i.e., inf-compactness is stronger than the restricted inf-compactness. The next result is MPVC version of Guo et al. (2014, Theorem 3.14)

**Theorem 11.3.3.** *Let $x^* \in \mathcal{O}(p^*)$; if $p - MPVC$-LICQ holds at $x^*$ and inf-compactness holds around $p^*$, then $\mathcal{O}(p)$ is nonempty and uniform compact around $p^*$.*

*Conversely, the nonemptiness and uniform compactness of $\mathcal{O}(p)$ around $p^*$ imply the inf-compactness around $p^*$ if there exist a sequence $p^k \to p^*$ such that $\mathbb{V}(p^k) \downarrow \mathbb{V}(p^*)$.*

*Proof.* Since the function $f$ is continuous, if

$$\emptyset \neq \{x \in \mathcal{W}(p) \mid f(x, p) \leqslant \beta\} \subseteq \Delta,$$

then the solution set is

$$\emptyset \neq \mathcal{O}(p) \subseteq \{x \in \mathcal{W}(p) \mid f(x, p) \leqslant \beta\} \subseteq \Delta.$$

Now to show uniform compactness of $\mathcal{O}(p)$ around $p^*$, it is sufficient to show that $\exists\ \delta_0 \in (0, \delta]$ such that the set $\{x \in \mathcal{W}(p) \mid f(x, p) \leqslant \alpha\} \neq \emptyset$ for each $p \in \mathbb{B}_{\delta_0}(p^*)$.

Now consider the system

$$\begin{aligned} g(x, p) &\leqslant 0, \\ h(x, p) &= 0, \end{aligned}$$

$$G_{I^*_{00} \cup I^*_{+0}}(x, p) = 0,$$
$$H_{I^*_{00} \cup I^*_{0+} \cup I^*_{0-}}(x, p) = 0.$$

Since $p - MPVC$-LICQ holds at $x^*$ for $p = p^*$, therefore by the implicit function theorem, there exist neighborhoods $V_1$ and $V_2$ of $p^*$ and $x^*$, respectively, and a smooth function $x(.) : V_1 \to V_2$ such that $x(p^*) = x^*$ and $x(p)$ satisfies the above system. Then $x(p) \in \mathcal{W}(p)$ and

$$f(x(p), p) \to f(x^*, p^*) < \alpha \quad \text{as} \quad p \to p^*.$$

Thus $\exists\, \delta_0 \in (0, \delta]$ such that the set $\{x \in \mathcal{W}(p) \mid f(x, p) \leqslant \alpha\} \neq \emptyset; \forall\, p \in \mathbb{B}_{\delta_0}(p^*)$.

Conversely, suppose inf-compactness does not hold around $p^*$, that is, for all $\alpha^k \downarrow \mathbb{V}(p^*)$, there exist $p^k \to p^*$ and $x^k \in \mathcal{W}(p^k)$ such that $f(x^k, p^k) \leqslant \alpha^k$ and $||x^k|| \to \infty$ as $k \to \infty$.

By the assumption $\alpha^k = \mathbb{V}(p^k)$ and by the nonemptiness and boundedness, let $x^k \in \mathcal{W}(p^k)$ such that $f(x^k, p^k) = \alpha^k$ and $\{x^k\}$ is bounded (by the uniform compactness of $\mathcal{O}(p)$ around $p^*$), which is a contradiction. Therefore inf-compactness holds around $p^*$. □

## 11.4 Subdifferential of the value function: enhanced Fritz John approach

This section is devoted to the analysis of limiting and singular mordukhovich subdifferential of value function using the enhanced optimality conditions. These enhanced optimality conditions are computationally significant over traditional in the sense that it provides information about a neighborhood of extremal points in terms of converging sequence towards it. Also they remain valid under weaker constraint qualifications, enable more accurate multiplier identification, and helps in sharper sensitivity analysis. This leads to better algorithmic performance, especially in nonsmooth or degenerate settings, and supports the design of more robust and reliable numerical methods for solving MPVCs.

Although a perturbed MPVC has been analyzed before in Izmailov and Solodov (2009) with smoothness assumptions. In Izmailov and Solodov (2009), Izmailov made a sensitivity analysis under MPVC-MFCQ and MPVC-LICQ. In this section, we are generalizing the work of Guo et al. (2014) for parametric MPVC with nonsmooth assumptions for the first time. Additionally, sharper upper bound for the limiting and singular subdifferentials based on the enhanced Fritz John condition under some weaker restrictions like restricted inf-compactness and quasinormality conditions are also estimated. For this section, we assume that all the involved functions $\{f, g, h, G, H\}$ are locally Lipschitz around the concerned point. First we write here the partial subdifferential of the lagrangian as

$$\partial_x \mathcal{L}^r(x, p, \lambda, \mu, \eta^G, \eta^H) = r\partial_x f(x, p) + \sum_{i=1}^{l} \lambda_i \partial_x g_i(x, p) + \sum_{i=1}^{m} \partial_x(\mu_i h_i)(x, p)$$

$$+\sum_{i=1}^{n}\partial_x(\eta_i^G G_i)(x,p)-\sum_{i=1}^{n}\partial_x(\eta_i^H H_i)(x,p);$$

$p-MPVC$ can be rewritten in more compact form

$$\begin{aligned} p-MPVC\ ; \quad & \min_{x\in C} f(x,p),\\ \text{such that} \quad & F(x,p)\in\Delta, \end{aligned} \tag{11.15}$$

where

$$F(x,p)=\left\{\begin{array}{c} g(x,p)\\ h(x,p)\\ G(x,p)\\ H(x,p)\end{array}\right\}\ ; \quad \Delta=\mathbb{R}_-^l\times\{0\}^m\times\Omega^n,$$

where $\Omega=\{(a,b)\mid b\geqslant 0,\ ab\leqslant 0\}$.

### 11.4.1 Subdifferential via enhanced Fritz John M-multipliers

In this subsection, an enhanced Fritz John-type optimality condition in terms of limiting subdifferential for nonsmooth case is derived. To obtain the main result, we use an established enhanced optimality condition from literature Guo et al. (2013b, Theorem 3.1, Corollaries 3.1, 3.2) for mathematical program with geometric constraints in Banach spaces.

**Theorem 11.4.1.** *Let $x^*\in\mathcal{W}(p^*)$ be a local minimizer of $p-MPVC$, then $\exists\ 0\neq(r,\lambda^*,\mu^*,\eta^{G^*},\eta^{H^*})$ with $r\geqslant 0$ such that*

1. $0\in r\partial_x f(x^*,p^*)+\sum_{i=1}^{l}\lambda_i^*\partial_x g_i(x^*,p^*)+\sum_{i=1}^{m}\partial_x(\mu_i^*h_i)(x^*,p^*)+\sum_{i=1}^{n}\partial_x(\eta_i^{G^*}G_i)(x^*,p^*)-\sum_{i=1}^{n}\partial_x(\eta_i^{H^*}H_i)(x^*,p^*)+\mathcal{N}_C(x^*)$,
   *or*
   $0\in\partial_x\mathcal{L}^r(x^*,p^*,\lambda^*,\mu^*,\eta^{G^*},\eta^{H^*})+\mathcal{N}_C(x^*)$,
   *where*

$$\begin{aligned} &\lambda_i^*\geqslant 0,\lambda_i^*=0\ ,\ \forall\, i\notin I_g^*,\\ &\mu_i^*\text{ is free},\forall\, i=1,\dots,m,\\ &\eta_i^{G^*}=0,\forall\, i\in I_{+-}^*\cup I_{0-}^*\cup I_{0+}^*,\\ &\eta_i^{G^*}\geqslant 0,\forall\, i\in I_{+0}^*\cup I_{00}^*,\\ &\eta_i^{H^*}=0,\forall\, i\in I_{+-}^*\cup I_{+0}^*,\\ &\eta_i^{H^*}\geqslant 0,\forall\, i\in I_{0-}^*,\\ &\eta_i^{G^*}\geqslant 0\text{ and }\eta_i^{G^*}\eta_i^{H^*}=0,\forall\, i\in I_{00}^*. \end{aligned} \tag{11.16}$$

**2.** *There is a sequence* $\{x^k\} \subset \mathcal{C}$ *converging to* $x^*$ *such that, for each* $k$

$$\begin{aligned} \lambda_i^* > 0 &\Rightarrow \lambda_i^* g_i(x^k, p^*) > 0, \\ \mu_i^* > 0 &\Rightarrow \mu_i^* h_i(x^k, p^*) > 0, \\ \eta_i^{G^*} \neq 0 &\Rightarrow \eta_i^{G^*} G_i(x^k, p^*) > 0, \\ \eta_i^{H^*} \neq 0 &\Rightarrow \eta_i^{H^*} H_i(x^k, p^*) < 0, \end{aligned} \tag{11.17}$$

*and* $\{g, h, G, H\}$ *are all differentiable with respect to* $x^*$ *at* $(x^k, p^*)$.

*Proof.* Since $x^* \in \mathcal{O}(p^*)$ is a local minimizer for problem (11.15), therefore by Guo et al. (2013b, Corollary 3.3), there exist a scalar $r \geqslant 0$ and a vector $\eta^*$ not all zero, such that the following conditions hold:

**(i)** $0 \in r\partial_x f(x^*, p^*) + \sum_{i=1}^{l+m+2n} \partial_x \langle \eta^*, e_i \rangle \langle F, e_i \rangle (x^*, p^*) + \mathcal{N}_{\mathcal{C}}(x^*)$.
**(ii)** $\eta^* \in \mathcal{N}_\Delta(F(x^*, p^*))$.
**(iii)** There is a sequence $\{x^k, y^k, \eta^k\} \subset \mathcal{C} \times \Delta \times \mathbb{R}^{l+m+2n}$ converges to $(x^*, F(x^*), \eta^*)$ such that for all $k$

$$f(x^k, p^*) < f(x^*, p^*),$$

$$\eta^k \in \mathcal{N}_\Delta(y^k), \tag{11.18}$$

$$\langle \eta^*, e_i \rangle \neq 0 \Rightarrow \langle \eta^*, e_i \rangle \langle F(x^k, p^*) - y^k, e_i \rangle > 0, \tag{11.19}$$

where $\{e_i \mid i = 1, ..., l + m + 2n\}$ is the orthogonal basis of $\mathbb{R}^{l+m+2n}$.

Let $\eta = (\lambda, \mu, \eta^G, \eta^H)$, and let $y = (y^1, y^2, y^3, y^4)$ with appropriate dimensional components corresponding to $(g, h, G, H)$. Now by the explicit expression of limiting normal cone $\mathcal{N}_\Delta$, since $\Delta = \mathbb{R}^l_- \times \{0\}^m \times \Omega^n$, therefore

$$\begin{aligned} \mathcal{N}_\Delta(F(x^*, p^*)) &= \mathcal{N}_{\mathbb{R}^l_-}(g(x^*, p^*)) \times \mathcal{N}_{\{0\}^m}(h(x^*, p^*)) \times \mathcal{N}_{\mathcal{C}^n}(G(x^*, p^*), H(x^*, p^*)) \\ &= \mathbb{R}^l_+ \times \mathbb{R} \times (d_1, d_2), \end{aligned}$$

where

$$\begin{aligned} d_1 = d_2 = 0, &\text{ if } H > 0,\ G < 0, \\ d_1 = 0,\ d_2 \geqslant 0, &\text{ if } H > 0,\ G = 0, \\ d_2 \geqslant 0,\ d_1 d_2 = 0, &\text{ if } H = 0 = G, \\ d_1 \leqslant 0,\ d_2 = 0, &\text{ if } H = 0,\ G < 0, \\ d_1 \in \mathbb{R},\ d_2 = 0, &\text{ if } H = 0,\ G > 0. \end{aligned}$$

Therefore using conditions (i), (ii), and expression of $\mathcal{N}_\Delta(F(x^*, p^*))$, we have

$$0 \in r\partial_x f(x^*, p^*) + \sum_{i=1}^{l} \lambda_i^* \partial_x g_i(x^*, p^*) + \sum_{i=1}^{m} \partial_x(\mu_i^* h_i)(x^*, p^*) + \sum_{i=1}^{n} \partial_x(\eta_i^{G^*} G_i)(x^*, p^*)$$

$$+\sum_{i=1}^{n}\partial_x(\eta_i^{H^*}H_i)(x^*,p^*)+\mathcal{N}_C(x^*),$$

and the multipliers $\lambda^*$, $\mu^*$, $\eta^{G^*}$ satisfy the restrictions in (11.16) and

$$\begin{aligned}\eta_i^{H^*}&=0,\forall\, i\in I_{+-}^*\cup I_{+0}^*\\ \eta_i^{H^*}&\leqslant 0,\forall\, i\in I_{0-}^*\\ \eta_i^{H^*}&\text{ is free},\forall\, i\in I_{0+}^*.\end{aligned}$$

If we replace $\eta_i^{H^*}$ with $-\eta_i^{H^*}$, then we obtain the (11.18) result directly.

Next we prove the second result of this theorem. Since we have $\{x^k\}\subseteq C$, $\{y^{1,k}\}\subseteq\mathbb{R}_-^l$, $\{y^{2,k}\}\subseteq\{0\}^m$, $\{(y^{3,k},y^{4,k})\}\subseteq\Omega^m$, and $\eta^k=\{(\lambda^k,\mu^k,\eta^{G^k},\eta^{H^k})\}\in\mathbb{R}^{l+m+2n}$, so from (iii), we have

$$\begin{aligned}\lambda_i^*>0&\Rightarrow\lambda_i^*(g_i(x^k,p^*)-y_i^{1,k})>0\\ \mu_i^*>0&\Rightarrow\mu_i^*(h_i(x^k,p^*)-y_i^{2,k})>0\\ \eta_i^{G^*}\neq 0&\Rightarrow\eta_i^{G^*}(G_i(x^k,p^*)-y_i^{3,k})>0\\ \eta_i^{H^*}\neq 0&\Rightarrow\eta_i^{H^*}(H_i(x^k,p^*)-y_i^{4,k})>0\end{aligned}$$

According to the nature of $\eta_i^{H^*}$ as analyzed before, we can replace it with $-\eta_i^{H^*}$, so

$$\eta_i^{H^*}\neq 0\;\Rightarrow\;\eta_i^{H^*}(H_i(x^k,p^*)-y_i^{4,k})<0.$$

Next we have to show $\{y_i^{1,k},y_i^{2,k},y_i^{3,k},y_i^{4,k}\}$ is zero for each sufficiently large $k$.

1. $y_i^{2,k}$ is obviously 0.
2. $y_i^{1,k}\leqslant 0$, because $y_i^{1,k}\in\mathbb{R}_-^l$. If $y_i^{1,k}<0$ for a subsequence $K_1\subseteq\{1,...m_1\}$, then $\mathcal{N}_{\mathbb{R}_-}(y_i^{1,k})=\{0\}$ for all $k\in K_1$. This implies $\lambda_i^k=0$, therefore $\lambda_i^k\to\lambda_i^*=0$ as $k\to\infty$ for all $k\in K_1$. Which contradicts $\lambda_i^*>0$, therefore $y_i^{1,k}\not<0$, $\Rightarrow$ $y_i^{1,k}=0$.
3. $\{(y_i^{3,k},y_i^{4,k})\}\subseteq\Omega^n$, that is, $y_i^{4,k}\geqslant 0$ and $y_i^{3,k}y_i^{4,k}\leqslant 0$.
   - If $y_i^{4,k}>0$, this implies either $y_i^{3,k}<0$ or $y_i^{3,k}=0$. Suppose $\{y_i^{3,k}\}<0$ for a subsequence $K_2\subseteq\{1,...,m\}$, then by the expression of $\mathcal{N}_\Omega$, we have

     $$\eta_i^{G^k}\to\eta_i^{G^*}=0,$$

     which is a contradiction of $\eta_i^{G^*}\neq 0$.
     If $\{y_i^{3,k}\}=0$ for a subsequence $K_3\subseteq\{1,...,m\}$, then we have

     $$\begin{aligned}\text{either }\;&\eta_i^{H^k}=0,\,\eta_i^{G^k}\geqslant 0\\ \text{or }\;&\eta_i^{G^k}\geqslant 0,\,\eta_i^{H^k}\eta_i^{G^k}=0,\end{aligned}$$

that is, either $\eta_i^{Gk} = 0$ or $\eta_i^{Hk} = 0$, hence

$$\text{either } \eta_i^{Gk} \to \eta_i^{G^*} = 0$$
$$\text{or } \eta_i^{Hk} \to \eta_i^{H^*} = 0,$$

which is a contradiction. Hence $y_i^{4,k} = 0$.

- If $\{y_i^{3,k}\} > 0$, then $y_i^{4,k} = 0$ for a subsequence $K_4 \subseteq \{1, ..., m\}$. This implies that either $\eta_i^{G^*} = 0$ or $\eta_i^{G^*} \geqslant 0$, $\eta_i^{G^*}\eta_i^{H^*} = 0$, that is, either $\eta_i^{G^*} = 0$ or $\eta_i^{H^*} = 0$, which is again a contradiction, because both of these multipliers are non zero. Hence $\{y_i^{3,k}\} \not> 0$.
- If $\{y_i^{3,k}\} < 0$, then either $\{y_i^{4,k}\} > 0$ or $\{y_i^{4,k}\} = 0$.
  **(a).** If $y_i^{4,k} > 0$, this implies $\eta_i^{Gk} \to \eta_i^{G^*} = 0$, which is a contradiction.
  **(b).** If $y_i^{4,k} = 0$, this implies that either $\eta_i^{G^*} = 0$ or $\eta_i^{H^*} = 0$, which is again a contradiction, hence $\{y_i^{3,k}\} \not< 0$. So by these last two points $y_i^{3,k} = 0$.

Finally, $(y_i^{1,k}, y_i^{2,k}, y_i^{3,k}, y_i^{4,k}) = (0, 0, 0, 0)$, i.e., all the required conditions of (11.19) holds. Now the differentiability of $(g, h, G, H)$ with respect to $x$ at $(x^k, p^*)$, can be represented similarly as in Guo et al. (2014, Theorem 4.1). □

Based on the above enhanced optimality condition, we define the following set of multipliers:

**Definition 11.4.1.** For given $r \geqslant 0$ and $x^* \in \mathcal{W}(p^*)$, $\mathcal{M}_M^r(x^*, p^*)$ denotes the set of vectors $(\lambda, \mu, \eta^G, \eta^H, \zeta)$ such that

**(i)** $0 \in \partial_{(x,p)}\mathscr{L}^r(x^*, p^*, \lambda, \mu, \eta^G, \eta^H) - (0, \zeta) + \mathcal{N}_C(x^*) \times \{0\}$
**(ii)**

$$\begin{aligned}
\lambda_i \geqslant 0, \ \ \lambda_i = 0, \forall\, i \notin I_g^* \\
\eta_i^G = 0, \forall\, i \in I_{+-}^* \cup I_{0-}^* \cup I_{0+}^* \\
\eta_i^G \geqslant 0, \forall\, i \in I_{+0}^* \cup I_{00}^* \\
\eta_i^H = 0, \forall\, i \in I_{+-}^* \cup I_{+0}^* \\
\eta_i^H \geqslant 0, \forall\, i \in I_{0-}^* \\
\eta_i^G \geqslant 0 \text{ and } \eta_i^G\eta_i^H = 0, i \in I_{00}^*
\end{aligned}$$

**(iii)** There exist a sequence $\{(x^k, p^k)\} \subseteq C \times \mathbb{R}^{n_2}$ converging to $(x^*, p^*)$ such that, for each $k$,

$$\begin{aligned}
\lambda_i > 0 &\Rightarrow \lambda_i g_i(x^k, p^k) > 0 \\
\mu_j > 0 &\Rightarrow \mu_j h_j(x^k, p^k) > 0 \\
\eta_l^G \neq 0 &\Rightarrow \eta_l^G G_l(x^k, p^k) > 0 \\
\eta_l^H \neq 0 &\Rightarrow \eta_l^H H_l(x^k, p^k) < 0
\end{aligned}$$

and $\{g, h, G, H\}$ are all differentiable at $(x^k, p^k)$.

Now to estimate the upper bounds, we define following constraint qualification:

**Definition 11.4.2.** Let $x^* \in \mathcal{W}(p^*)$, then MPVC-M-quasinormality holds at $x^* \in \mathcal{W}(p^*)$ if $(\lambda, \mu, \eta^G, \eta^H, 0) \in \mathcal{M}_M^0(x^*, p^*) \implies (\lambda, \mu, \eta^G, \eta^H) = \{0\}$.

**Lemma 11.4.1.** *Let $x^* \in \mathcal{W}(p^*)$ and $(x^*, p^*)$ is MPVC-M-quasinormal for the constraint region $gph\mathcal{W}$, then the following upper estimates for the Fréchet subdifferential of value function holds:*

$$\hat{\partial}\mathbb{V}(p^*) \subseteq \{\zeta \mid (\lambda, \mu, \eta^G, \eta^H, \zeta) \in \mathcal{M}_M^1(x^*, p^*)\}.$$

*Proof.* Let $\zeta \in \hat{\partial}\mathbb{V}(p^*)$, then by the definition of Fréchet subdifferential, for an arbitrary $\epsilon > 0$, $\exists$ a $\delta_\epsilon > 0$ such that

$$\mathbb{V}(p) - \mathbb{V}(p^*) \geqslant \zeta^T(p - p^*) - \epsilon||p - p^*|| \ , \quad \forall\, p \in \mathbb{B}_{\delta_\epsilon}(p^*),$$

we have $f(x, p) \geqslant \mathbb{V}(p)$ for all $x \in \mathcal{W}(p)$, and hence

$$f(x, p) - \zeta^T(p - p^*) + \epsilon||p - p^*|| \geqslant f(x^*, p^*) \ , \quad \forall\, x \in \mathcal{W}(p) \text{ and } \forall\, p \in \mathbb{B}_{\delta_\epsilon}(p^*).$$

Thus $(x^*, p^*)$ is a locally optimal solution of the optimization problem:

$$\begin{aligned} &\min f(x, p) - \zeta^T(p - p^*) + \epsilon||p - p^*|| \ = \ \mathcal{F}(x, p) \\ &\text{such that} \quad g(x, p) \leqslant 0\,, \quad h(x, p) = 0 \\ &\qquad\qquad\quad H(x, p) \geqslant 0\,, \quad G(x, p)H(x, p) \leqslant 0 \\ &\qquad\qquad\qquad\qquad (x, p) \in \mathcal{C} \times \mathbb{R}^{n_2}. \end{aligned}$$

By Theorem 11.4.1 and the MPVC-M-quasinormality assumption, there exist a vector $(\lambda, \mu, \eta^G, \eta^H)$ such that the following conditions hold:

$$\begin{aligned} 0 \in r\partial_{(x,p)}\mathcal{F}(x^*, p^*) &+ \sum_{i=1}^{l} \partial_{(x,p)}\lambda_i g_i(x^*, p^*) + \sum_{i=1}^{m} \partial_{(x,p)}\mu_i h_i(x^*, p^*) + \sum_{i=1}^{n} \partial_{(x,p)}\eta_i^G G_i(x^*, p^*) \\ &- \sum_{i=1}^{n} \partial_{(x,p)}\eta_i^H H_i(x^*, p^*) + \mathcal{N}_{\mathcal{C}\times\mathbb{R}^{n_2}}(x^*, p^*). \end{aligned}$$

Since MPVC-M-quasinormality holds, therefore $r > 0$, this implies

$$\begin{aligned} 0 \in \partial_{(x,p)} f(x^*, p^*) &+ \sum_{i=1}^{l} \partial_{(x,p)}\lambda_i g_i(x^*, p^*) + \sum_{i=1}^{m} \partial_{(x,p)}\mu_i h_i(x^*, p^*) + \sum_{i=1}^{n} \partial_{(x,p)}\eta_i^G G_i(x^*, p^*) \\ &- \sum_{i=1}^{n} \partial_{(x,p)}\eta_i^H H_i(x^*, p^*) - (0, \zeta) + \mathcal{N}_{\mathcal{C}}(x^*) \times \{0\} + \epsilon(0, \bar{\mathbb{B}})^T \\ \implies 0 \in \partial_{(x,p)}&\mathcal{L}^1(x^*, p^*, \lambda, \mu, \eta^G, \eta^H) - (0, \zeta) + \mathcal{N}_{\mathcal{C}}(x^*) \times \{0\} + \epsilon(0, \bar{\mathbb{B}})^T, \end{aligned}$$

where multiplier $(\lambda, \mu, \eta^G, \eta^H)$ satisfies the restrictions in Eqs. (11.16) and (11.17), and the functions $\{g, h, G, H\}$ are all differentiable at $(x^k, p^k)$. Hence required result holds. □

Next we derive the tighter upper estimates for the limiting and singular subdifferentials in terms of MPVC-M-quasinormal multipliers. Since the proof is same as in Guo et al. (2013b, Proposition 6.2), Ye and Zhang (2013, Lemma 1), Guo et al. (2014, Theorem 4.7, Corollary 4.8), respectively, we are giving only statement here.

**Lemma 11.4.2.** *If a vector $(x^*, p^*)$ is MPVC-M-quasinormal for the constraint region $gph\mathcal{W}$, then there is a neighborhood $V$ of $(x^*, p^*)$ such that all vectors $(x, p) \in gph\mathcal{W} \cap V$ are MPVC-M-quasinormal.*

**Theorem 11.4.2.** *Assume that the restricted inf-compactness holds for $p - MPVC$ around $p^*$, then the value function $\mathbb{V}(p)$ is lower semicontinuous at $p^*$. Suppose further that for each $x^* \in \mathcal{O}(p^*)$, $(x^*, p^*)$ is MPVC M-quasinormal for the constraint region $gph\mathcal{W}$. Then*

$$\partial\mathbb{V}(p^*) \subseteq \bigcup_{x^* \in \mathcal{O}(p^*)} \{\zeta \mid (\lambda, \mu, \eta^G, \eta^H, \zeta) \in \mathcal{M}_M^1(x^*, p^*)\}$$

$$\partial^\infty\mathbb{V}(p^*) \subseteq \bigcup_{x^* \in \mathcal{O}(p^*)} \{\zeta \mid (\lambda, \mu, \eta^G, \eta^H, \zeta) \in \mathcal{M}_M^0(x^*, p^*)\}$$

**Corollary 11.4.1.** *Assume that the restricted inf-compactness holds for $p - MPVC$ around $p^*$. Suppose that for each $x^* \in \mathcal{O}(p^*)$, $(p^*, x^*)$ is MPVC M-quasinormal for the constraint region $gph\mathcal{W}$. If*

$$\bigcup_{x^* \in \mathcal{O}(p^*)} \{\zeta \mid (\lambda, \mu, \eta^G, \eta^H, \zeta) \in \mathcal{M}_M^0(x^*, p^*)\} = \{0\},$$

*then the value function $\mathbb{V}$ is Lipschitzian around $p^*$ with*

$$\emptyset \neq \partial\mathbb{V}(p^*) \subseteq \bigcup_{x^* \in \mathcal{O}(p^*)} \{\zeta \mid (\lambda, \mu, \eta^G, \eta^H, \zeta) \in \mathcal{M}_M^1(x^*, p^*)\}.$$

*In addition to the above assumption, if*

$$\bigcup_{x^* \in \mathcal{O}(p^*)} \{\zeta \mid (\lambda, \mu, \eta^G, \eta^H, \zeta) \in \mathcal{M}_M^1(x^*, p^*)\} = \{\zeta\},$$

*then $\mathbb{V}$ is strictly differentiable at $p^*$ and $\nabla\mathbb{V}(p^*) = \zeta$.*

Now we illustrate the above Corollary in following example:

**Example 11.4.1.**

$$\min\ x_1^2 + x_2^2$$
$$x_1 \leqslant 0$$

$$x_2 + p \geqslant 0$$
$$(x_2 - x_1 + p)(x_2 + p) \leqslant 0;$$

clearly $(0, 0)$ is the optimal solution for parameter $p > 0$. We can find the singular and nonsingular set of multipliers by solving the following system:

$$\lambda \begin{pmatrix} 1 \\ 0 \end{pmatrix} + \eta^G \begin{pmatrix} -1 \\ 1 \end{pmatrix} - \eta^H \begin{pmatrix} 0 \\ 1 \end{pmatrix} = \begin{pmatrix} 0 \\ 0 \end{pmatrix},$$

then $\lambda = \eta^G = \eta^H$ and $\mathcal{M}^r_M(x^*, p^*) = \{\lambda(1, 1, 1) \mid \lambda \geqslant 0\}$ are the sets of singular and nonsingular multipliers for $r = 0, 1$, respectively. Now there is a sequence $\{(x^k, p^k)\} \to (x^*, p^*)$ and any multiplier $\lambda(1, 1, 1)$ with $\lambda > 0$; we have

$$\lambda x_1^k > 0\,,\ \lambda(x_2^k - x_1^k + p^k) > 0\,,\ \lambda(x_2^k + p^k) < 0.$$

It is possible only for $\lambda = 0$. Then clearly set of enhanced singular and nonsingular M-multipliers are $\mathcal{M}^r_M(x^*, p^*) = \{(0, 0, 0)\}$ for $r = 0, 1$, which are contained in $\mathcal{M}^r_M(x^*, p^*)$. Then the MPVC-M-quasinormality holds at $x^* \in \mathcal{W}_p$. Now we apply Corollary 11.4.1 to show the strict differentiability of value function, for this we need value of $\zeta$. We use Lemma 11.4.1 to calculate $\zeta$.

$$\begin{pmatrix} 0 \\ 0 \\ \zeta \end{pmatrix} = \begin{pmatrix} 0 \\ 0 \\ 0 \end{pmatrix} + \lambda \begin{pmatrix} 1 \\ 0 \\ o \end{pmatrix} + \eta^G \begin{pmatrix} -1 \\ 1 \\ 1 \end{pmatrix} - \eta^H \begin{pmatrix} 0 \\ 1 \\ 1 \end{pmatrix}.$$

$$\zeta = \eta^G - \eta^H.$$

So by the Corollary 11.4.1, we have

$$\{\lambda - \eta^H \mid (\lambda, \eta^G, \eta^H) \in \mathcal{M}^0_M(x^*, p^*)\} = \{\lambda - \eta^H \mid (\lambda, \eta^G, \eta^H) \in \mathcal{M}^1_M(x^*, p^*)\} = \{0\}.$$

Therefore the value function $\mathbb{V}$ is strictly differentiable with $\nabla\mathbb{V}(p^*) = 0$.

### 11.4.2 Subdifferential via enhanced Fritz John C-multiplier

This subsection is again devoted to the study of subdifferential of the value function in terms of enhanced C-multiplier. First we define the enhanced C-multiplier type Fritz John optimality condition. For this we need the following auxiliary lemmas:

**Lemma 11.4.3.** (Ye and Zhang, 2013, Proposition 2)

**(i)** *Let $f : \mathbb{R}^n \to \mathbb{R}$ be Lipschitz near a point $x_0$ and $g : \mathbb{R}^n \to \mathbb{R} \cup \{+\infty\}$ be lower semicontinuous and finite at $x_0$. Let $\alpha, \beta$ be nonnegative scalars. Then*

$$\partial(\alpha f + \beta g)(x_0) \subset \alpha\partial f(x_0) + \beta\partial g(x_0).$$

**(ii)** *Let $f : \mathbb{R}^n \to \mathbb{R} \cup \{+\infty\}$ be lower semicontinuous near $x_0$ and $g : \mathbb{R}^n \to \mathbb{R}$ be Lipschitz near $x_0$. Assume that $\partial^F g(x_0) \neq \emptyset$ for all $x$ near $x_0$. Then*

$$\partial(f-g)(x_0) \subset \partial f(x_0) - \partial g(x_0).$$

**Lemma 11.4.4.** (see Mordukhovich and Shao, 1996, Theorem 7.5, 7.6) *Let*

$$g(x) = \max\{g_i(x) \mid i = 1, ..., m\},$$

*where $g_i : \mathbb{R}^n \to \mathbb{R} \cup \{\infty\}$, and $I(\bar{x}) = \{i \mid g_i(\bar{x}) = g(\bar{x})\}$. Let $g_i, i = 1, ..., m$, be Lipschitzian around $\bar{x}$. Then $g$ is Lipschitzian around $\bar{x}$ and*

$$\partial g(\bar{x}) \subseteq \bigcup \left\{ \partial \left( \sum_{i \in I(\bar{x})} \lambda_i g_i \right)(\bar{x}) \mid \sum_{i \in I(\bar{x})} \lambda_i = 1, \lambda_i \geqslant 0, i \in I(\bar{x}) \right\}.$$

*Let $f(x) = \min\{f_i(x) \mid i = 1, ..., m\}$, where $f_i : \mathbb{R}^n \to \mathbb{R} \cup \{\infty\}$, and $J(\bar{x}) = \{i \mid f_i(\bar{x}) = f(\bar{x})\}$. Assume that $f_i$ is lower semicontinuous near $\bar{x}$ for $i \in J(\bar{x})$ and lower semicontinuous at $\bar{x}$ for $i \notin J(\bar{x})$. Then $f$ is lower semicontinuous near $\bar{x}$ and*

$$\partial f(\bar{x}) \subseteq \bigcup \{\partial f_i(\bar{x}) \mid i \in J(\bar{x})\}.$$

**Theorem 11.4.3.** *If $x^*$ is a local minimizer of $p - MPVC$, then $\exists\ (r, \lambda^*, \mu^*, \eta^{G^*}, \eta^{H^*}) \neq 0$ with $r \geqslant 0$ such that*

1. $0 \in \partial_x \mathscr{L}^r(x^*, p^*, \lambda^*, \mu^*, \eta^{G^*}, \eta^{H^*}) + \mathcal{N}_C(x^*)$
   or
   $0 \in r\partial_x f(x^*, p^*) + \sum_{i=1}^{l} \lambda_i^* \partial_x g_i(x^*, p^*) + \sum_{i=1}^{m} \partial_x(\mu_i^* h_i)(x^*, p^*) + \sum_{i=1}^{n} \partial_x(\eta_i^{G^*} G_i)(x^*, p^*) - \sum_{i=1}^{n} \partial_x(\eta_i^{H^*} H_i)(x^*, p^*) + \mathcal{N}_C(x^*)$
2. $\lambda_i^* \geqslant 0, \quad \lambda_i^* = 0\ \forall\, i \notin I_g^*$
   $\eta_i^{H^*} = 0, \forall\, i \in I_+^*, \eta_i^{G^*} = 0, \forall\, i \in I_{0+}^* \cup I_{0-}^* \cup I_{+-}^*$
   *that is* $\eta_i^G G_i(x^*, p^*) = 0$ *and* $\eta_i^H H_i(x^*, p^*) = 0$
3. *There is a sequence $\{x^k\} \subseteq C$ converging to $x^*$ such that for each $k$,*

$$\lambda_i^* > 0 \implies \lambda_i^* g_i(x^k, p^*) > 0$$
$$\mu_j^* \neq 0 \implies \mu_j^* h_j(x^k, p^*) > 0$$
$$\eta_i^{G^*} \neq 0 \implies \eta_i^{G^*} \max\{0, -H_l(x^k, p^*), \min(G_l(x^k, p^*), H_l(x^k, p^*))\} > 0$$
$$\eta_i^{H^*} \neq 0 \implies \eta_i^{H^*} \max\{0, -H_l(x^k, p^*), \min(G_l(x^k, p^*), H_l(x^k, p^*))\} > 0$$

*and $h(., p^*), g(., p^*), \max\{0, -H_l, \min(g_l, H_l)\}(., p^*)$ are all differentiable at $x^k$.*

*Proof.* Feasible region of $p - MPVC$ can be rewritten as

$$\mathcal{W}(p) = \{x \in C \mid g(x,p) \leqslant 0, h(x,p) = 0, \max\{0, -H(x,p), \min(G(x,p), H(x,p))\} = 0\}.$$

Now by Guo et al. (2013b, Corollary 3.4) and Ye and Zhang (2013, Theorem 3.1), ∃ nonzero vectors $(r, \lambda, \mu, \xi)$ with $r \geqslant 0$ such that the following holds:

**1.**

$$0 \in r\partial_x f(x^*, p^*) + \sum_{i=1}^{l} \lambda_i \partial_x g_i(x^*, p^*) + \sum_{j=1}^{m} \partial_x(\mu_j h_j)(x^*, p^*)$$
$$+ \sum_{l=1}^{n} \partial_x[\xi_l \max(0, -H_l, \min(G_l, H_l))](x^*, p^*) + \mathcal{N}_C(x^*).$$

**2.** There is a sequence $\{x^k\} \subseteq C$ converging to $x^*$ such that for all $k$,

$$\begin{aligned}
\lambda_i > 0 &\Longrightarrow \lambda_i g_i(x^k, p^*) > 0\\
\mu_j \neq 0 &\Longrightarrow \mu_j h_j(x^k, p^*) > 0\\
\xi_l \neq 0 &\Longrightarrow \xi_l \max\{0, -H_l(x^k, p^*), \min(G_l(x^k, p^*), H_l(x^k, p^*))\} > 0
\end{aligned}$$

and $h(., p^*), g(., p^*), \max\{0, -H_l, \min(G_l, H_l)\}(., p^*)$ are all differentiable at $x^k$.

Next we proceed to calculate the subdifferential of $\xi_l \max(0, -H_l, \min(G_l, H_l))(x^*, p^*)$.

**Step I.** If $\xi_l \geqslant 0$, then by using Lemma 11.4.3 and 11.4.4, we have

$$\partial_x[\xi_l \max(0, -H_l, \min(G_l, H_l))](x^*, p^*) \subseteq \partial_x \max(0, -\xi_l H_l, \xi_l \min(G_l, H_l))(x^*, p^*)$$
$$= \begin{cases} \partial_x(0)(x^*, p^*)\,; & i \in I^*_{+-}\\ \alpha_1 \partial_x(\xi_l G_l)(x^*, p^*)\,; & i \in I^*_{+0}\\ \alpha_1 \partial_x(-\xi_l H_l)(x^*, p^*)\,; & i \in I^*_{0-}\\ (\alpha_1 \partial_x(-\xi_l H_l) + \alpha_2 \partial_x(\xi_l H_l))(x^*, p^*)\,; & i \in I^*_{0+}\\ [(\alpha_1 \partial_x(-\xi_l H_l) + \alpha_2(\partial_x(\xi_l H_l) \cup \partial_x(\xi_l G_l))](x^*, p^*)\,; & i \in I^*_{00} \end{cases}$$

where in each cases $\sum \alpha_i = 1$ and $0 \leqslant \alpha_i \leqslant 1$.

**Step II.** If $\xi_l < 0$, then

$$\partial_x[\xi_l \max(0, -H_l, \min(G_l, H_l))](x^*, p^*) = \partial_x \min(0, -\xi_l H_l, \xi_l \min(G_l, H_l))(x^*, p^*)$$
$$= \begin{cases} \partial_x(0)(x^*, p^*)\,; & i \in I^*_{+-}\\ \partial_x(\xi_l G_l)(x^*, p^*)\,; & i \in I^*_{+0}\\ \partial_x(-\xi_l H_l)(x^*, p^*)\,; & i \in I^*_{0-}\\ (\partial_x(-\xi_l H_l) \cup \partial_x(\xi_l H_l))(x^*, p^*)\,; & i \in I^*_{0+}\\ (\partial_x(-\xi_l H_l) \cup \partial_x(\xi_l H_l) \cup \partial_x(\xi_l G_l))(x^*, p^*)\,; & i \in I^*_{00} \end{cases}$$

Now by defining the multipliers $\eta^{G^*}, \eta^{H^*}$ in terms of $\{\xi, \alpha_i\}$, we can obtain all the desired restrictions. □

**Remark 11.4.1.** C-stationarity for MPEC has topological relevance (Dorsch et al., 2012), and the similar topological relevance for MPVC is T-stationarity. This result cannot be obtained by using clark subdifferential (pointed out in Hoheisel et al., 2012), as in the case of MPEC for C-stationarity (Scheel and Scholtes, 2000). But our C-stationarity is the adaptation of the result Guo et al. (2014, Theorem 4.11).

Similar to $\mathcal{M}^r_M(x^*, p^*)$, we can define the set of all enhanced Fritz John C-multipliers.

**Definition 11.4.3.** For $r \geqslant 0$, $\mathcal{M}^r_C(x^*, p^*)$ denote the set of vectors $(\lambda, \mu, \eta^G, \eta^H, \zeta)$ at $x^* \in \mathcal{W}(p^*)$ such that

1. $0 \in \partial_{(x,p)} \mathscr{L}^r(x^*, p^*, \lambda, \mu, \eta^G, \eta^H) - (0, \zeta) + \mathcal{N}_C(x^*) \times \{0\}$.
2. $\lambda_i \geqslant 0, \lambda_i = 0, (i \in I^*_g)$
   $\eta^H_i = 0, \forall\, i \in I^*_{+}, \eta^G_i = 0, \forall\, i \in I^*_{0+} \cup I^*_{0-} \cup I^*_{+-}$
   that is $\eta^G_i G_i(x^*, p^*) = 0$ and $\eta^H_i H_i(x^*, p^*) = 0$.
3. There is a sequence $\{x^k, p^k\} \subseteq C \times \mathbb{R}^{n_2}$ converging to $(x^*, p^*)$ such that for each $k$,

$$
\begin{aligned}
\lambda_i > 0 &\implies \lambda_i g_i(x^k, p^k) > 0\\
\mu_j \neq 0 &\implies \mu_j h_j(x^k, p^k) > 0\\
\eta^G_i \neq 0 &\implies \eta^G_i \max\{0, -H_l(x^k, p^k), \min(G_l(x^k, p^k), H_l(x^k, p^k))\} > 0\\
\eta^H_i \neq 0 &\implies \eta^H_i \max\{0, -H_l(x^k, p^k), \min(G_l(x^k, p^k), H_l(x^k, p^k))\} > 0
\end{aligned}
$$

and $g, h, \max(0, -H_l, \min(G_l, H_l))$ are all differentiable at $(x^k, p^k)$.

Similarly, we provide upper bounds for the limiting and singular subdifferential of value function in terms of MPVC-C-quasinormal multipliers. Again the proofs are skipped here as they are similar to Guo et al. (2014, Theorem 4.15, Corollary 4.16), respectively.

**Definition 11.4.4.** We say that the MPVC-C-quasinormality holds at $(x^*, p^*)$ for the region $gph\mathcal{W}$ if $(\lambda, \mu, \eta^G, \eta^H, 0) \in \mathcal{M}^0_C(x^*, p^*) \implies (\lambda, \mu, \eta^G, \eta^H) = 0$.

**Theorem 11.4.4.** *Assume that the restricted inf-compactness holds for $p - MPVC$ around $p^*$. Then the value function $\mathbb{V}(p)$ is lower semicontinuous at $p^*$. Suppose further that for each $x^* \in \mathcal{O}(p^*)$, $(x^*, p^*)$ is MPVC-C-quasinormal for the constraint region $gph\mathcal{W}$. Then*

$$
\partial \mathbb{V}(p^*) \subseteq \bigcup_{x^* \in \mathcal{O}(p^*)} \{\zeta \mid (\lambda, \mu, \eta^G, \eta^H, \zeta) \in \mathcal{M}^1_C(x^*, p^*)\}
$$

$$
\partial^\infty \mathbb{V}(p^*) \subseteq \bigcup_{x^* \in \mathcal{O}(p^*)} \{\zeta \mid (\lambda, \mu, \eta^G, \eta^H, \zeta) \in \mathcal{M}^0_C(x^*, p^*)\}
$$

**Corollary 11.4.2.** *Assume that the restricted inf-compactness holds for $p - MPVC$ around $p^*$. Suppose that for each $x^* \in \mathcal{O}(p^*)$, $(x^*, p^*)$ is MPVC-C-quasinormal for the constraint region $gph\mathcal{W}$. If*

$$
\bigcup_{x^* \in \mathcal{O}(p^*)} \{\zeta \mid (\lambda, \mu, \eta^G, \eta^H, \zeta) \in \mathcal{M}^0_C(x^*, p^*)\} = \{0\}
$$

*Then the value function* $\mathbb{V}$ *is Lipschitz around* $p^*$ *with*

$$\emptyset \neq \partial \mathbb{V}(p^*) \subseteq \bigcup_{x^* \in \mathcal{O}(p^*)} \{\zeta \mid (\lambda, \mu, \eta^G, \eta^H, \zeta) \in \mathcal{M}_C^1(x^*, p^*)\}.$$

*In addition to the above assumption, if*

$$\bigcup_{x^* \in \mathcal{O}(p^*)} \{\zeta \mid (\lambda, \mu, \eta^G, \eta^H, \zeta) \in \mathcal{M}_C^0(x^*, p^*)\} = \{\zeta\},$$

*then* $\mathbb{V}$ *is strictly differentiable at* $p^*$ *and* $\nabla \mathbb{V}(p^*) = \zeta$.

## 11.5 Conclusion

Under the MPVC-RCR regularity and the restricted inf-compactness assumption, we have established a min–max type formula for the directional derivative of the value function in every direction, for a class of parametric mathematical programs with vanishing constraints (p-MPVC). This formula is derived through a detailed analysis of several auxiliary results, particularly involving modified constraint regions, such as the tightened MPVC and piecewise MPVC, under the frameworks of MPVC-R and MPVC-RCR regularity.

In the second part of our study, we investigated a nonsmooth parametric MPVC and derived, for the first time, enhanced Fritz John-type optimality conditions utilizing the tools of limiting subdifferentials. This development leads us to introduce new notions of enhanced-M and enhanced-C multipliers, which in turn define MPVC-M and MPVC-C quasinormality, respectively. Furthermore, using these enhanced multipliers, we obtained sharper upper estimates for both, the limiting and singular subdifferentials of the value function.

## References

Achtziger, W., 2004. On optimality conditions and primal-dual methods for the detection of singular optima. In: Cinquini, C., Rovati, M., Venini, P., Nascimbene, R. (Eds.), Proceedings of the FifthWorld Congress of Structural and Multidisciplinary Optimization (WCSMO-5), vol. 73. Schönenfeld & Ziegler, Italy, pp. 1–6.

Achtziger, W., Kanzow, C., 2008. Mathematical programs with vanishing constraints: optimality conditions and constraint qualification. Mathematical Programming 114, 69–99.

Achtziger, W., Hoheisel, T., Kanzow, C., 2013. A smoothing-regularization approach to mathematical programs with vanishing constraints. Computational Optimization and Applications 55, 733–767.

Bendsøe, M.P., Sigmund, O., 2003. Topology Optimization Theory, Methods and Applications, 2nd ed. Springer, Heidelberg, Germany.

Bonnas, J.F., Shapiro, A., 2000. Perturbation Analysis of Optimization Problems. Springer, New York.

Cheng, G.D., Guo, X., 1997. $\epsilon$-relaxed approach in structural topology optimization. Structural Optimization 13, 258–266.

Dorsch, D., Shikhman, O., Stein, V., 2012. Mathematical programs with vanishing constraints: critical point theory. Journal of Global Optimization 52, 591–605.

Gauvin, J., 1979. The generalized gradient of a marginal function in mathematical programming 4 (4), 458–463.

Gauvin, J., Janin, R., 1990. Directional derivative of the value function in parametric optimization. Annals of Operations Research 27, 237–252.
Guo, L., Lin, G.H., Ye, J.J., Zhang, J., 2012. Stability analysis for parametric mathematical programs with geometric constraints and its applications. SIAM Journal on Optimization 22, 1151–1176.
Guo, L., Lin, G.H., Ye, J.J., 2013a. Second-order optimality conditions for mathematical programs with equilibrium constraint. Journal of Optimization Theory and Applications 158, 33–64.
Guo, L., Ye, J.J., Zhang, J., 2013b. Mathematical programs with geometric constraints in Banach spaces: enhanced optimality, exact penalty, and sensitivity. SIAM Journal on Optimization 23, 2295–2319.
Guo, L., Lin, G.H., Ye, J.J., Zhang, J., 2014. Sensitivity analysis of the value function for parametric mathematical programs with equilibrium constraints. SIAM Journal on Optimization 24 (3), 1206–1237.
Hogan, W., 1973. Directional derivative for extremal-value function with applications to the completely convex case. Operations Research 21 (1), 188–209.
Hoheisel, T., Kanzow, C., 2007. First and second-order optimality conditions for mathematical programs with vanishing constraints. Applications of Mathematics 52, 495–514.
Hoheisel, T., Kanzow, C., 2008. Stationary conditions for mathematical programs with vanishing constraints using weak constraint qualifications. Journal of Mathematical Analysis and Applications 337, 292–310.
Hoheisel, T., Kanzow, C., 2009. On the Abadie and Guignard constraint qualification for mathematical programs with vanishing constraints. Optimization 58, 431–448.
Hoheisel, T., Kanzow, C., Schwartz, A., 2012. Convergence of a local regularization approach for mathematical programmes with complementarity for vanishing constraints. Optimization Methods & Software 27 (3), 483–512.
Hu, Q., Wang, J., Chen, Y., Zhu, Z., 2017. On an $l_1$ exact penalty result for mathematical programs with vanishing constraints. Optimization Letters 11 (3), 641–653.
Hu, X.M., Ralph, D., 2002. A note on sensitivity of value function of mathematical programs with complementarity constraints. Mathematical Programming 93, 265–279.
Izmailov, A.F., Solodov, M.V., 2009. Mathematical programs with vanishing constraints: optimality conditions, sensitivity and a relaxation method. Journal of Optimization Theory and Applications 142, 501–532.
Jabr, R.A., 2012. Solution to economic dispatching with disjoint feasible regions via semidefinite programming. IEEE Transactions on Power Systems 27 (1), 572–573.
Janin, R., 1984. Directional derivative of the marginal function in nonlinear programming. Mathematical Programming Studies 21, 110–126.
Khare, A., Nath, T., 2019. Enhanced Fritz John stationarity, new constraint qualifications and local error bound for mathematical programs with vanishing constraints. Journal of Mathematical Analysis and Applications 472 (1), 1042–1077.
Lucet, Y., Ye, J.J., 2001. Sensitivity analysis of the value function for optimization problems with variational inequality constraints. SIAM Journal on Control and Optimization 40, 699–723.
Lucet, Y., Ye, J.J., 2002. Erratum: sensitivity analysis of the value function for optimization problems with variational inequality constraints. SIAM Journal on Control and Optimization 41, 1315–1319.
Michael, N.J., Kirches, C., Sager, S., 2013. On perspective functions and vanishing constraints in mixed integer nonlinear optimal control. In: Facets of Combinatorial Optimization, pp. 387–417.
Minchenko, L., Stakhovski, S., 2011a. Parametric nonlinear programming problems under the relaxed constant rank condition. SIAM Journal on Optimization 21 (1), 314–332.
Mordukhovich, B.S., Nam, N.M., 2005. Variational stability and marginal functions via generalized differentiation. Mathematics of Operations Research 30, 800–816.
Mordukhovich, B.S., Nam, N.M., 2009. Subgradients of marginal functions in parametric mathematical programming. Mathematical Programming 116, 369–396.
Mordukhovich, B.S., Shao, Y.H., 1996. Nonsmooth sequential analysis in Asplund space. Transactions of the American Mathematical Society 348, 1235–1280.
Nath, T., Khare, A., 2019. On an exact penalty result and new constraint qualifications for mathematical programs with vanishing constraints. Yugoslav Journal of Operations Research. Available at: http://www.yujor.fon.bg.ac.rs/index.php/yujor/article/view/637.
Rockafellar, R.T., 1984. Directional differentiability of the optimal value function in a nonlinear programming problem. Mathematical Programming Studies 21, 213–226.

Rockafellar, R.T., Wets, R.J.-B., 1998. Variational Analysis. Springer, Berlin.
Scheel, H., Scholtes, S., 2000. Mathematical programs with complementarity constraints: stationarity, optimality, and sensitivity. Mathematics of Operations Research 25 (1), 1–22.
Shapiro, A., 1995. Directional differentiability of the optimal value function in convex semi-infinite programming. Mathematical Programming 70, 149–157.
Ye, J.J., Zhang, J., 2013. Enhanced Karush-Kuhn-Tucker condition and weaker constraint qualifications. Mathematical Programming 139, 353–381.

# 12

# Optimizing IFS parameters of rational fractal linear splines with a memory-efficient quasi-Newton method for time series forecasting

Sana Abdulla and K. Mahipal Reddy
*Department of Mathematics, VIT-AP University, Amaravati, Andhra Pradesh, India*

## 12.1 Introduction

Fractal geometry, first introduced by Mandelbrot, laid the foundation for understanding self-similar and irregular structures in nature. Building upon this concept, the mathematical framework of fractals and their applications was extensively explored in Barnsley (2012). Interpolation, a classical technique used to estimate intermediate data points from a known set, takes on a new dimension when applied through fractal principles. The concept of fractal interpolation functions (FIFs), introduced in Barnsley and Harrington (1989), allows for the generation of new data points that not only interpolate the original dataset but also preserve self-similarity, as further elaborated in Barnsley (1986). A crucial aspect in constructing FIFs is the appropriate determination of the iterated function system (IFS) parameters, which directly influence the geometry, smoothness, and complexity of the resulting fractal curve. These parameters—particularly the vertical scaling factors—play a key role in shaping the behavior of the interpolation. Therefore the selection process must be carried out with careful analysis, as improper parameter choices can significantly degrade the accuracy and visual integrity of the interpolated curve.

Fractal interpolation has garnered significant attention due to its capability to reconstruct complex, self-similar data patterns. Nonetheless, identifying appropriate parameters for the iterated function system (IFS) remains a nontrivial task. Conventional methods often rely on heuristic or manual tuning, which is not only time-intensive but also prone to inconsistencies and subjectivity. To streamline this process, an explicit fractal interpolation algorithm utilizing affine transformations was proposed by the authors in Li and Li (2008), wherein the vertical scaling factor ($\alpha$) is estimated through a locally defined expression. Although this method enhances computational efficiency, it may lead to oversimplified representations, potentially overlooking fine fractal structures and causing deviations in the resulting interpolation. In another study, the authors in Raubitzek and Neubauer

Advanced Computational and Mathematical Approaches in Applied Differential Equations
https://doi.org/10.1016/B978-0-44-345743-2.00023-2

(2021) implemented a fractal interpolation-based framework to enhance neural network predictions for complex time series, incorporating the Hurst parameter into the model. However, since the Hurst parameter is highly sensitive to noise and fluctuations in data, relying on it for estimating $\alpha$ can introduce instability, ultimately affecting interpolation accuracy. To overcome such limitations, a rational cubic spline fractal interpolation function was introduced in Reddy and Vijender (2023), leveraging a rational IFS framework. This technique allows for adjustable $\alpha$ values and shape parameters within each subinterval, offering better adherence to boundary conditions and improved control over the interpolation process. Nonetheless, this added adaptability increases algorithmic complexity, necessitating precise tuning and contributing to longer computation times. In the realm of time series prediction, the authors in Zhai (2015) employed fractal theory combined with wavelet analysis to estimate $\alpha$ within an IFS context, showcasing the practicality of fractal methods in real-world data modeling. Similarly, in Zhai et al. (2011), a fractal interpolation algorithm tailored for self-affine signal reconstruction was developed, where genetic algorithms (GA) were used to optimize $\alpha$ for enhanced signal fidelity. Despite yielding high-quality results, the multivariable nature of the fitness function introduced additional complexity, making it harder to attain globally optimal solutions. Moreover, minor alterations in IFS parameters can significantly influence the behavior of fractal interpolation curves, challenging the stability and reproducibility of these models across varying datasets. To mitigate this, optimization-based approaches are increasingly adopted to automate the selection of suitable IFS parameters, aiming to enhance robustness. However, this parameter sensitivity continues to pose limitations on the broader applicability and generalization of fractal interpolation methods across diverse domains. Recent advances in fractal interpolation have demonstrated the potential of combining iterated function system (IFS)-based models with optimization and machine learning techniques. In our previous works, we explored the use of clustering algorithms like k-means and self-organizing maps for segmenting fractal garment patterns generated by IFS, enhancing pattern complexity analysis and classification accuracy (see Abdulla and K, 2024). Building upon this, we developed a fractal differential evolution (FDE) algorithm to optimize the parameters of rational fractal cubic (RFC) splines, significantly improving their adaptability and precision, and further integrated these optimized splines into a neural network framework for extrapolative modeling (see Abdulla and Reddy, 2025a). In another study, we extended this approach by incorporating modified optimization strategies—namely fractal Nelder–Mead, fractal particle swarm optimization, and fractal differential evolution—to fine-tune vertical scaling parameters in nonaffine fractal interpolation (NAFI), and evaluated their effectiveness in time-series prediction using neural networks, ARIMA, and Nu-SVR models (see Abdulla and Reddy, 2025b). These studies collectively highlight the synergy between fractal modeling, evolutionary optimization, and machine learning in enhancing interpolation accuracy, and predictive capabilities.

Rational fractal linear (RFL) splines are a class of fractal interpolation functions (FIFs) that integrate rational functions with the structural richness of fractal geometry (Chand et al., 2015). Unlike traditional fractal splines, which are typically constructed using polynomial bases (e.g., linear or cubic), RFL splines employ rational functions—expressed as ratios of polynomials—as their base. The rational fractal cubic (RFC) spline exhibits key distinctions from conventional interpolation methods, including: (i) its formulation based on iterated function system (IFS) theory, which leads to a functional equation and reveals self-similarity at finer scales, (ii) the generation of the interpolant through an iterative process rather than a closed-form analytical expression, and (iii) the incorporation of scaling factors that allow enhanced flexibility in selecting the interpolant, in contrast to the uniqueness associated with classical interpolants In this paper, we focus on the formulation of an RFL spline using optimized IFS parameters. To achieve this, we developed a memory-efficient quasi-Newton optimization strategy named the fractal limited-memory Broyden–Fletcher–Goldfarb–Shanno algorithm for box constraints (FL-BFGS-B) to optimize the parameters of the iterated function system (IFS). Subsequently, we illustrate an application of the optimized RFL spline for forecasting a time series model, employing the ARIMA model to predict temperature data.

## 12.2 Preliminaries

Consider a given set of interpolation data points $\{(x_i, y_i) : i \in \mathbb{N}_N\}$ with strictly increasing abscissae, where $\mathbb{N}_N$ represents the set of the first $N$ natural numbers. Let $d_i$ denote the derivative value at $x_i$, either provided or computed using standard estimation techniques. Define $h_i$ as the local step size $x_{i+1} - x_i$ for $i \in \mathbb{N}_{N-1}$, and introduce the positive shape-controlling parameters $r_i$ and $t_i$. The $C^1$-continuous linear-linear rational fractal interpolation function is formulated based on the iterated function system $\{(L_i(x), f(L_i(x)))\}_{i=1}^{N-1}$, where $f(x)$ satisfies the conditions $f(x_i) = y_i$ for all $i \in \mathbb{N}_N$. The affine transformation $L_i(x)$ is given by

$$a_i = \frac{h_i}{x_N - x_1}, \quad b_i = \frac{x_i x_N - x_{i+1} x_1}{x_N - x_1}. \tag{12.1}$$

The function $f(L_i(x))$ is expressed as

$$f(L_i(x)) = \alpha_i y_i + \frac{p_i(\theta)}{q_i(\theta)}, \quad x \in I, \tag{12.2}$$

where $I = [x_1, x_N]$, and $\theta$ is defined as

$$\theta = \frac{x - x_1}{x_N - x_1}. \tag{12.3}$$

The numerator $p_i(\theta)$ and the denominator $q_i(\theta)$ are expressed as

$$p_i(\theta) = r_i(y_i - \alpha_i y_1)(1-\theta) + t_i(y_{i+1} - \alpha_i y_N)\theta, \quad q_i(\theta) = r_i(1-\theta) + t_i\theta. \tag{12.4}$$

Given the dataset $\{(x_i, y_i) : i \in \mathbb{N}_N\}$, the derivatives at $x_i$ are represented by $d_i$. The mesh spacing is taken as $h_i = x_{i+1} - x_i$, for $i \in \mathbb{N}_{N-1}$. The parameters $r_i$ and $t_i$ are selected to ensure stability and smoothness of the fractal interpolation function. To maintain the positivity of the denominator, constraints $r_i > 0$ and $t_i > 0$ are imposed. For further details on the construction of linear-linear fractal interpolation functions, readers may refer to Chand and Reddy (2018); Chand et al. (2014); Reddy et al. (2020).

## 12.3 Methodology

Fractal interpolation has emerged as a robust method for modeling data that exhibits irregular, complex, or self-similar structures. The rational fractal linear (RFL) interpolation function enhances classical fractal interpolation by incorporating scaling parameters $\alpha$ and shape-controlling parameters $r$ and $t$, allowing a high degree of flexibility in the shape of the interpolant. A central feature of this approach is the nonuniqueness of the parameters $\alpha$, $r$, and $t$, which means that a wide variety of interpolating curves can be generated from the same data set by varying these parameters. Fig. 12.1a illustrates this by showing RFL splines generated with different combinations of $\alpha$, $r$, and $t$, resulting in visibly different curve shapes. In contrast, Fig. 12.1b shows the spline generated with a fixed set of parameters, emphasizing the influence of these choices. While this parameter-dependency offers flexibility, it also introduces the challenge of determining the most appropriate set of parameters for a given interpolation task. This raises a fundamental research question: *Given the data, how should we select* $\alpha$, $r$, *and* $t$ *to obtain the best possible interpolating function under a chosen criterion?* To address this, we propose an optimization-based approach, where the quality of the interpolant is captured by a suitably defined objective function, and we employ the quasi-Newton L-BFGS-B algorithm to determine the optimal parameters within specified bounds. This strategy enables automated and effective selection of parameters, aiming to produce the most suitable RFL spline for the given data.

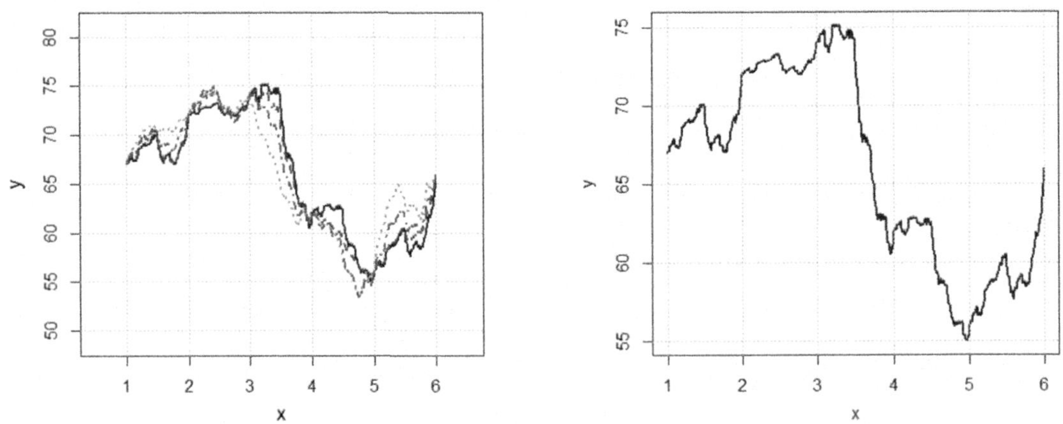


**(a)** Different combinations of $\alpha, r, t$.

**(b)** A single fixed set of $\alpha, r, t$.

**FIGURE 12.1** Comparison of RFL splines: (a) using different parameter combinations and (b) using a fixed set of parameters.

### 12.3.1 Target fractal curve formation

Here we delineate an example to show how a target RFL spline is formulated.

**Step 1:** Fix the data points: $\{(x_i, y_i)\}_{i=1}^{6} = \{(1, 67), (2, 72), (3, 74), (4, 62), (5, 56), (6, 66)\}$. Calculate the step sizes $h_i$ and the divided differences $\Delta_i$ using $h_i = x_{i+1} - x_i$, $\Delta_i = \frac{y_{i+1}-y_i}{h_i}$, for $i \in \mathbb{N}_5$. Using the given data, we find $h_1 = 1.0$, $\Delta_1 = 5.0$; $h_2 = 1.0$, $\Delta_2 = 2.0$; $h_3 = 1.0$, $\Delta_3 = -12.0$; $h_4 = 1.0$, $\Delta_4 = -6.0$; and $h_5 = 1.0$, $\Delta_5 = 10.0$. In addition, the values $r_c = \{3, 3, 3, 3, 3\}$, $t_c = \{1, 1, 1, 1, 1\}$, and $\alpha_c = \{0.24, 0.10, 0.40, 0.24, 0.26\}$ are chosen. Here $r_c$ and $t_c$ are the target shape parameters, and $\alpha_c$ is the target scaling factor. These serve as the IFS parameters for the target fractal curve.

**Step 2:** For each segment $i$, the parameters $a_i$ and $b_i$ are calculated using Eqs. (12.1). The computed values are as follows: $a_1 = 0.2$, $b_1 = 0.8$; $a_2 = 0.2$, $b_2 = 1.8$; $a_3 = 0.2$, $b_3 = 2.8$; $a_4 = 0.2$, $b_4 = 3.8$; and $a_5 = 0.2$, $b_5 = 4.8$.

**Step 3:** Arithmetic mean method: The three-point difference approximation at internal grids is given as Chand et al. (2014)

$$d_i = \begin{cases} 0 & \text{if } \Delta_{i-1} = 0 \text{ or } \Delta_i = 0 \\ \frac{h_i \Delta_{i-1} + h_{i-1} \Delta_i}{h_i + h_{i-1}} & \text{otherwise, for } i = 2, 3, \ldots, N-1, \end{cases}$$

and at the endpoints, the derivatives are approximated as

$$d_1 = \begin{cases} 0, & \text{if } \Delta_1 = 0 \text{ or } \operatorname{sgn}(D_1^*) \neq \operatorname{sgn}(\Delta_1), \\ D_1^* = \Delta_1 + \frac{(\Delta_1 - \Delta_2) h_1}{h_1 + h_2}, & \text{otherwise.} \end{cases}$$

$$d_n = \begin{cases} 0, & \text{if } \Delta_{N-1} = 0 \text{ or } \operatorname{sgn}(D_N^*) \neq \operatorname{sgn}(\Delta_{N-1}), \\ D_n^* = \Delta_{N-1} + \frac{(\Delta_{n-1} - \Delta_{n-2}) h_{N-1}}{h_{N-1} + h_{N-2}}, & \text{otherwise.} \end{cases}$$

Using this, we find $d_1 = 6.5$, $d_2 = 3.5$, $d_3 = -5.0$, $d_4 = -9.0$, $d_5 = 2.0$, and $d_6 = 18.0$. For each segment $i \in \mathbb{N}_4$, $T_{1_i}$, and $T_{2_i}$ are computed as follows:

$$T1_i = (y_i - \alpha_i \cdot y_1) \cdot r_i,$$
$$T2_i = (y_{i+1} - \alpha_i \cdot y_N) \cdot t_i.$$

The $T_{j_i}$ for $j, i = 1, 2$ are given in Table 12.1.

**Table 12.1** Values of $T_{j_i}$ for $j = 1, 2$ and $i = 1, 2, 3, 4, 5$.

| $T_{j_i}$ | $i = 1$ | $i = 2$ | $i = 3$ | $i = 4$ | $i = 5$ |
|---|---|---|---|---|---|
| $j = 1$ | 152.760000 | 195.900000 | 141.600000 | 137.760000 | 115.740000 |
| $j = 2$ | 56.160000 | 67.400000 | 35.600000 | 40.160000 | 48.840000 |

**Step 4:** For each subinterval $[x_i, x_{i+1}]$, where $i = 1, 2, \ldots, N-1$, define the affine map $L_i(x) = a_i x + b_i$, where $a_i = \frac{h_i}{h}$, $b_i = \frac{x_i x_N - x_{i+1} x_1}{h}$, and $h = x_N - x_1$. For a given point $x \in [x_1, x_N]$, the normalized parameter is defined as $\theta = \frac{x - x_1}{x_N - x_1}$, with $0 \leq$

$\theta \le 1$. Using this $\theta$, compute the intermediate function $\pi_i(\theta) = T_{1i}(1-\theta) + T_{2i}\theta$, where $T_{1i} = (y_i - \alpha_i y_1)r_i$ and $T_{2i} = (y_{i+1} - \alpha_i y_N)t_i$. The corresponding scaling denominator is given by $q_i(\theta) = r_i(1-\theta) + t_i\theta$. Finally, the interpolated value at the transformed point is computed as $f_i(x) = \alpha_i y + \frac{\pi_i(\theta)}{q_i(\theta)}$, see Table 12.2.

**Table 12.2** Computed values of $\theta_i$, $P_i$, $q_i$, $L_i$, and $f_i$ for each segment in Iteration 1.

| Segment | $\theta_i$ | $P_i$ | $q_i$ | $L_i$ | $f_i$ |
|---|---|---|---|---|---|
| | | **Segment 1** | | | |
| 1 | 0.0000 | 152.7600 | 3.0000 | 1.0000 | 67.0000 |
| 2 | 0.2000 | 133.4400 | 2.6000 | 1.2000 | 68.6031 |
| 3 | 0.4000 | 114.1200 | 2.2000 | 1.4000 | 69.6327 |
| 4 | 0.6000 | 94.8000 | 1.8000 | 1.6000 | 67.5467 |
| 5 | 0.8000 | 75.4800 | 1.4000 | 1.8000 | 67.3543 |
| 6 | 1.0000 | 56.1600 | 1.0000 | 2.0000 | 72.0000 |
| | | **Segment 2** | | | |
| 1 | 0.0000 | 195.9000 | 3.0000 | 2.0000 | 72.0000 |
| 2 | 0.2000 | 170.2000 | 2.6000 | 2.2000 | 72.6615 |
| 3 | 0.4000 | 144.5000 | 2.2000 | 2.4000 | 73.0818 |
| 4 | 0.6000 | 118.8000 | 1.8000 | 2.6000 | 72.2000 |
| 5 | 0.8000 | 93.1000 | 1.4000 | 2.8000 | 72.1000 |
| 6 | 1.0000 | 67.4000 | 1.0000 | 3.0000 | 74.0000 |
| | | **Segment 3** | | | |
| 1 | 0.0000 | 141.6000 | 3.0000 | 3.0000 | 74.0000 |
| 2 | 0.2000 | 120.4000 | 2.6000 | 3.2000 | 75.1077 |
| 3 | 0.4000 | 99.2000 | 2.2000 | 3.4000 | 74.6909 |
| 4 | 0.6000 | 78.0000 | 1.8000 | 3.6000 | 68.1333 |
| 5 | 0.8000 | 56.8000 | 1.4000 | 3.8000 | 62.9714 |
| 6 | 1.0000 | 35.6000 | 1.0000 | 4.0000 | 62.0000 |
| | | **Segment 4** | | | |
| 1 | 0.0000 | 137.7600 | 3.0000 | 4.0000 | 62.0000 |
| 2 | 0.2000 | 118.2400 | 2.6000 | 4.2000 | 62.7569 |
| 3 | 0.4000 | 98.7200 | 2.2000 | 4.4000 | 62.6327 |
| 4 | 0.6000 | 79.2000 | 1.8000 | 4.6000 | 58.8800 |
| 5 | 0.8000 | 59.6800 | 1.4000 | 4.8000 | 56.0686 |
| 6 | 1.0000 | 40.1600 | 1.0000 | 5.0000 | 56.0000 |
| | | **Segment 5** | | | |
| 1 | 0.0000 | 115.7400 | 3.0000 | 5.0000 | 56.0000 |
| 2 | 0.2000 | 102.3600 | 2.6000 | 5.2000 | 58.0892 |
| 3 | 0.4000 | 88.9800 | 2.2000 | 5.4000 | 59.6855 |
| 4 | 0.6000 | 75.6000 | 1.8000 | 5.6000 | 58.1200 |
| 5 | 0.8000 | 62.2200 | 1.4000 | 5.8000 | 59.0029 |
| 6 | 1.0000 | 48.8400 | 1.0000 | 6.0000 | 66.0000 |

### 12.3.2 Optimization problem for fine-tuning scale and shape parameters

The fractal interpolated curve $\mathcal{G}$ is defined as the attractor of an iterated function system (IFS). The objective is to optimize the scale and shape parameters such that the approximated fractal curve is as close as possible to the target curve. The optimization problem is formulated as minimizing the mean squared error (MSE) between the target fractal curve $\mathcal{G}(\vec{\alpha}_c, \vec{r}_c, \vec{t}_c)$, which is defined by the fixed target parameters $\vec{\alpha}_c$, $\vec{r}_c$, $\vec{t}_c$, and the approximated fractal curve $\mathcal{G}^*(\vec{\alpha}_j, \vec{r}_j, \vec{t}_j)$, where the parameters $\vec{\alpha}_j$, $\vec{r}_j$, $\vec{t}_j$ are the variables to be optimized. The objective function is given by the MSE:

$$E(\vec{\alpha}_j, \vec{r}_j, \vec{t}_j) = \frac{1}{n}\sum_{i=1}^{n}\left(\mathcal{G}(\vec{\alpha}_c, \vec{r}_c, \vec{t}_c) - \mathcal{G}^*(\vec{\alpha}_j, \vec{r}_j, \vec{t}_j)\right)^2,$$

which is subject to the following constraints:

$$|\alpha_{ji}| < a_i, \quad i \in \mathbb{N}_{N-1}, \quad j \in \mathbb{N}_N, \quad r_{ji} > 0, \quad t_{ji} > 0, \quad \forall i, j.$$

Here, $\alpha_{ji}$, $r_{ji}$, and $t_{ji}$ represent the scaling factors and shape parameters for each segment, while $a_i$ are upper bounds for the scaling factors. The optimization aims to find the values of $\alpha_{ji}$, $r_{ji}$, and $t_{ji}$ that minimize the MSE, thereby achieving the best approximation of the target curve while respecting the constraints on the parameters.

### 12.3.3 FL-BFGS-B algorithm for optimizing the IFS parameters

To estimate the optimal IFS parameters of the RFL spline, we employ the limited-memory Broyden–Fletcher–Goldfarb–Shanno with Box constraints (L-BFGS-B) algorithm. This method is particularly effective for large-scale optimization problems with bound constraints. It approximates the inverse Hessian matrix using limited memory, making it computationally efficient. The objective function to be minimized is the mean squared error between the actual and interpolated values. Gradients are computed analytically, and parameter updates are iteratively performed within the defined bounds. The proposed memory-efficient quasi-Newton method differs from conventional optimization techniques primarily in how it approximates and updates the Hessian matrix. Traditional quasi-Newton methods, such as BFGS or DFP, require storing and manipulating full dense matrices, leading to substantial memory and computational overhead, especially for large-scale problems. In contrast, our approach employs a limited-memory update strategy that avoids the need to explicitly form or store the full Hessian or its inverse. Instead, it uses a compact representation of low-rank updates based on gradient differences from recent iterations. This significantly reduces the memory requirements while preserving the superlinear convergence properties of quasi-Newton methods. Additionally, this strategy is particularly well-suited for real-time or resource-constrained environments, such as fractal spline parameter optimization over large datasets, where conventional methods become impractical due to their storage complexity. The following algorithm outlines the step-by-step procedure used to implement this optimization technique.

**Algorithm 1** FL-BFGS-B algorithm for optimizing IFS parameters.

**Input:** Initial parameters $\alpha^{(0)}$, $r^{(0)}$, $t^{(0)}$, data points $\{(x_i, y_i)\}_{i\in\mathcal{N}}$, maximum iterations `iter`, bounds for $\alpha$: $L_\alpha$ and $U_\alpha$ (lower and upper bounds for $\alpha$), and tolerance `tol`.
**Initialize:** Set iteration counter $k = 0$, choose initial estimates $\alpha^{(k)}$, $r^{(k)}$, $t^{(k)}$ where $\alpha^{(k)} \in [L_\alpha, U_\alpha]$, and compute initial objective function and gradient.

Define $y_i$ as the target RFL data points, and $\tilde{y}_i$ as the approximated RFL spline values for each $i \in \mathcal{N}$, where $N = |\mathcal{N}|$ is the total number of data points.
**while** $\|\nabla f_k\| >$ `tol` and $k <$ `iter` **do**
    Compute gradient of the objective function (mean squared error):

$$\nabla f(\alpha, r, t) = \frac{\partial}{\partial(\alpha, r, t)}\left(\frac{1}{N}\sum_{i\in\mathcal{N}}(y_i - \tilde{y}_i)^2\right)$$

    Approximate the inverse Hessian $H_k$ using the L-BFGS two-loop recursion. Let $q = \nabla f_k$
    **for** $i = k-1$ to $k-m$ **do**
        Compute memory values: $\rho_i = \frac{1}{s_i^T y_i}$, where $s_i = x_{i+1} - x_i$ and $y_i = \nabla f_{i+1} - \nabla f_i$
        $\theta_i = \rho_i s_i^T q$
        Update: $q \leftarrow q - \theta_i y_i$
    **end for**
    Apply initial inverse Hessian approximation:

$$H_0 = \frac{s_{k-1}^T y_{k-1}}{y_{k-1}^T y_{k-1}} I, \quad z = H_0 q$$

    **for** $i = k-m$ to $k-1$ **do**
        $\gamma_i = \rho_i y_i^T z$
        Update: $z \leftarrow z + s_i(\theta_i - \gamma_i)$
    **end for**
    Set search direction: $d_k = -z$
    Perform line search to determine optimal step size $\lambda_k$ along direction $d_k$
    Update parameters: $x_{k+1} = x_k + \lambda_k d_k$
    Apply box constraints: $x_{k+1} = \max(L, \min(U, x_{k+1}))$
    Update memory: $s_k = x_{k+1} - x_k$, $y_k = \nabla f_{k+1} - \nabla f_k$
    Update iteration count: $k \leftarrow k + 1$
**end while**

**Output:** Optimized parameters $\alpha^*, r^*, t^*$
**Notes:**

- $H_k$: Inverse Hessian approximation at iteration $k$, estimated using the L-BFGS update strategy.
- $s_i$: Difference in successive iterates, $s_i = x_{i+1} - x_i$.
- $y_i$: Difference in successive gradients, $y_i = \nabla f_{i+1} - \nabla f_i$.
- $z$: Initialized as $z = H_0 q$, where $H_0$ is the scaled identity matrix.
- Termination criteria: Algorithm stops if gradient norm $\|\nabla f_k\| <$ `tol` or if iteration limit `iter` is reached.

## 12.4 Results and discussion

In this study, we employed the L-BFGS-B optimization algorithm to determine the optimal parameters $(\alpha, r, t)$ of a rational fractal interpolation function. The built-in `airquality` dataset in R was used for demonstration. This dataset records daily air quality measures in New York between May and September 1973. For the interpolation task, we selected the `Temperature` and `Day` columns and extracted the first six records. The resulting set of data points used for interpolation was

$$\{(x_i, y_i)\}_{i=1}^{6} = \{(1, 67), (2, 72), (3, 74), (4, 62), (5, 56), (6, 66)\}.$$

To approximate the given dataset with a rational fractal spline, the L-BFGS-B algorithm was initialized with feasible bounds and parameter estimates. The optimization aimed to minimize the mean squared error (MSE) between the actual and interpolated values. Over multiple iterations, the parameters were updated based on gradient evaluations and inverse Hessian approximations, adhering to the Box constraints set for each variable. Visual results of the interpolation at successive optimization steps are depicted in Figs. 12.3a to 12.3d. These plots highlight how the fractal curve progressively improves in approximating the underlying data pattern. The optimized parameter values at each iteration are summarized in Table 12.3. It is evident that the optimization procedure fine-tuned the $\alpha$, $r$, and $t$ values to improve interpolation accuracy.

**Table 12.3** Optimized parameters $\alpha$, $r$, and $t$ across iterations.

| Iteration | Optimized $\alpha$ | | | | |
|---|---|---|---|---|---|
| 1 | 0.2179 | 0.1813 | 0.2397 | 0.2276 | 0.1744 |
| 2 | 0.1766 | 0.2240 | 0.1663 | 0.2450 | 0.1967 |
| 3 | 0.2325 | 0.1586 | 0.2491 | 0.2490 | 0.2056 |
| 4 | 0.1864 | 0.1699 | 0.1685 | 0.2242 | 0.1549 |

| Iteration | Optimized $r$ | | | | |
|---|---|---|---|---|---|
| 1 | 3.4568 | 2.6903 | 2.7253 | 3.4664 | 2.8463 |
| 2 | 2.8706 | 3.2905 | 2.9308 | 2.9411 | 3.4716 |
| 3 | 2.6783 | 3.1648 | 2.6187 | 2.5483 | 2.7834 |
| 4 | 2.5926 | 2.9497 | 2.9892 | 2.8446 | 2.5006 |

| Iteration | Optimized $t$ | | | | |
|---|---|---|---|---|---|
| 1 | 0.9216 | 0.8852 | 1.1465 | 0.8444 | 1.0144 |
| 2 | 0.9786 | 1.1235 | 0.8929 | 1.1045 | 0.8508 |
| 3 | 1.1670 | 0.9778 | 1.0704 | 1.1230 | 1.0831 |
| 4 | 1.0547 | 1.1211 | 1.1754 | 1.0086 | 0.9148 |

The convergence of the algorithm was confirmed by the significant reduction in the mean squared error (MSE) across iterations. At the fifth iteration, the MSE reached a value of $MSE = 3.365807 \times 10^{-29}$, which is approximately 0.0000000000000000000000000000000000003365807 in decimal form. This extremely low error demonstrates the efficacy of the L-BFGS-B optimization in tuning the fractal interpolation parameters and achieving a close fit to the original dataset. Fig. 12.3a illustrates both the target RFL and its approximation obtained using the FL-BFGS-B algorithm corresponding to the first level of iteration in the RFL interpolation process. Similarly, Fig. 12.3b presents the comparison between the target and the approximated RFL for the second iteration. Fig. 12.3c depicts this comparison for the third iterated points, while Fig. 12.3d demonstrates the results for the fourth iteration. In each case, the FL-BFGS-B algorithm is employed to approximate the RFL with increasing refinement through successive iterations. Also the optimized IFS parameter values for each iteration of the RFL are shown in the Table 12.3. As illustrated in Fig. 12.2, the mean squared error (MSE) reduces progressively over the five iterations, confirming the effectiveness of the optimization strategy.

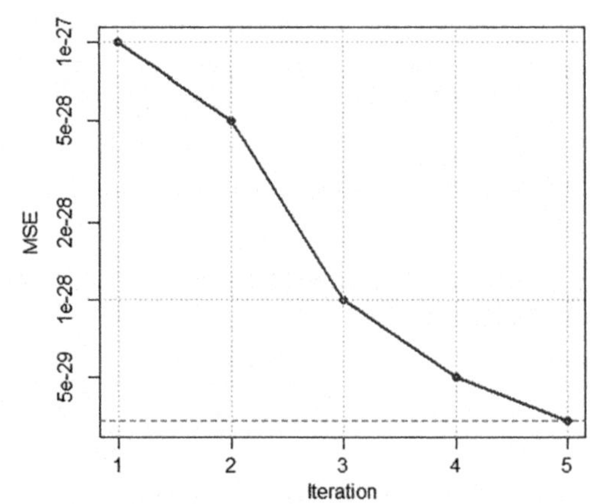


**FIGURE 12.2** Reduction of mean squared error (MSE) over five iterations of the optimization process. The MSE decreases consistently, reaching a minimum value of $3.365807 \times 10^{-29}$, indicating convergence.

## 12.5 Application

Optimizing the parameters of the iterated function system (IFS) is critical, because these parameters govern the shape, self-affinity, and local variability of the resulting fractal interpolation function (FIF). In time series modeling, especially when the data exhibit irregularities or self-similar patterns, classical models often fail to accurately capture both the global trends and fine local details. An optimized fractal spline—generated by tuning the IFS parameters—produces a more accurate and structurally faithful representation of the observed time series.

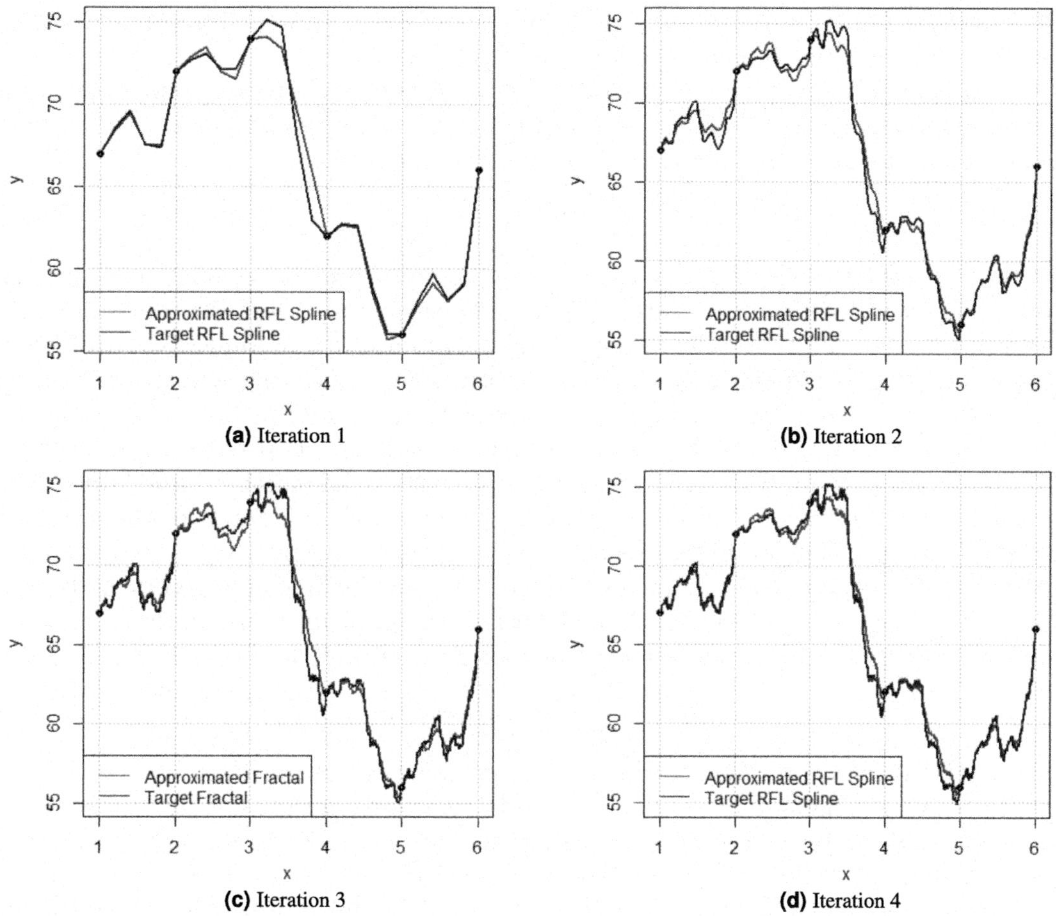


**FIGURE 12.3** Optimized RFL for different iteration of the fractal interpolation.

This high-quality approximation serves as a preprocessed version of the data, reducing noise and enhancing important dynamic features. By fitting the fractal spline to the historical data through parameter optimization, we obtain a model that reflects the intrinsic geometry of the time series. This optimized representation can then be used as an input to statistical forecasting models (e.g., ARIMA), thereby improving their ability to extrapolate future points. In essence, accurate forecasting depends on the quality of the input model, and the optimized IFS-based fractal spline enhances this quality by preserving meaningful data characteristics while filtering out noninformative variations. Fractal interpolation enhances the structure of raw time series data by smoothing noise while preserving essential trends and local variations. This results in a cleaner and more informative signal, which allows ARIMA to model patterns more accurately. As a result, forecasting accuracy improves, because ARIMA operates on a well-structured, denoised version of the data that better reflects its underlying dynamics.

The fractal series was modeled using different ARIMA processes for various iterations. Each model was selected based on optimal fit criteria, ensuring stationarity and minimizing residual variance. The general form of an ARIMA model is

$$(1-\phi_1 B-\phi_2 B^2-\cdots-\phi_p B^p)(1-B)^d Y_t=(1+\theta_1 B+\theta_2 B^2+\cdots+\theta_q B^q)\varepsilon_t, \tag{12.5}$$

where $B$ is the backward shift operator, $\phi_i$ are the autoregressive coefficients, $\theta_i$ are the moving average coefficients, $d$ is the differencing order, and $\varepsilon_t$ represents a white noise error term. In an ARIMA model, $p$ represents the order of the autoregressive (AR) term, $d$ is the degree of differencing, and $q$ is the order of the moving average (MA) term. The parameter $p$ indicates the number of past values, $d$ indicates how many times the data is differenced, and $q$ is the number of past forecast errors used. We employed the `auto.arima` function in R programming to identify the most suitable ARIMA model for our data by automatically selecting the optimal values of $p$, $d$, and $q$ based on model selection criteria. Now we predict points for the Fig. 12.3a, which is the optimized RFL spline, which is generated from first iteration points; and the predicted points can be seen in the Fig. 12.5(a). The ARIMA(0,1,1) model was fitted, implying a first-order differencing step with a single moving average term. The model equation is

$$(1-B)Y_t=(1+0.6413B)\varepsilon_t, \tag{12.6}$$

where $B$ is the backward shift operator, and $\varepsilon_t$ represents white noise.

The residual variance is $\sigma^2=4.967$. The model selection criteria values are AIC = 114.53, AICc = 115.07, and BIC = 116.96, indicating an acceptable fit. Fig. 12.4a delineates the residual plot, ACF, PACF, and Q-Q plot for the ARIMA(0,1,1).

Now we do predict points for the Fig. 12.3b, which is the optimized RFL spline, which is generated from second iteration points; and the predicted points can be seen in the Fig. 12.5(b). The ARIMA(2,1,3) model, incorporating two autoregressive and three moving average terms, is given by

$$(1+0.4269B+0.3320B^2)(1-B)Y_t=(1+1.4172B+1.4118B^2+0.7839B^3)\varepsilon_t. \tag{12.7}$$

The residual variance is $\sigma^2=0.1767$, with model selection values AIC = 148.34, AICc = 149.05, and BIC = 165.31. Fig. 12.4b delineates the residual plot, ACF, PACF, and Q-Q plot for the ARIMA(2,1,3).

Now we predict points for the Fig. 12.3c, which is the optimized RFL spline, which is generated from third iteration points; and the predicted points can be seen in Fig. 12.5(c). The ARIMA(5,2,4) model consists of five autoregressive and four moving average terms with second-order differencing. The model equation is

$$(1-0.1000B+0.0042B^2+0.1407B^3-0.2213B^4-0.5377B^5)(1-B)^2Y_t, \tag{12.8}$$

$$=(1-0.4169B-0.2419B^2-0.0136B^3+0.3179B^4)\varepsilon_t. \tag{12.9}$$

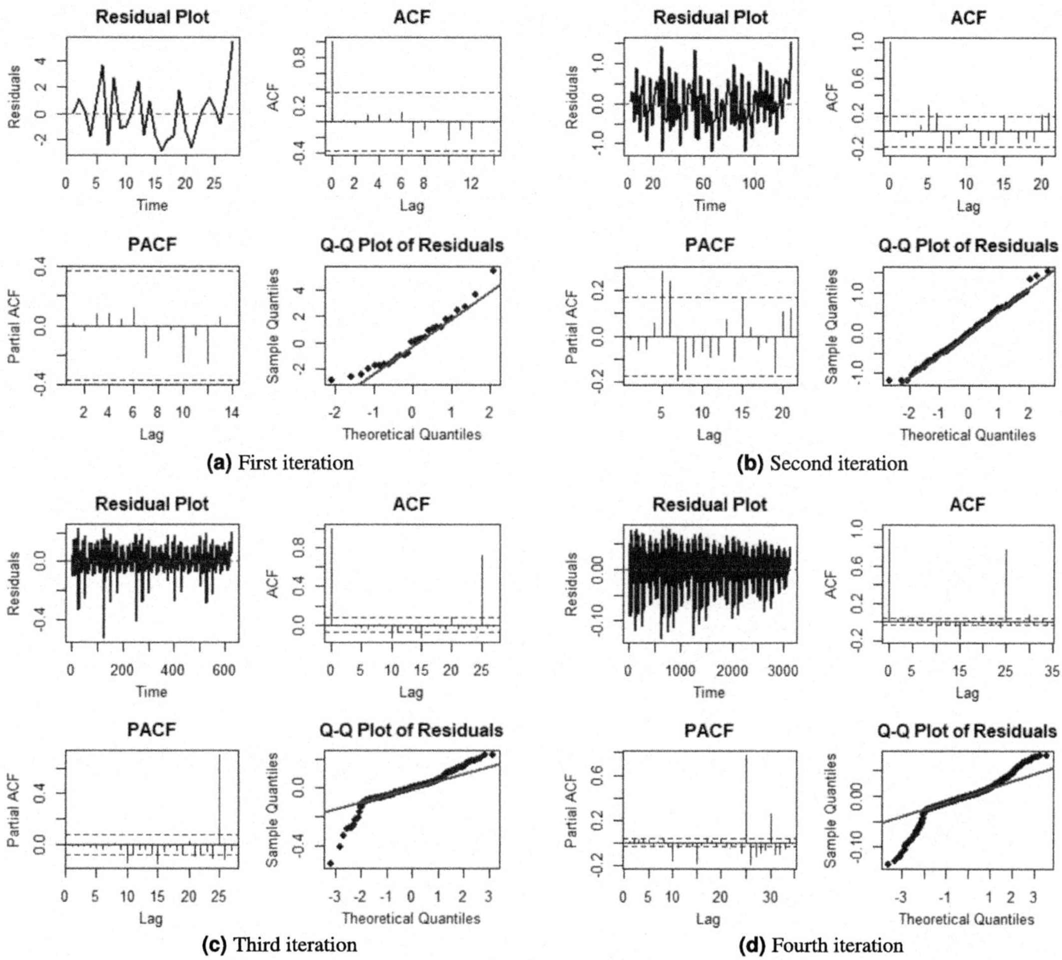


**(a)** First iteration **(b)** Second iteration **(c)** Third iteration **(d)** Fourth iteration

**FIGURE 12.4** ARIMA forecasts for the newly generated points in successive iterations of the RFL interpolation.

The residual variance is $\sigma^2 = 0.006773$. Model selection values are log-likelihood = 676.88, AIC = −1333.77, AICc = −1333.41, and BIC = −1289.40. Fig. 12.4c delineates the residual plot, ACF, PACF, and Q-Q plot for the ARIMA(5,2,4).

Now we predict points for the Fig. 12.3d, which is the optimized RFL spline, which is generated from fourth iteration points; the predicted points can be seen in Fig. 12.5(d). The ARIMA(5,2,0) model contains five autoregressive terms and second-order differencing:

$$(1 + 0.3099B + 0.3734B^2 + 0.3214B^3 + 0.2197B^4 - 0.3387B^5)(1 - B)^2 Y_t = \varepsilon_t. \qquad (12.10)$$

The residual variance is $\sigma^2 = 0.0005406$. The model selection values are log-likelihood = 7327.8, AIC = −14643.61, AICc = −14643.58, and BIC = −14607.33. Fig. 12.4d delineates

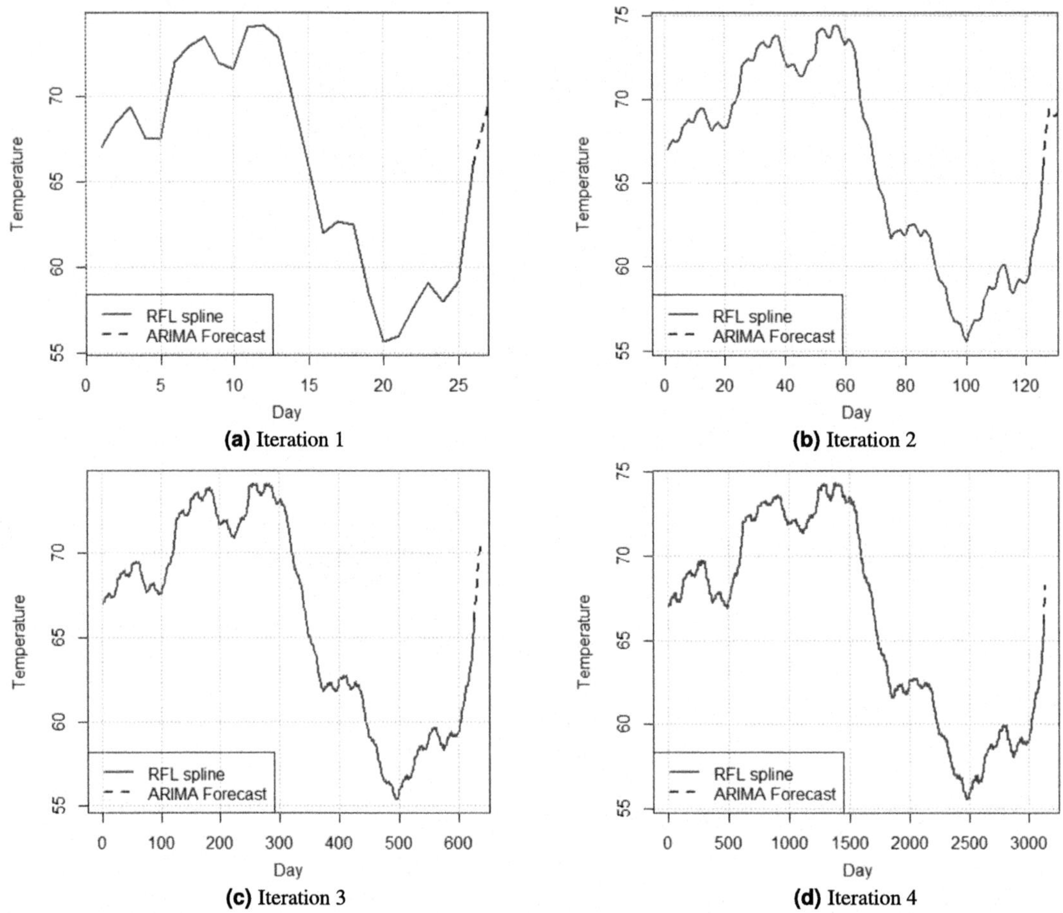


**FIGURE 12.5** ARIMA model plots over four successive iterations of the RFL.

the residual plot, ACF, PACF, and Q-Q plot for the ARIMA(5,2,0). Table 12.4 presents the parameters, estimated values, and standard errors for the ARIMA(0,1,1) model. Similarly, Table 12.5 provides the parameters, estimated values, and standard errors for the ARIMA(2,1,3) model. Table 12.6 displays the corresponding information for the ARIMA(5,2,4) model, while Table 12.7 summarizes the results for the ARIMA(5,2,0) model.

**Table 12.4** Estimated ARIMA(0,1,1) parameters for Iteration 1.

| Parameter | Estimate | Standard error |
|---|---|---|
| MA(1) ($\theta_1$) | 0.6413 | 0.1851 |

**Table 12.5** Estimated ARIMA(2,1,3) parameters for Iteration 2.

| Parameter | Estimate | Standard error |
|---|---|---|
| AR(1) ($\phi_1$) | −0.4269 | 0.1260 |
| AR(2) ($\phi_2$) | −0.3320 | 0.1254 |
| MA(1) ($\theta_1$) | 1.4172 | 0.0987 |
| MA(2) ($\theta_2$) | 1.4118 | 0.0980 |
| MA(3) ($\theta_3$) | 0.7839 | 0.0849 |

**Table 12.6** Estimated ARIMA(5,2,4) parameters for Iteration 3.

| Parameter | Estimate | Standard Error |
|---|---|---|
| AR(1) ($\phi_1$) | 0.1000 | 0.0752 |
| AR(2) ($\phi_2$) | −0.0042 | 0.0590 |
| AR(3) ($\phi_3$) | −0.1407 | 0.0903 |
| AR(4) ($\phi_4$) | 0.2213 | 0.0708 |
| AR(5) ($\phi_5$) | 0.5377 | 0.0477 |
| MA(1) ($\theta_1$) | −0.4169 | 0.0937 |
| MA(2) ($\theta_2$) | −0.2419 | 0.0713 |
| MA(3) ($\theta_3$) | 0.0136 | 0.1469 |
| MA(4) ($\theta_4$) | −0.3179 | 0.0877 |

**Table 12.7** Estimated ARIMA(5,2,0) parameters for Iteration 4.

| Parameter | Estimate | Standard Error |
|---|---|---|
| AR(1) ($\phi_1$) | −0.3099 | 0.0169 |
| AR(2) ($\phi_2$) | −0.3734 | 0.0174 |
| AR(3) ($\phi_3$) | −0.3214 | 0.0177 |
| AR(4) ($\phi_4$) | −0.2197 | 0.0174 |
| AR(5) ($\phi_5$) | 0.3387 | 0.0169 |

## 12.6 Conclusion

In this study, we proposed an optimization framework for **rational fractal linear (RFL)** splines to enhance modeling and forecasting in time series analysis. The use of fractal interpolation allowed us to generate flexible, self-similar curves that retain the inherent irregularities of real-world data. By optimizing the iterated function system (IFS) parameters—$\alpha$, $r$, and $t$—we gained precise control over the structure and smoothness of the interpolants. To efficiently perform this optimization, we employed the **limited-memory BFGS with Box constraints (L-BFGS-B)** algorithm, a quasi-Newton method known for fast convergence in large-scale problems while handling bound constraints. Iterative optimization led to a progressive reduction in the mean squared error (MSE), eventually reaching values as low as $10^{-29}$. We validated the performance of the optimized RFL splines using temperature time series data from the `airquality` dataset. Forecasting was conducted using the **ARIMA** model applied to iterated points generated from the fractal interpolants. Visual results over multiple iterations confirmed enhanced smoothness and continuity, while effectively preserving the data's original trends and improving predictive quality. This successful integration of RFL spline optimization with ARIMA forecasting opens avenues for further research. Future work will explore the incorporation of advanced forecasting models, such as LSTM networks, Prophet, and hybrid deep learning techniques. We also aim to extend the framework to more complex, high-dimensional datasets across domains like finance, environment, and biomedicine, while investigating adaptive parameter control, sensitivity analysis, and real-time optimization strategies.

## References

Abdulla, S., K, M.R., 2024. Machine learning clustering techniques for segmenting ifs-generated garment patterns. In: 2024 International Conference on Intelligent Computing and Emerging Communication Technologies (ICEC), pp. 1–6.

Abdulla, S., Reddy, K.M., 2025a. Optimizing the neural network and iterated function system parameters for fractal approximation using a modified evolutionary algorithm. Scientific Reports 15.

Abdulla, S., Reddy, K.M., 2025b. Time-series prediction with optimized non-affine fractal interpolations using neural network. The European Physical Journal Special Topics.

Barnsley, M.F., 1986. Constructive approximation fractal functions and interpolation. Constructive Approximation 2, 303–329.

Barnsley, M.F., 2012. Fractals Everywhere, 2 edition. Dover Publications, New York.

Barnsley, M.F., Harrington, A.N., 1989. The calculus of fractal interpolation functions. Journal of Approximation Theory 57, 14–34.

Chand, A.K., Reddy, K.M., 2018. Constrained fractal interpolation functions with variable scaling. Siberian Electronic Mathematical Reports 15, 60–73.

Chand, A.K., Vijender, N., Navascués, M.A., 2014. Shape preservation of scientific data through rational fractal splines. Calcolo 51, 329–362.

Chand, A.K.B., Viswanathan, P., Reddy, K.M., 2015. Towards a more general type of univariate constrained interpolation with fractal splines. Fractals 23 (04), 1550040.

Li, X.-F., Li, X.-F., 2008. An explicit fractal interpolation algorithm for reconstruction of seismic data. Chinese Physics Letters 25 (3), 1157–1159.

Raubitzek, S., Neubauer, T., 2021. A fractal interpolation approach to improve neural network predictions for difficult time series data. Expert Systems with Applications 169.

Reddy, K.M., Chand, A.K., Viswanathan, P., 2020. Data visualization by rational fractal function based on function values. Journal of Analysis 28, 261–277.

Reddy, K.M., Vijender, N., 2023. A fractal model for constrained curve and surface interpolation. The European Physical Journal Special Topics 232, 1015–1025.

Zhai, M.Y., 2015. A new method for short-term load forecasting based on fractal interpretation and wavelet analysis. International Journal of Electrical Power & Energy Systems 69, 241–245.

Zhai, M.Y., FernÁndez-MartÍnez, J.L., Rector, J.W., 2011. A new fractal interpolation algorithm and its applications to self-affine signal reconstruction. Fractals 19, 355–365.

# 13

# Structure and arboricity of $(k-r,k)$-trees

M. VijayaLakshmi and V. Raja
*Department of Mathematics, School of Advanced Sciences, VIT-AP University, Inavolu, Beside AP Secretariat, Amaravati, AP, India*

## 13.1 Introduction

The family of $k$-trees is a well-studied generalization of trees in graph theory, introduced as a recursive construction starting from a complete graph $K_k$ and repeatedly adding vertices adjacent to a $k$-clique. These graphs are of particular interest, because they are exactly the $k$-connected, triangulated, and $K_{k+2}$-free graphs, as established in Patil (1986). $k$-trees have appeared in diverse contexts, including perfect elimination schemes, sparse graph representations, and as a basis for studying chordal graphs (Cho et al., 1986; Dmitriev, 1982; Patil, 1984, 1986).

Several extensions of $k$-trees have been explored in the literature. Harary and Palmer (1968) introduced acyclic simplicial complexes, which later led Beineke and Pippert (1977) and Gionfriddo (1982) to investigate $(j,k)$-trees as higher-dimensional generalizations. Patil (1984, 1986) studied $k$-trees in depth, providing characterizations and structural results. Borowiecki and Patil (2017) examined coloring properties of $(k-r,k)$-trees, revealing their rich combinatorial behavior.

In this work, the special case $j=k-r$, with $1 \leq r \leq k$, is considered, leading to the definition of a new family of graphs, called $(k-r,k)$-trees. This construction naturally generalizes $k$-trees (when $r=1$) and encompasses other interesting graph families; for example, $(1,2)$-trees correspond to maximal outerplanar graphs, while $(2,3)$-trees include certain maximal planar graphs. The study of $(k-r,k)$-trees is motivated by their structural resemblance to $k$-trees, while exhibiting distinct properties in terms of connectivity, minimal separators, and clique structure.

## 13.2 Structure of $(k-r,k)$-trees

**Definition 13.2.1.** For integers $r$ and $k$ $(1 \leq r \leq k)$, we define the class of $\mathcal{T}_{k,r}$ (hereafter referred to as $\mathcal{T}_{k,r}$) through the following recursive steps:

1. The initial $\mathcal{T}_{k,r}$ is defined as the complete graph $K_{k-r+1}$.


Advanced Computational and Mathematical Approaches in Applied Differential Equations
https://doi.org/10.1016/B978-0-44-345743-2.00024-4

2. Starting with $\mathcal{T}_{k,r}$ a $G$ of order $p = k + (i - 1)r + 1$ $(i > 0)$, we can build an expanded tree of order $p + r$. This is achieved by introducing a set of $r$ new, mutually adjacent vertices and ensuring each is joined to every vertex of a preexisting $(k - r + 1)$-clique within $G$.

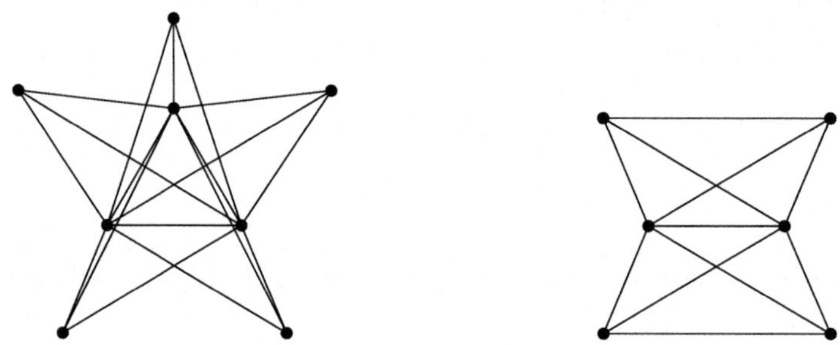

(a) $(1, 3)$-tree of order 8 (b) $(2, 3)$-tree of order 6

**FIGURE 13.1** Illustrations of $\mathcal{T}_{k,r}$ where $k = 3$: (a) A $(1, 3)$-tree of order 8, generated by appending $r = 1$ pendant vertices to a $K_3$ core; (b) A $(2, 3)$-tree of order 6, formed by joining $r = 2$ mutually adjacent vertices to a $K_2$ base clique.

**Remark 13.2.2.**

1. $r = 1$; the class of $(k - 1, k)$-trees are equivalent to the well-known family of $k$-trees (see, Cho et al., 1986; Dmitriev, 1982; Patil, 1984, 1986). Furthermore, notice that $(1, 2)$-trees are just 2-trees, which extend the familiar concept of the maximal outerplanar graphs. Notably, the $(1, 2)$-tree class exclusively consists of maximal are the only $(1, 2)$-trees, which are outerplanar.
2. $r = k$; Indeed, the category of $\mathcal{T}_{k,r}$ represents the same structural group as $(0, k)$-trees, which are defined by their $K_{k+1}$ block components. Notice that $(0, 1)$-trees are just the usual trees.
3. $(1, 3)$-trees are the special kind of 3-connected, planar graphs (see Fig. 13.1(a)).
4. $(2, 3)$-trees are the special kind of maximal planar graphs. In fact, the planar $(2, 3)$-trees are equivalent to the familiar class of triangulated, maximal planar graphs (see Fig. 13.1(b)).

**Proposition 1.** *For any graph $G \in \mathcal{T}_{k,r}$ with $1 \le r \le k$, let $p = |V(G)|$ satisfy the condition $p \ge k - r + 1$. The edge count of $G$ is given by $|E(G)| = p\left(k - \binom{r-1}{2}\right) - \frac{(k+1)(k-r+1)}{2}$.*

*Proof.* From Definition 13.2.1, consider the edge count $|E(G)|$ for a graph where the number of vertices $|V(G)| = k + (i - 1)r + 1$ and $i \in \{0, 1, 2, \ldots\}$ can be readily computed to be $|E(G)| = \binom{k-r+1}{2} + i\left(\binom{r}{2} + r(k - r + 1)\right) = p\left(k - \binom{r-1}{2}\right) - \frac{(k+1)(k-r+1)}{2}$. □

For $n \geq 0$, a graph $G$ is $n$-degenerate if each of its induced subgraphs contains a vertex with degree $d(v) \leq n$. Following Harary (1969), we use $\kappa(G)$ and $\lambda(G)$ for the vertex and edge-connectivity of $G$. The properties below follow directly from Definition 13.2.1:

**Proposition 2.** *Whenever $k \geq r \geq 1$ and the vertex count of a graph $G \in \mathcal{T}_{k,r}$ reaches at least $k+r+1$, the following results hold:*

1. $\kappa(G) = k - r + 1$.
2. $\lambda(G) = k = \delta(G)$.
3. *$G$ is triangulated, $k$-degenerate and $K_{k+2}$-free.*

## 13.3 Properties

Notice that a single nonpendant vertex (*i.e.*, $K_1$), disconnects a nontrivial tree. This fact is generalized to a clique $K_{k-r+1}$ of a $\mathcal{T}_{k,r}$, $G$ of order $\geq k+r+1$, due to (1) of Proposition 2. Such cliques, we call the *minimal separators* of $G$, because the removal of such cliques from $G$ results in a disconnected graph. The following result is obvious:

**Proposition 3.** *Suppose $G$ is a $\mathcal{T}_{k,r}$ such that $|V(G)| \geq k+r+1$. For every nonadjacent vertex pair $\{u,v\}$, there is a subgraph $K_{k-r+1} \subset G$ where $u \sim K_{k-r+1}$ (i.e., $u$ is complete to the clique). Additionally, any path connecting $u$ to $v$ is necessarily intercepted by this $K_{k-r+1}$ structure.*

Given $G \in \mathcal{T}_{k,r}$, let its cardinality be $p \geq k+1$; a vertex $v \in V(G)$ is designated as a *pendant vertex* if its membership is restricted to a unique $(k+1)$-clique. It follows naturally that every vertex in a $(k-r,k)$-tree of order $k+1$ satisfies this criterion. To facilitate our subsequent analysis, we now identify a fundamental structural property of this graph family.

**Proposition 4.** *For any graph $G \in \mathcal{T}_{k,r}$ with vertex count $|V(G)| \geq k+r+1$, there exists a chain of $(k+1)$-cliques $\{K_1, K_2, \ldots, K_j\}$ $(j \geq 2)$ such that each consecutive intersection $K_i \cap K_{i+1}$ $(1 \leq i < j)$ constitutes a $(k-r+1)$-clique that acts as a minimal separator for $G$.*

The above proposition leads the following definition:

**Definition 13.3.1.** In a $\mathcal{T}_{k,r}$ graph $G$ where $|V(G)| \geq k+r+1$, a clique $K_{k+1}$ is defined as an *end-clique* provided that $K_{k+1} \cap (G - K_{k+1})$ is a unique minimal $K_{k-r+1}$ whose deletion increases the number of connected components.

The following attribute of $\mathcal{T}_{k,r}$ is easily verified:

**Proposition 5.** *For cases where $p \geq k+r+1$, let $G$ represent a $\mathcal{T}_{k,r}$ graph of order $p$.*

1. *$G$ contains at least two end-cliques: $K^i_{k+1}$ and $K^j_{k+1}$, each containing exactly $r$ mutually adjacent pendant vertices.*
2. *For a $(k-r,k)$-tree $G$ of order $p$, let $t$ represent the count of its pendant vertices. The following bounds apply:*

$$2r \leq t \leq p - k + r - 1. \tag{13.1}$$

Subsequently, we present several structural results for $\mathcal{T}_{k,r}$ pertaining to their vertex degrees.

**Proposition 6.** *Given $G \in \mathcal{T}_{k,r}$ with vertex count $p = k + ir + 1$ ($i \geq 0$), the number of vertices satisfying $\deg(v) \leq 2k - r$ is $k + 1$ or greater.*

*Proof.* Suppose $G \in \mathcal{T}_{k,r}$ has cardinality $p$. Invoking the results of Proposition 1, the edge count is given by $|E(G)| = \frac{1}{2}[p(2k - r + 1) - (k + 1)(k - r + 1)]$. Using the Handshaking Lemma, we determine the sum of all vertex degrees in $V(G)$:

$$\sum_{v \in V(G)} \deg(v) = 2|E(G)| = p(2k - r + 1) - (k + 1)(k - r + 1). \tag{13.2}$$

Assume $t$ represents the count of vertices in $G$ whose degrees satisfy $\deg(v) \leq 2k - r$. Since every vertex in such a tree has a degree of at least $k$ (per Proposition 2), and the remaining $(p - t)$ vertices have a minimum degree of $2k - r + 1$, we obtain the following bound:

$$\sum_{v \in V(G)} \deg(v) \geq tk + (p - t)(2k - r + 1). \tag{13.3}$$

By comparing Eqs. (13.2) and (13.3), it follows that

$$p(2k - r + 1) - (k + 1)(k - r + 1) \geq tk + (p - t)(2k - r + 1). \tag{13.4}$$

Rearranging the terms to isolate $t$:

$$tk - t(2k - r + 1) \leq -(k + 1)(k - r + 1). \tag{13.5}$$

This simplification results in $t(k - r + 1) \geq (k + 1)(k - r + 1)$, which clearly implies $t \geq k + 1$. Consequently, $G$ must possess at least $k + 1$ vertices with degrees bounded by $2k - r$. □

## 13.4 Characterizations

**Definition 13.4.1.** In a graph $G$, a vertex $v$ is classified as a *simplicial vertex* provided its neighborhood $N(v)$ forms a complete subgraph (clique).

Based on the recursive structure of Definition 13.2.1, the following result offers a characterization of $\mathcal{T}_{k,r}$ using the existence of simplicial vertices, mirroring the perspective in Borowiecki and Patil (2017):

**Proposition 7.** *Consider a graph $G$ with order $|V(G)| \geq k + r + 1$ ($k \geq r \geq 1$). Then $G$ belongs to the $\mathcal{T}_{k,r}$ family if and only if there exists a set of $r$ mutually adjacent simplicial vertices $\{u_1, u_2, \ldots, u_r\}$ such that each $u_i$ has $\deg(u_i) = k$, and the induced subgraph $G - \{u_1, \ldots, u_r\}$ remains a $(k - r, k)$-tree.*

In simpler terms, $G$ is reducible to a complete graph $K_{k-r+1}$ through the iterative deletion of $r$-cliques, where each deleted clique consists of pendant vertices of degree $k$.

Following directly from Proposition 7, we establish the following property:

**Corollary 13.4.2.** *An embedding of the join $K_{k-r+1} \vee 2K_r$ is guaranteed within the structure of every $\mathcal{T}_{k,r}$ graph possessing at least $k+r+1$ vertices.*

**Proposition 8.** *Consider a connected graph G with vertex count $p \geq k+r+1$. Then G is categorized as a $\mathcal{T}_{k,r}$ if and only if there exists a finite collection $\mathcal{S}$ of distinct $(k+1)$-cliques whose union constitutes G. Furthermore, for every vertex $v \in V(G)$, there is a unique chain of cliques $(K^a_{k+1}, K^b_{k+1}) \in \mathcal{S} \times \mathcal{S}$, there exists a unique finite sequence of cliques: $K^1_{k+1}, K^2_{k+1}, \ldots, K^j_{k+1}$, each in $\mathcal{S}$ so that $K^1_{k+1} = K^a_{k+1}$, $K^j_{k+1} = K^b_{k+1}$ and $(K^i_{k+1} \cap K^{i+1}_{k+1}) = K_{k-r+1}$ for $1 \leq i \leq j-1$, is a minimal disconnecting clique of G.*

*Proof.* The forward implication is immediate, because of Proposition 4. The backward implication follows directly by using induction on $|\mathcal{S}|$. □

## 13.5 Arboricity

The *edge-arboricity*, often denoted simply as the *arboricity* $\gamma(G)$, represents the minimum number of forests required to partition the edge set $E(G)$ of a nontrivial graph $G$ (Harary, 1969). A landmark result by Nash-Williams (1961) provides a precise characterization of this parameter:

$$\gamma(G) = \max\left\{\left\lceil \frac{|E(H)|}{|V(H)|-1} \right\rceil : H \subseteq G, |V(H)| \geq 2\right\}, \tag{13.6}$$

where the maximum is taken over every nontrivial induced subgraph $H$ of $G$, and $\lceil x \rceil$ indicates the ceiling function (the smallest integer not less than $x$). Notably, for a complete graph $K_k$, the arboricity is $\gamma(K_k) = \lceil k/2 \rceil$. In the theorem that follows, we establish the specific bounds for the arboricity of the $(k-r,k)$-tree family.

**Theorem 13.5.1.** *For any graph G belonging to the $\mathcal{T}_{k,r}$ family $(1 \leq r \leq k)$, the following bounds are established for the arboricity $\gamma(G)$:*

$$\left\lceil \frac{k-r+1}{2} \right\rceil \leq \gamma(G) \leq \left\lceil \frac{2k-r+1}{2} \right\rceil. \tag{13.7}$$

*Proof.* Let $G$ be a $\mathcal{T}_{k,r}$ of order $p = k+(i-1)r+1$, where $i \in \{0,1,2,\ldots\}$. We evaluate the arboricity based on the construction index $i$.

**Case 1:** $i = 0$. Here, the graph order is $p = k-r+1$, which implies $G \cong K_{k-r+1}$. The arboricity of a complete graph $K_n$ is $\lceil n/2 \rceil$; thus $\gamma(G) = \lceil (k-r+1)/2 \rceil$. This confirms that the lower bound is sharp.

**Case 2:** $i = 1$. In this instance, $p = k+1$ and $G \cong K_{k+1}$. Consequently, $\gamma(G) = \lceil (k+1)/2 \rceil$, which strictly exceeds the lower bound $\lceil (k-r+1)/2 \rceil$ for $r > 0$.

**Case 3:** $i \geq 2$. For orders $p \geq k+2$, Corollary 13.4.2 implies $G$ contains $K_{k-r+1} \vee 2K_r$ as a subgraph, which contains $K_{k+1}$. Thus $\gamma(G) \geq \gamma(K_{k+1})$. Applying the Nash-Williams formula, we observe that the ratio $\frac{|E(H)|}{|V(H)|-1}$ is maximized when $H = G$ due to the uniform density of the $(k-r,k)$-tree structure. Setting $X = \frac{|E(G)|}{p-1}$, we obtain

$$X = \frac{1}{p-1}\left[\frac{p(2k-r+1)-(k+1)(k-r+1)}{2}\right]. \tag{13.8}$$

Algebraic simplification yields

$$X = \frac{2k-r+1}{2} - \frac{k(k-r)}{2(p-1)}. \tag{13.9}$$

We now analyze the ceiling $\lceil X \rceil$ relative to the order $p$:

1. If $p \leq \frac{k(k-r)}{2} + 1$, then $\gamma(G) = \lceil X \rceil$ remains strictly below the upper bound $\lceil \frac{2k-r+1}{2} \rceil$.
2. If $p > \frac{k(k-r)}{2} + 1$, the fractional reduction becomes sufficiently small such that $\gamma(G) = \lceil \frac{2k-r+1}{2} \rceil$.

This completes the proof. □

## 13.6 Conclusion

The work in Borowiecki and Patil (2017) introduced a conjecture regarding the chromatic index and explored the chromatic polynomials and partition numbers associated with $\mathcal{T}_{k,r}$ $(k \geq r \geq 1)$. Furthermore, the specific subclass of $(k-1,k)$-trees with order $p \geq k+1$ has been shown to coincide with standard $k$-trees, representing the family of triangulated, $k$-connected, and $K_{k+2}$-free graphs (Patil, 1986). Various other structural properties and unique characterizations for these graphs were further developed in Patil (1984). Our current study establishes that any $\mathcal{T}_{k,r}$ of order $p \geq k+r+1$ is inherently triangulated, $k$-degenerate, $(k-r+1)$-connected, and contains no $K_{k+2}$ subgraph.

Based on these findings, we propose the following direction for future investigation:

**Open Problem.** Consider a $(p,q)$-graph $G$ where $p \geq k+r+1$ and $1 \leq r \leq k$. If $G$ is triangulated and $(k-r+1)$-connected, and satisfies either the condition that it is $K_{k+2}$-free or its edge count is exactly

$$q = \frac{p(2k-r+1)-(k+1)(k-r+1)}{2},$$

can it be concluded that $G$ is a $\mathcal{T}_{k,r}$?

Additionally, it remains a valuable pursuit to identify further characterizations of $\mathcal{T}_{k,r}$ by examining the interplay between connectivity, $k$-degeneracy, and chromatic properties.

# References

Beineke, L.W., Pippert, R.E., 1977. On the structure of $(m,n)$-trees. In: Proceedings of the 8th Southeastern Conference on Combinatorics, Graph Theory and Computing, pp. 75–80.
Borowiecki, M., Patil, H.P., 2017. Colourings of $(k-r,k)$-trees. Opuscula Mathematica 37 (4), 491–500.
Cho, C.Y., Li, N.Z., Xu, S.J., 1986. On $q$-trees. Journal of Graph Theory 10, 129–136.
Dmitriev, I.G., 1982. Characterization of $k$-trees (in Russian). Sbornik Trudov Instituta Matematiki, Sibirskoe Otdelenie Akademii Nauk USSR 38, 9–18.
Gionfriddo, M., 1982. Characterizations and properties of the $(m,n)$-trees. Journal of Combinatorics, Information & System Sciences 7 (4), 297–302.
Harary, F., 1969. Graph Theory. Addison-Wesley, Reading, MA.
Harary, F., Palmer, E.M., 1968. On acyclic simplicial complexes. Mathematika 15, 115–122.
Nash-Williams, C.St.J.A., 1961. Edge-disjoint spanning trees of finite graphs. Journal of the London Mathematical Society 36 (1), 445–450.
Patil, H.P., 1984. Studies on $k$-trees and Some Related Topics. Doctoral Thesis. University of Warsaw, Poland.
Patil, H.P., 1986. On the structure of $k$-trees. Journal of Combinatorics, Information & System Sciences 11 (2–4), 57–64.

# 14

# Optimal control of a tumor-immune model considering targeted chemotherapy

P.T. Sowndarrajan[a,b] and T. Dilleshwar Rao[b]
[a]*School of Advanced Sciences, VIT-AP University, Amaravati, India* [b]*School of Advanced Sciences, Vellore Institute of Technology, Chennai, India*

## 14.1 Introduction

Cancer is a disease, which can start everywhere in the human body. It is a leading cause of death worldwide. It accounts for nearly one in six deaths (World Health Organization, 2020). Cancer or malignant tumors can affect any part of the body and also spread to other organs. Through the International Agency for Research on Cancer (IARC), World Health Organization (WHO) classifies tobacco use, alcohol consumption, unhealthy diet, physical inactivity, and air pollution are the causing agents of cancer and other noncommunicable diseases. Between 30–50% of cancer burden can be prevented or reduced by implementing appropriate treatment and prevention strategies.

An estimated 400 000 children and adolescents from 0–19 years old develop cancer each year (Steliarova-Foucher et al., 2017). These childhood cancers include leukemias, brain cancers, lymphomas, and solid tumors, such as neuroblastoma and Wilms tumors. It cannot be prevented or identified through the screening. But, it can be cured with generic medicines and treatment, including surgery and radiotherapy. However, very few cancers are caused by environmental or lifestyle factors. Children need continued physical and cognitive growth and nutritional status, which requires a dedicated, multidisciplinary team (World Health Organization, 2021). Nowadays, the prevention of such cancer is an important health challenge around the world. Many researchers show that it can be treated effectively with certain prevention measures and controlled through earlier detection. Our immune system plays a major role in limiting the development of these cancerous cells. If cancer is identified early and treated appropriately, it has a high chance of being cured. It results in less morbidity and less treatment expense. A proper cancer diagnosis is required for effective treatment. It usually includes surgery, radiotherapy, chemotherapy, hormonal treatments, and targeted biological therapies. The first step is to determine the treatment goal which cures cancer cells and improves the individual's life.

**Advanced Computational and Mathematical Approaches in Applied Differential Equations**
**https://doi.org/10.1016/B978-0-44-345743-2.00025-6**

Among the treatment types, chemotherapy is one of the finest treatments that help to eliminate or control tumor growth through drug injection. This drug treatment also harms the normal cells that divide rapidly. It results in the most common harmful side effects: mouth sores, nausea, hair loss, etc. Recently, many mathematicians started focusing on tumor therapy mathematical models that play a significant role in better understanding tumor prevention and its treatment. To avoid the chemotherapy side effects, oncologists established a treatment called targeted chemotherapy. It targets the changes in tumor cells that help them to grow, divide, and spread (Glassman and Balthasar, 2014; Weiner, 2015). Many researchers have proposed mathematical models in the literature on tumor systems. Kuznetsov et al. (1994) analyzed a tumor model of the cytotoxic T lymphocyte response and estimated the parameter's value with experimental data. de Pillis et al. (2005) proposed a mathematical model of cell-mediated tumor-immune interactions with natural killer (NK) and $CD8^+$ T cells. They established variable sensitivity analysis and predict the patient's response to treatment. During the last few years, several mathematical models have been developed to describe the tumor-immune interaction (Liu et al., 2021; Li and Xiao, 2022; Pinho et al., 2013; Chan and Yu, 2012). The model developed by Dhar and Gupta (2021) focuses on a tumor-immune system with antibodies effect and efficacy of the mAb drug. They showed the effect of tumor interaction and antibodies with various values of drug efficacy. Based on the tumor-immune model, some researchers established a chemotherapy model to simulate the effects of the drugs. Kirschner and Panetta (1998) investigated a tumor-immune model with therapy and explored the elimination of tumors.

In this paper, we consider a tumor-immune mathematical model with targeted chemotherapy and control intervention to reduce the tumor cell population in the communities. In the literature, some researchers investigated on the tumor growth models with optimal control strategies (de Pillis and Radunskaya, 2001; Rocha et al., 2018; Alkama et al., 2018; Engelhart et al., 2011; Fister and Panetta, 2000, 2003; Dehingia et al., 2021; Rihan, 2021). de Pillis et al. (2007) investigated a tumor-immune chemotherapy model with linear and quadratic control in the objective functional. They also provided numerical results of the effects of those control strategies. Pang et al. (2016) proposed and analyzed an optimal strategy for a tumor model with a combination of immunotherapy and chemotherapy. In Le et al. (2021), the authors develop a data-driven mathematical model to study the interactions between immune cells and the primary bone tumor microenvironment. They investigated the effect of various treatments on the dynamics of tumor cells. Sharma and Samanta (2016) analyzed a quadratic optimal control for a tumor-immune model with chemotherapeutic and immunotherapeutic drugs. The authors in Dehingia et al. (2022), proposed a cancer-obesity-treatment model with the control parameters: IL-2 therapy, ACI therapy, and nutritional diet. They analyzed the existence and stability of the equilibrium points and demonstrated an optimal cancer treatment. However, there are currently no mathematical models to study the optimal control analysis for the progression of tumor-immune dynamics with targeted chemotherapy, to the best of our knowledge. Taking these studies as motivation, in this work, we investigate the optimal control problem for such tumor-immune mathematical model.

## 14.2 Mathematical model

In this section, we present a mathematical model for tumor-immune dynamics with targeted chemotherapy (Liu and Liu, 2017). Thus the system of nonlinear ordinary differential equations is as follows:

$$\begin{cases} \dfrac{dT(t)}{dt} = aT(t)(1-bT(t)) - c_1 N(t)T(t) - k_T M(t)T(t), \\ \dfrac{dN(t)}{dt} = \alpha_1 + \dfrac{gT(t)N(t)}{s+T(t)} - \mu N(t) - pT(t)N(t) - k_N(1-\eta)M(t)N(t), \\ \dfrac{dC(t)}{dt} = \alpha_2 - \beta C(t) - k_C(1-\eta)M(t)C(t), \\ \dfrac{dM(t)}{dt} = -\gamma M(t) + V_M - K_T T(t)M(t), \end{cases} \tag{14.1}$$

with nonnegative initial conditions $T(0)$, $N(0)$, $C(0)$, and $M(0)$. The considered model is of interactions between tumor cell populations (T), effector immune cell population (N), circulating lymphocyte population (C), and chemotherapeutic drugs (M). All parameters in (14.1) are positive constants, which are shown in Table 14.1.

Liu and Liu (2017) investigated the mathematical model (14.1) of tumor-immune dynamics with drug concentration. This is an extension of de Pillis et al. (2007) model by considering the effect of targeted chemotherapy. In Liu and Liu (2017), the authors investigated the qualitative analysis and illustrated some numerical simulations on the constructed new mathematical model. Also the authors compared the targeted chemotherapy model with regular chemotherapy model. They confirm that the tumor-immune model with targeted chemotherapy helps to kill tumor cells. Moreover, they concluded that using targeted chemotherapy drug, we can eliminate a small size tumor, but not a large size tumor. Therefore to understand the progression in which the tumor cell population to be controlled, we implement the optimal control theory. We include a control intervention in our tumor-immune model (14.1). The control mechanism is incorporated in (14.1) by replacing the constant chemotherapy treatment rate with $u(t)$. The idea behind this is to determine optimal strategy to minimize the tumor cell population and the cost administered.

Therefore in this paper, we extend the tumor-immune model of Liu and Liu (2017) by incorporating the control term in the drug concentration. The fact in this considered optimal control problem is that the implemented control intervention helps to reduced the tumor burden efficiently with cost-effectiveness of the treatment. The paper is structured as follows: In Section 14.3, we formulate an optimal control problem and investigate the existence of an optimal solution that minimize the functional. Then using Pontryagin's Maximum Principle (PMP), we derive characterization of the optimal control. We perform some numerical simulations to our problem to understand the effect of control interventions in Section 14.4.

**Table 14.1** Symbols and description of parameters.

| Symbol | Description |
|---|---|
| $T$ | Tumor cell population |
| $N$ | Effector immune cell population |
| $C$ | Circulating lymphocyte population |
| $M$ | Concentration of chemotherapeutic drug |
| $t$ | Time |
| $T(0)$ | Initial tumor population |
| $N(0)$ | Initial immune population |
| $C(0)$ | Initial circulating lymphocyte population |
| $M(0)$ | Initial drug concentration |
| $a$ | Tumor growth rate |
| $b$ | $\frac{1}{b}$ is tumor carrying capacity |
| $c_1$ | Fractional tumor cells killed by effector cells |
| $k_T$ | Fractional tumor cells killed by chemotherapy |
| $\alpha_1$ | Constant source of effector cells |
| $g$ | Maximum effector cells recruitment rate by tumor cells |
| $s$ | Steepness coefficient of the effector cells recruitment |
| $\mu$ | Death rate of effector cells |
| $p$ | Effector cells inactivation rate by tumor cells |
| $k_N$ | Fractional effectors cells killed by chemotherapy |
| $K_T$ | Combination rate of chemotherapy drug with tumor cells |
| $\alpha_2$ | Constant source of circulating lymphocytes |
| $\gamma$ | Rate of chemotherapy drug decay |
| $\beta$ | Death rate of circulating lymphocytes |
| $k_C$ | Fractional circulating lymphocytes killed by chemotherapy |
| $\eta$ | Efficacy of chemotherapy |
| $V_M$ | Chemotherapy treatment |

## 14.3 Optimal control model

In this section, we begin by formulating an optimal control problem with vaccination and treatment as control interventions. Then we prove existence of an optimal control, and we characterize it through Pontryagin's maximum principle.

### 14.3.1 The total cost functional

Our main goal is to decrease the density of tumor cells and the cost of implementing the control intervention. Therefore we consider the total cost functional $J$ to minimize, as follows:

$$J[u(\cdot)] = \int_0^{t_f} [A_1 T + B_1 u^2(t)]dt \longrightarrow \min, \tag{14.2}$$

subject to the model control system

$$\begin{cases} \dfrac{dT}{dt} = aT(1-bT) - c_1NT - k_TMT, \\ \dfrac{dN}{dt} = \alpha_1 + \dfrac{gTN}{s+T} - \mu N - pTN - k_N(1-\eta)MN, \\ \dfrac{dC}{dt} = \alpha_2 - \beta C - k_C(1-\eta)MC, \\ \dfrac{dM}{dt} = -\gamma M + u - K_TTM, \end{cases} \tag{14.3}$$

with fixed initial conditions

$$T(0) = T_0 > 0,\ N(0) = N_0 > 0,\ C(0) = C_0 > 0,\ M(0) = M_0 > 0. \tag{14.4}$$

Here $t_f$ is the fixed terminal time. The detailed report of the components in the cost functional (14.2) is as follows:

- **(i)** The cost induced by tumor effect itself is $\int_0^{t_f} A_1T(t)dt$, which is proportional to the density of tumor cells $T$. The coefficient $A_1$ represents the positive weight constant of the tumor cells.
- **(ii)** Chemotherapy to tumor individuals. Here we consider the constant drug injection rate $V_M$ as the control variable $u(t)$, which measures the treatment intensity. The cost $\int_0^{t_f} B_1u^2(t)dt$ is involved in providing chemo treatment. We consider a nonlinearity of order two to reflect the severity of the side effects of the drugs imposed. The coefficient $B_1$ represents the positive weight constant associated with treatment to keep balance in the size.

Thus the Lagrangian function $L$ for our optimal control problem is given by

$$L(T, N, C, M, u) = A_1T + B_1u^2. \tag{14.5}$$

The control variable $u(t)$ of our optimal control problem involve the following admissible control set:

$$U_{ad} = \{u|u(t) \text{ is Lebesgue measurable on } [0, t_f] : 0 \le u(t) \le u_{max}\},$$

where $u_{max}$ is a fixed positive constant.

### 14.3.2 Existence of optimal control

In this subsection, we prove that there exists an optimal control $u^*$ which minimizes the cost functional $J$ in finite time.

**Theorem 14.3.1.** *There exists an optimal control $u^*$ in $U_{ad}$ such that $J(u^*) = \min\{J(u) : u \in U_{ad}\}$, solution to the optimal control problem (14.2)–(14.4).*

*Proof.* To prove existence of solution, we need to satisfy the following conditions:

**(1)** The admissible set of controls $U_{ad}$ and the state solutions of (14.3) is nonempty.
**(2)** The control set $U_{ad}$ is closed, convex, and the state system can be expressed as a linear function of the control variables with coefficients that depend on time and state variables.
**(3)** The integrand $L$ in the cost functional (14.2) is convex on the control set $U_{ad}$.
**(4)** There exists $c > 0$ and $q > 1$ such that the objective functional $L(T, u)$ satisfies

$$L(T, u) \geq c(|u|^2)^{q/2}.$$

Therefore the results follows directly from Fleming and Rishel (1975). □

## 14.3.3 Characterization of optimal control functions

Now we derive necessary optimality condition using Pontryagin's maximum principle (PMP) (Lenhart and Workman, 2007; Pontryagin et al., 1964).

**Theorem 14.3.2.** *Let $u^*$ be the optimal control of problem (14.2)–(14.4) and $T^*$, $N^*$, $C^*$, $M^*$ the corresponding optimal state trajectories satisfying (14.3)–(14.4). Then there exists an adjoint variable $\lambda = (\lambda_1, \lambda_2, \lambda_3, \lambda_4) \in \mathbb{R}^4$ that satisfies the following equations:*

$$\begin{cases} \dfrac{d\lambda_1}{dt} = \lambda_1[2abT - a + c_1N + k_TM] + \lambda_2\left[pN - \dfrac{sgN}{(s+T)^2}\right] + \lambda_4K_TM - A_1, \\ \dfrac{d\lambda_2}{dt} = \lambda_1c_1T + \lambda_2\left[\mu + pT + k_N(1-\eta)M - \dfrac{gT}{s+T}\right], \\ \dfrac{d\lambda_3}{dt} = \lambda_3[\beta + k_C(1-\eta)M], \\ \dfrac{d\lambda_4}{dt} = \lambda_1k_T + \lambda_2[k_N(1-\eta)N] + \lambda_3k_C(1-\eta)C + \lambda_4[\gamma + K_TT], \end{cases} \tag{14.6}$$

*with transversality conditions*

$$\lambda_i(t_f) = 0, \quad i = 1, \ldots, 4. \tag{14.7}$$

*Moreover, the optimal control $u^*$ is given as*

$$u = \min\left\{\max\left\{0, \left(\frac{-\lambda_4}{2B_1}\right)\right\}, u_{max}\right\}. \tag{14.8}$$

*Proof.* We define the Hamiltonian function as follows:

$$\begin{aligned} H(T, N, C, M, u, \lambda) = {} & A_1T + B_1u^2 \\ & + \lambda_1\left(aT(1-bT) - c_1NT - k_TMT\right) \\ & + \lambda_2\left(\alpha_1 + g\frac{T}{s+T}N - \mu N - pTN - k_N(1-\eta)MN\right) \\ & + \lambda_3\left(\alpha_2 - \beta C - k_C(1-\eta)MC\right) \end{aligned}$$

$$+\lambda_4(-\gamma M + u - K_T T M).$$

From PMP, there exist functions $\lambda_1, \ldots, \lambda_4$ that satisfy the adjoint equations

$$\frac{d\lambda_1}{dt} = -\frac{\partial H}{\partial T}, \quad \frac{d\lambda_2}{dt} = -\frac{\partial H}{\partial N}, \quad \frac{d\lambda_3}{dt} = -\frac{\partial H}{\partial C}, \quad \frac{d\lambda_4}{dt} = -\frac{\partial H}{\partial M},$$

evaluated at the optimal controls and corresponding state variables, subject to the transversality conditions $\lambda_i(t_f) = 0$, $i = 1, \ldots, 4$. Therefore we obtain the adjoint system (14.6) and terminal conditions (14.7). Finally,

$$\frac{\partial H}{\partial u} = 2B_1 u^* \lambda_4 = 0;$$

we obtain from the minimality condition of PMP that (14.8) holds. □

Concluding, the optimality conditions consist of the state system (14.3) with given initial conditions (14.4), the adjoint system (14.6) with transversality conditions (14.7), and the optimal control functions (14.8). In the next Section 14.4, we solve numerically the obtained optimality system:

$$\begin{cases} \dfrac{dT}{dt} = aT(1-bT) - c_1 NT - k_T MT, \\ \dfrac{dN}{dt} = \alpha_1 + \dfrac{gTN}{s+T} - \mu N - pTN - k_N(1-\eta)MN, \\ \dfrac{dC}{dt} = \alpha_2 - \beta C - k_C(1-\eta)MC, \\ \dfrac{dM}{dt} = -\gamma M + u - K_T TM, \\ \dfrac{d\lambda_1}{dt} = \lambda_1[2abT - a + c_1 N + k_T M] + \lambda_2\left[pN - \dfrac{sgN}{(s+T)^2}\right] + \lambda_4 K_T M - A_1, \\ \dfrac{d\lambda_2}{dt} = \lambda_1 c_1 T + \lambda_2\left[\mu + pT + k_N(1-\eta)M - \dfrac{gT}{s+T}\right] \\ \dfrac{d\lambda_3}{dt} = \lambda_3[\beta + k_C(1-\eta)M], \\ \dfrac{d\lambda_4}{dt} = \lambda_1 k_T + \lambda_2[k_N(1-\eta)N] + \lambda_3 k_C(1-\eta)C + \lambda_4[\gamma + K_T T], \\ T(0) = T_0, \; N(0) = N_0, \; C(0) = C_0, \; M(0) = M_0, \\ \lambda_i(t_f) = 0, \; i = 1, \ldots, 4, \end{cases} \tag{14.9}$$

with

$$u = \min\left\{\max\left\{0, \left(\frac{-\lambda_4}{2B_1}\right)\right\}, u_{max}\right\}. \tag{14.10}$$

## 14.4 Numerical results and discussion

In this section, we use a fourth-order Runge–Kutta algorithm to perform the numerical simulation of the optimal system (14.9) with (14.10). Choosing the initial conditions for the states and the initial guesses for the controls, the state system (14.3) is solved forward in time using a fourth-order Runge–Kutta scheme. Using the current iteration solution of the state Eq. (14.3) and the transversality conditions (14.7), the adjoint system (14.6) is solved backwards in time by the fourth-order Runge–Kutta scheme. We repeat the iteration process by updating the controls using the state and adjoint values. This process will continue until the values of the state, adjoint, and controls converge. The initial values of unknowns we have used are given by $T(0) = 10^5$, $N(0) = 3 \times 10^5$, $C(0) = 6.25 \times 10^{10}$, and $M(0) = 0.5$. Population profiles and control intervention are plotted for a time period of $t_f = 120$ days. We use the set of parameter values as in the Table 14.2 to determine the numerical simulation of the optimality system.

**Table 14.2** Values of the parameters used to illustrate the endemic situation and optimal control.

| Parameter | Value | Units |
|---|---|---|
| $a$ | 0.00431 | per day |
| $b$ | $1.02 \times 10^{-9}$ | per day |
| $c_1$ | $6.41 \times 10^{-11}$ | per day |
| $k_T$ | 0.08 | per day |
| $\alpha_1$ | 12000 | per day |
| $g$ | 0.0125 | persons per day |
| $s$ | 20,200,000 | persons per day |
| $\mu$ | 0.0412 | per day |
| $p$ | $3.42 \times 10^{-6}$ | – |
| $k_N$ | $2 \times 10^{-11}$ | – |
| $K_T$ | $3.2 \times 10^{-9}$ | – |
| $\alpha_2$ | $7.50 \times 10^8$ | – |
| $\beta$ | 0.012 | – |
| $k_C$ | $6 \times 10^{-11}$ | – |
| $\gamma$ | 0.9 | – |
| $\eta$ | 0.8 | – |
| $A_1$ | 10 | – |
| $B_1$ | 50 | – |
| $u_{max}$ | 1.0 | – |

Note that here the positive weights in the objective functional (14.2) are assumed. Before studying the scenario under optimal control, we consider the targeted chemotherapy case without optimal control ($u = 0$) using the parameter values given in Table 14.2. The dynamics of our model, tumor cell population, effector-immune cell population, circulating lymphocyte population, and drug chemotherapy are respectively depicted in Fig. 14.1a–d.

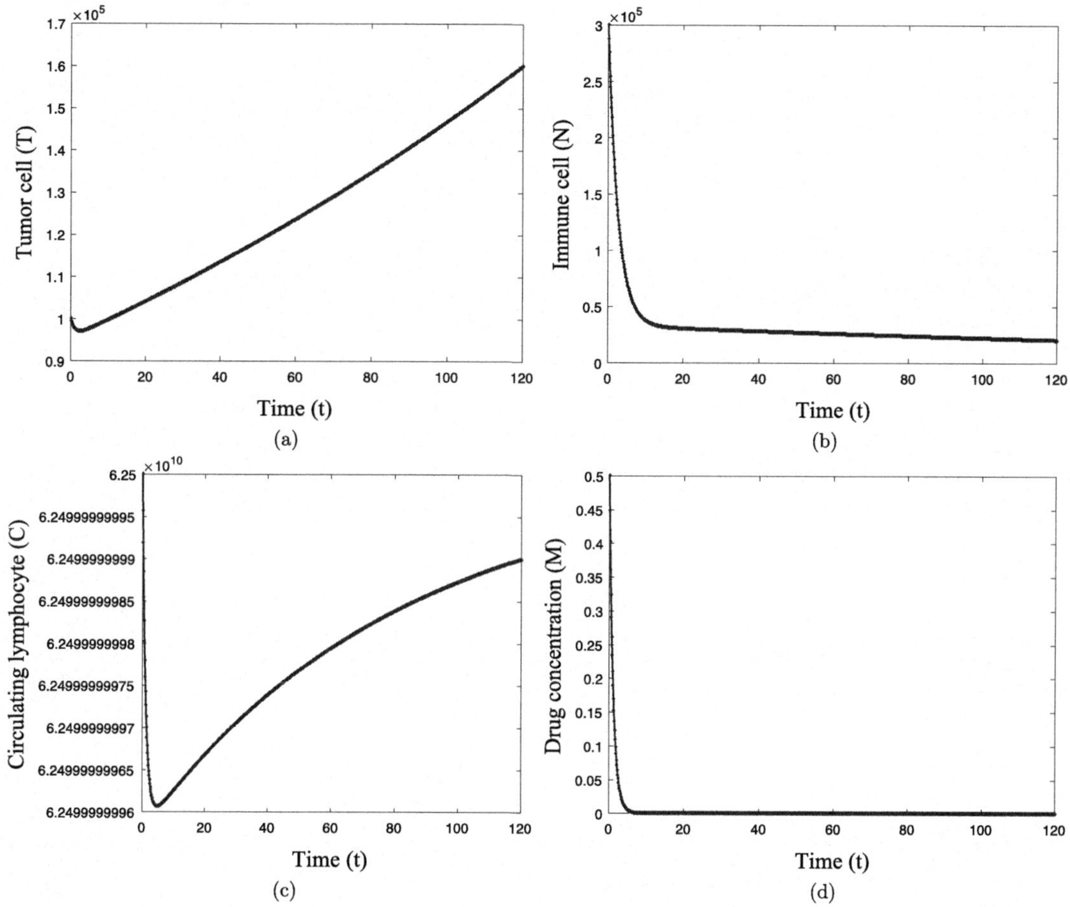


**FIGURE 14.1** Profiles of (a) Tumor cell, (b) Effector-immune cell, (c) Circulating lymphocyte and (d) Drug concentration without control, i.e., $u = 0$.

In Fig. 14.1, (a) and (b) represent the dynamics of tumor cell and immune cell population, respectively. It tells that if the tumor cell grows, then the immune cell will not remain itself. Similarly, in Fig. 14.1, (c)–(d), respectively, represent the dynamics of circulating lymphocyte population and drug concentration using the same parameter values. The concentration of drugs decreases to zero in the absence of control term in our system. It means that drug level is reduced drastically when there is large density of tumor cells.

Using the same parameters value as in Table 14.2, we present the profiles of individual population with and without control strategy in Fig. 14.2, 14.3, 14.4, and 14.5. Our aim in proposing this optimal control problem is to decrease the tumor cell population and the total cost administered during the chemotherapy treatment using the implementation of control strategy. First the dynamics of tumor cell (T) with and without control are respectively depicted in Fig. 14.2(a) and Fig. 14.2(b). In these figures, we observed that the

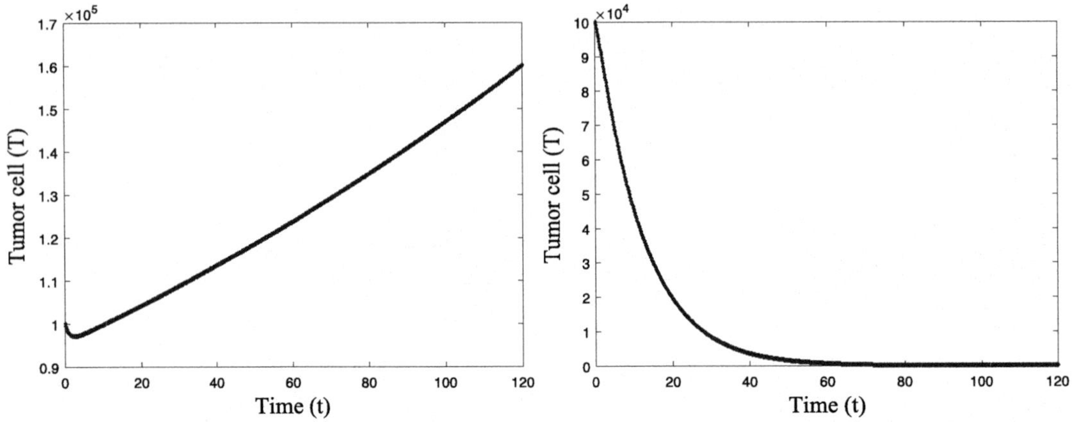


(a) T with optimal control ($u = 0$) (b) T with optimal control ($u \neq 0$)

**FIGURE 14.2** Tumor cell (T) profiles.

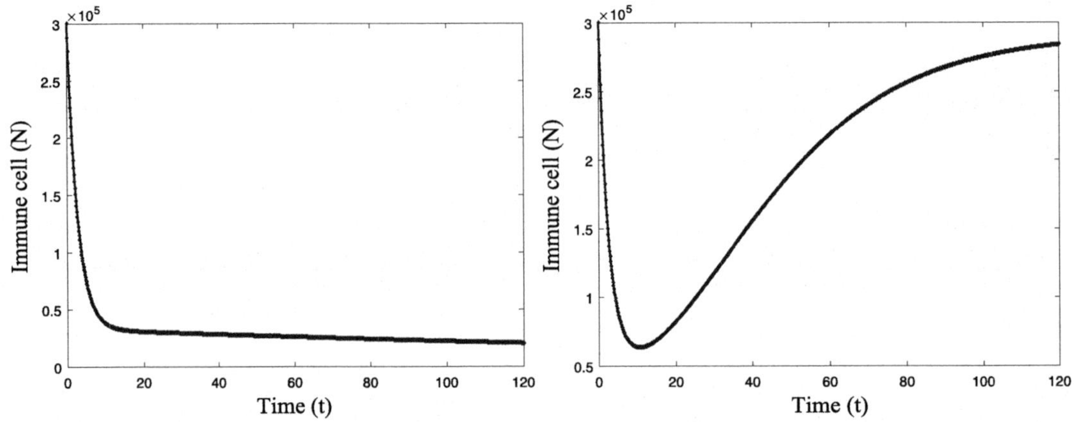


(a) N without optimal control ($u = 0$) (b) N with optimal control ($u \neq 0$)

**FIGURE 14.3** Effector-immune cell (N) profiles.

density of tumor cell decreases gradually with the presence of control variable than the system without control for increased in chemotherapy treatment. In Fig. 14.3, we present the profiles of effector-immune cell (N) without control (see, Fig. 14.3(a)) and with control variable (see, Fig. 14.3(b)). From these figures, we notice that the immune cell increases over time with the impact of control than compared without control variable. This helps in realizing the importance of control term $u(t)$ in our system.

Furthermore, we present the dynamics of circulating lymphocyte with and without control in Fig. 14.4. From Fig. 14.4(a) ($u(t) = 0$) and Fig. 14.4(b) ($u(t) \neq 0$), we observed that the lymphocyte population remain same from initial density. It means that lymphocyte count rebound to normal level after recover from the infection. Therefore in Fig. 14.5, no

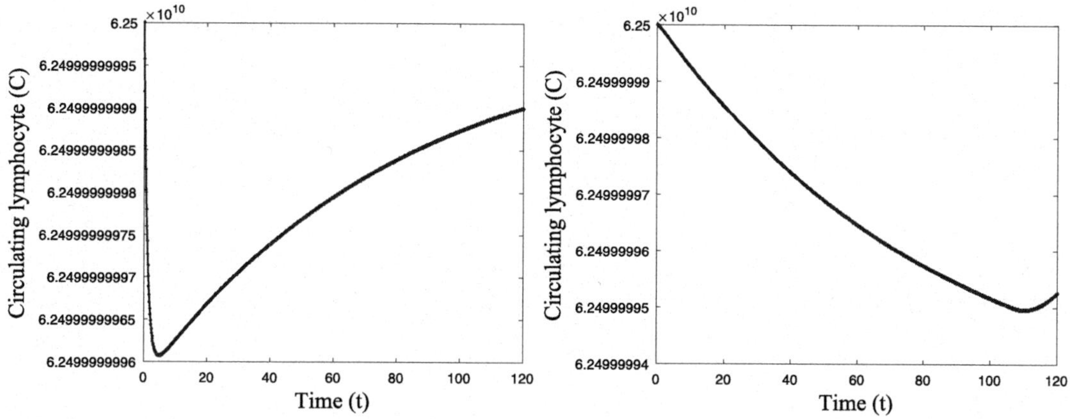


(a) C without optimal control ($u = 0$)

(b) C with optimal control ($u \neq 0$)

**FIGURE 14.4** Circulating lymphocyte (C) profiles.

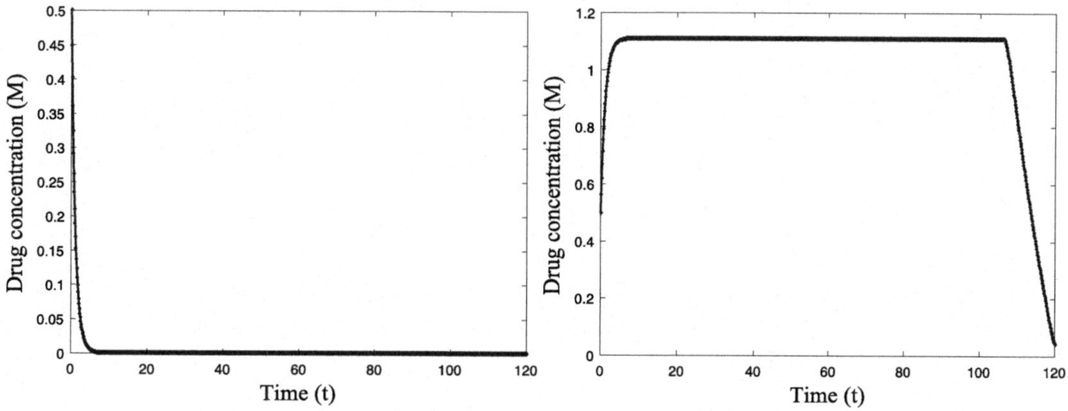


(a) M without optimal control ($u = 0$)

(b) M with optimal control ($u \neq 0$)

**FIGURE 14.5** Drug concentration (M) profiles.

morphological change is observed in the circulating lymphocyte population among with control and absence of control in our system. Generally, lymphocyte level indicates that either too high or too low level may show significance of infection or mild illness in patient. Next, in Fig. 14.5, we present the dynamics of drug concentration with and without control term. This drug concentration helps to reduce the growth of cancer cell in the individual population. In Fig. 14.5a, we noticed that when there is absence of control term ($u(t) = 0$) in our system, the concentration of drug drastically decreases to zero, even the tumor cell population is increase in larger over time. However, the presence of control variable ($u(t) \neq 0$) in the drug concentration increases gradually in the initial time and it continues its drug level in reducing the cancer cell over certain period of time, see Fig. 14.5b. Then, we understand that the drug treatment rapidly decreases in the stipulated time and then goes to zero.

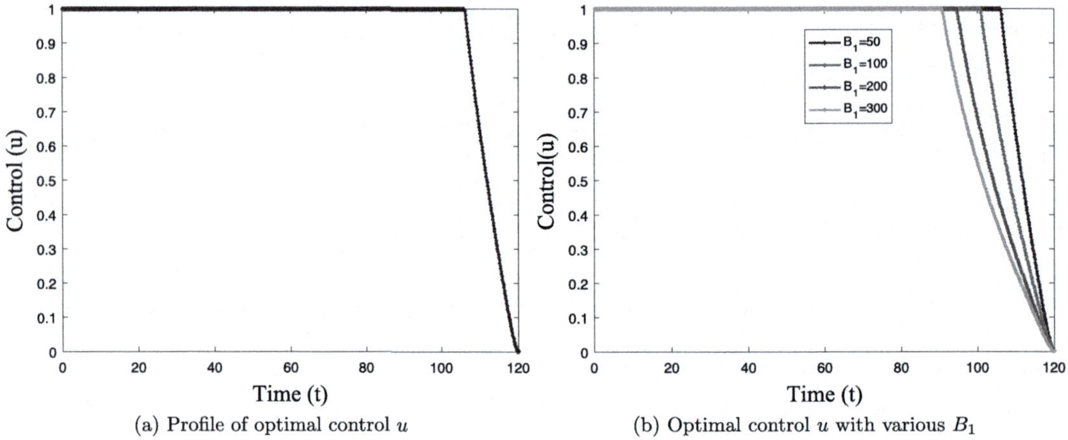


**FIGURE 14.6** Optimal control profiles.

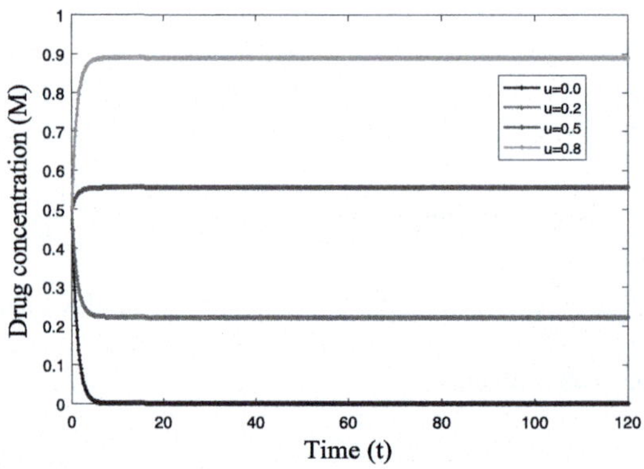


**FIGURE 14.7** Drug concentration (M) with different constant treatment rate.

Furthermore, we also perform some numerical simulations with different weights in $B_1$ for the control variable $u$ and given in Fig. 14.6. Therefore from Fig. 14.6b, it is observed that if the weight $B_1$ increases, then the amount of control policy $u$ decreases. Additionally, various computational results were carried out with different treatment rate in the drug concentration and depicted in Fig. 14.7. It shows that an increase in the constant treatment rate increases the drug concentration and remains maximum level intake which helps in controlling the cancer cell growth among the population.

## 14.5 Conclusion

We investigated an optimal control problem for a tumor-immune model with targeted chemotherapy. The model considered here, an extension of Liu and Liu (2017), consists of four compartments which describes the interaction of tumor cell, effector-immune cell, circulating lymphocyte, and drug chemotherapy. By introducing the targeted chemotherapy, Liu et al. claimed that we can eliminate only a small size tumor but not a tumor in large size. Therefore the main goal in this work is to minimize the tumor cell population and the cost of drug associated. For this, we formulated a optimal control problem and the cost functional to be minimized. We implemented the optimal control theory and incorporated the control intervention by replacing the constant treatment rate $V_M$ with variable $u(t)$. We established the existence of an optimal solution and its characterization of optimal control by PMP for the proposed control problem. With the qualitative analysis in Liu and Liu (2017) and the effect of control strategy, we perform some numerical simulations to illustrate the theoretical findings. Finally, we concluded that the implemented control strategy with targeted chemotherapy is more effective than the regular chemotherapy in reducing the tumor cell and its cost burden.

## Declaration of competing interest

The author has no conflict of interests to declare that are relevant to the content of this article.

## Data availability

The datasets generated during and/or analyzed during the current study are available from the corresponding author on reasonable request.

## References

Alkama, M., Larrache, A., Rachik, M., Elmouki, I., 2018. Optimal duration and dosage of BCG intravesical immunotherapy: a free final time optimal control approach. Mathematical Methods in the Applied Sciences 41, 2209–2219.

Chan, B.S., Yu, P., 2012. Bifurcation analysis in a model of cytotoxic T-lymphocyte response to viral infections. Nonlinear Analysis: Real World Applications 13, 64–77.

de Pillis, L.G., Radunskaya, A.E., 2001. A mathematical tumor model with immune resistance and drug therapy: an optimal control approach. Journal of Theoretical Medicine 3, 79.

de Pillis, L.G., Radunskaya, A.E., Wiseman, C.L., 2005. A validated mathematical model of cell-mediated immune response to tumor growth. Cancer Research 65, 7950–7958.

de Pillis, L.G., Gu, W., Fister, K.R., Head, T., Maples, K., Murugan, A., 2007. Chemotherapy for tumors: an analysis of the dynamics and a study of quadratic and linear optimal controls. Mathematical Biosciences 209, 292–315.

Dehingia, K., Sarmah, H.K., Hosseini, K., Sadri, K., Salahshour, S., Park, C., 2021. An optimal control problem of immuno-chemotherapy in presence of gene therapy. AIMS Mathematics 6, 11530–11549.

Dehingia, K., Yao, S.W., Sadri, K., Das, A., Sarmah, H.K., Zeb, A., Inc, M., 2022. A study on cancer-obesity-treatment model with quadratic optimal control approach for better outcomes. Results in Physics, 105963.

Dhar, B., Gupta, P.K., 2021. A numerical approach of tumor-immune model with B cells and monoclonal antibody drug by multi-step differential transformation method. Mathematical Methods in the Applied Sciences 44, 4058–4070.

Engelhart, M., Lebiedz, D., Sager, S., 2011. Optimal control for selected cancer chemotherapy ODE models: a view on the potential of optimal schedules and choice of objective function. Mathematical Biosciences 229, 123–134.

Fister, K.R., Panetta, J.C., 2000. Optimal control applied to cell-cycle specific cancer chemotherapy. SIAM Journal on Applied Mathematics 60 (3), 1059.

Fister, K.R., Panetta, J.C., 2003. Optimal control applied to competing chemotherapeutic cell-kill strategies. SIAM Journal on Applied Mathematics 63 (6), 1954.

Fleming, W.H., Rishel, R.W., 1975. Deterministic and Stochastic Optimal Control. Springer, New York.

Glassman, P.M., Balthasar, J.P., 2014. Mechanistic considerations for the use of mono- clonal antibodies for cancer therapy. Cancer Biology & Medicine 11, 20–33.

Kirschner, D., Panetta, J.C., 1998. Modeling immunotherapy of the tumor–immune interaction. Journal of Mathematical Biology 37, 235–252.

Kuznetsov, V.A., Makalkin, I.A., Taylor, M.A., Perelson, A.S., 1994. Nonlinear dynamics of immunogenic tumors: parameter estimation and global bifurcation analysis. Bulletin of Mathematical Biology 56, 295–321.

Le, T., Su, S., Kirshtein, A., Shahriyari, L., 2021. Data-driven mathematical model of osteosarcoma. Cancers 13, 2367.

Lenhart, S.M., Workman, J.T., 2007. Optimal Control Applied to Biological Models. CRC Press, Boca Raton, FL.

Li, Q., Xiao, Y., 2022. Modeling the virus-induced tumor-specific immune response with delay in tumor virotherapy. Communications in Nonlinear Science and Numerical Simulation 108, 106196.

Liu, H.Y., Yang, H.L., Yang, L.G., 2021. Dynamics analysis in a tumor-immune system with chemotherapy. Chinese Physics B 30, 058201.

Liu, P., Liu, X., 2017. Dynamics of a tumor-immune model considering targeted chemotherapy. Chaos, Solitons and Fractals 98, 7–13.

Pang, L., Zhao, Z., Song, X., 2016. Cost-effectiveness analysis of optimal strategy for tumor treatment. Chaos, Solitons and Fractals 87, 293–301.

Pinho, S.T.R., Bacelar, F.S., Andrade, R.F.S., Freedman, H.I., 2013. A mathematical model for the effect of anti-angiogenic therapy in the treatment of cancer tumors by chemotherapy. Nonlinear Analysis: Real World Applications 14, 815–828.

Pontryagin, L.S., Boltyanskii, V.G., Gamkrelidze, R.V., Mishchenko, E.F., 1964. The Mathematical Theory of Optimal Processes. A Pergamon Press Book, the Macmillan Co., New York.

Rihan, F.A., 2021. Delay differential equations of tumor-immune system with treatment and control. In: Delay Differential Equations and Applications to Biology. Springer, Singapore, pp. 167–189.

Rocha, A.M.A., Costa, M.F.P., Fernandes, E.M., 2018. On a multiobjective optimal control of a tumor growth model with immune response and drug therapies. International Transactions in Operational Research 25, 269–294.

Sharma, S., Samanta, G., 2016. Analysis of the dynamics of a tumor–immune system with chemotherapy and immunotherapy and quadratic optimal control. Differential Equations and Dynamical Systems 24, 149–171.

Steliarova-Foucher, E., Colombet, M., Ries, L.A.G., 2017. International incidence of childhood cancer, 2001-10: a population-based registry study. Lancet Oncology 18, 719–731.

Weiner, G.J., 2015. Building better monoclonal antibody-based therapeutics. Nature Reviews. Cancer 15, 361–370.

World Health Organization, 2020. Assessing national capacity for the prevention and control of noncommunicable diseases: report of the 2019 global survey. World Health Organization, Geneva.

World Health Organization, 2021. CureAll framework: WHO global initiative for childhood cancer: increasing access, advancing quality, saving lives. World Health Organization. https://apps.who.int/iris/handle/10665/347370.

# 15

# Quantitative analysis of structural vibration problems by using modified physics-informed neural networks


Shivani Saini[a], Ramesh Kumar Vats[a], and Arup Kumar Sahoo[b]

[a]*Department of Mathematics and Scientific Computing, National Institute of Technology, Hamirpur, Himachal Pradesh, India* [b]*Hatter Department of Marine Technologies, University of Haifa, Haifa, Israel*


## 15.1 Introduction

Deep learning (DL) has demonstrated remarkable effectiveness for handling problems that involve high dimensionality and strong nonlinearity. As a result, it has achieved significant success across a wide range of applications, including speech recognition, image classification, natural language processing, and high-dimensional data approximate modeling (Chakraverty et al., 2025; Rao et al., 2024a; Saini et al., 2025; Kumar and Vats, 2025). Because of its strong learning capability and adaptability, it has arisen as a promising approach for handling complex dynamical systems governed by differential equations, integral equations, or algebraic equations that are often difficult to solve using traditional numerical and analytical methods. Moreover, it offers an effective alternative for scientific and engineering simulations. In recent years, the utilization of neural networks to approximate solutions of differential equations has grown rapidly. Earlier approaches were mostly based on trial solutions that automatically satisfied the initial conditions (ICs) and boundary conditions (BCs) (Lagaris et al., 1998). While this idea was conceptually attractive, such artificial neural network (ANN) algorithms had several drawbacks, such as the trial functions often needed to be redesigned for each new problem, training being sensitive to the scaling of variables, and the models lacking mechanisms to explicitly enforce the governing physics throughout the domain of the system. As a result, these methods often exhibit limited effectiveness when applied for complex geometries, nonlinear behavior, or higher-dimensional systems.

To address the limitations, Raissi et al. (2019) introduced the physics-informed neural network (PINN) methodology, enabling the solution process to remain consistent with the underlying physics. PINNs embed the underlying ODE and PDE constraints in the loss function, thereby guiding the network to approximate solutions that both fulfill available data and satisfy the governing differential equations. This integration of


**Advanced Computational and Mathematical Approaches in Applied Differential Equations**
**https://doi.org/10.1016/B978-0-44-345743-2.00026-8**

physics with DL significantly improves accuracy. Another advantage of PINNs is that they are mesh-free, which enables them to provide a closed form of solution. It also mitigates the curse of dimensionality that often affects numerical techniques, including the finite difference method and finite element methods. Over the past few years, PINNs and their variants have been successfully applied across a broad areas, such as scientific (Sahoo and Klein, 2026; Kumar et al., 2025) and engineering (Yang et al., 2020; Sahoo and Klein, 2025), fluid mechanics (Kumar et al., 2023), solid and structural mechanics (Hu et al., 2024), geophysics (Sahoo et al., 2025b; Yadav et al., 2025), system health monitoring (Chang et al., 2018), and structural vibration and dynamics (Sahoo et al., 2025a). Early developments demonstrated that PINNs can approximate modal responses, wave propagation, and vibration characteristics in continuous structures without relying on mesh-based numerical discretization. These pioneering works led to the way for extending PINNs to more challenging structural vibration problems, such as forced responses, foundation interactions, damping effects, and parameter-dependent systems.

In structural system identification, recent DL approaches have sought to enhance reliability by structurally incorporating governing equations into neural network architectures. Building on this philosophy, Lai et al. (2021) introduced a physics-informed neural ODE framework that couples a known structural dynamics system with a learned nonlinear discrepancy term, thereby maintaining physical consistency while capturing unmodeled behaviors. Extending this direction, Chatterjee et al. (2024) implemented PINNs in MATLAB® for forward and inverse vibration problems and highlighted that standard formulations suffer from scaling-related convergence issues. To mitigate these limitations, they proposed a modified loss strategy that substantially improved accuracy without additional computational overhead. Further advancing the application of PINNs in structural vibration, Kapoor et al. (2023) investigated Euler–Bernoulli and Timoshenko beam systems on a Winkler foundation, demonstrating that PINNs can simultaneously handle both the forward and inverse vibration analyses, even under noisy data. Complementing these PINN studies with classical perspectives, Sunday et al. (2016) developed a sixth-order collocation-based numerical scheme for damped free vibrations in mass-spring systems, establishing its stability and convergence as an important benchmark for evaluating neural network solvers.

In the context of fluid structure interaction, Cheng et al. (2021) and Li et al. (2025) developed PINN frameworks to analyze vortex-induced vibrations of cylinder structures by embedding both the Navier–Stokes equations and the structural dynamics into a unified loss formulation. Their approaches were able to capture the flow field and hydrodynamic forces with good accuracy, while reducing computational cost compared with traditional computational fluid dynamics methods. Building on these ideas, Chen et al. (2023) extended the methodology to perform second-order analyses of slender steel beam columns using adaptive loss weighting and transfer learning, achieving notable improvements in data efficiency and predictive accuracy. Well-established numerical schemes, for example, finite element, finite difference, and various spectral methods continue to be robust tools for vibration analysis (Delic, 2025; Rao et al., 2024b). However, these approaches often depend on mesh generation, boundary discretization, and substantial computational resources, which become especially demanding for nonlinear or high-dimensional systems. In contrast, PINNs provide a mesh-free alternative that incorporates the governing physical laws within the learning process, making them an attractive option for complex vibration problems.

Despite their advantage, applying PINNs to structural analysis in practical settings remains challenging. Their performance can reduce in accuracy for systems exhibiting strong nonlinearity, high-frequency oscillations, or stiffness in the vibration system. PINNs have sometimes converged slowly, become unstable, or produced incorrect phase and amplitude in dynamic systems. Furthermore, if the data and physics parts of the loss function are not weighted properly, the network may struggle to learn the actual vibration characteristics. These limitations have motivated the development of improved variants, such as gradient-enhanced physics-informed neural networks (G-PINNs) (Yu et al., 2022), adaptive loss weighting schemes, and domain decomposition approaches. G-PINNs, in particular, incorporate derivative information into the loss formulation, leading to enhanced stability, accuracy, and robustness in dynamic simulations.

Following are the outcomes of this chapter:

1. G-PINNs is implemented for the vibration systems, such as free and forced vibrations of mass-spring systems and longitudinal vibrations of rods.
2. The performance of PINNs and G-PINNs is compared based on training convergence, error accuracy, stability, and the reliability of the predicted response.
3. The results show that including gradient information improves accuracy and the physical consistency of the predicted vibration system.

This chapter is organized as follows: Section 15.2 provides the preliminaries, including the mathematical formulations of the vibration problems and the theoretical foundations of ANNs, PINNs, and G-PINNs. Section 15.3 presents the numerical experiments along with comparative analyses and performance evaluations of the proposed models. Section 15.4 presents the final summary of findings and describes several potential directions for continued research.

## 15.2 Preliminaries

This section summarizes the governing equations of vibration systems and simultaneously introduces ANN, PINN, and G-PINN based frameworks for learning their responses while respecting the underlying physical laws.

### 15.2.1 A mass–spring system with a single degree of freedom

In a mass-spring single degree of freedom system, the mass $m$ is connected to a spring of stiffness $k$, and a viscous damping component is $c$. Let $u(t)$ represent the displacement of the mass measured from its equilibrium position. Applying Newton's second law, the equation of motion is defined as Rao (2019)

$$m\ddot{u}(t) + c\dot{u}(t) + ku(t) = F(t), \tag{15.1}$$

where $F(t)$ is an external force with respect to time.

**(i) An undamped vibration mass-spring free system**
If the system is neither externally forced nor damped ($F(t) = 0$, $c = 0$), then Eq. (15.1) reduces to

$$m\ddot{u} + ku = 0. \tag{15.2}$$

The inherent characteristics of the system are described by the natural circular frequency, natural frequency, and vibration period:

$$\omega_n = \sqrt{\frac{k}{m}}, \qquad f = \frac{\omega_n}{2\pi}, \qquad T = \frac{2\pi}{\omega_n}, \qquad n = 1, 2, 3, \ldots.$$

The displacement response can be written in the general form

$$u(t) = A_1 \cos(\omega_n t) + A_2 \sin(\omega_n t) = A \sin(\omega_n t + \phi), \tag{15.3}$$

where $A$ is the amplitude and $\phi$ phase are depend on the ICs:

$$A = \sqrt{u_0^2 + \left(\frac{\dot{u}_0}{\omega_n}\right)^2}, \qquad \phi = \arctan\left(\frac{u_0\omega_n}{\dot{u}_0}\right).$$

**(ii) An undamped vibration mass-spring force system**
In the force an undamped vibration mass-spring system has $F(t) = F_0 \sin(\omega t)$ and $c = 0$, Eq. (15.1) becomes

$$\ddot{u} + \omega_n^2 u = \frac{F_0}{m} \sin(\omega t). \tag{15.4}$$

The general solution is

$$u(t) = A_1 \cos(\omega_n t) + A_2 \sin(\omega_n t) + \frac{F_0/m}{\omega_n^2 - \omega^2} \sin(\omega t). \tag{15.5}$$

Here the final term corresponds to the stead-state dynamic response of the system. As the excitation frequency $\omega$ moves closer to the natural frequency $\omega_n$, the vibration amplitude rises rapidly, leading to the well-known phenomenon of resonance.

### 15.2.2 Longitudinal vibration of rod

A uniform elastic rod represents a distributed-parameter vibration system where the deformation varies continuously along its length. Unlike single degree of freedom models, both space and time influence the response. Considering a length of a rod $L$, cross-sectional area $A$, density $\rho$, and Young's modulus $E$, the displacement $u(x,t)$ is formulated by

$$\frac{\partial^2 u}{\partial t^2} = c^2 \frac{\partial^2 u}{\partial x^2}, \qquad c = \sqrt{\frac{E}{\rho}}, \tag{15.6}$$

where c is the axial wave-propagation speed. The BCs are

$$u(0,t) = 0, \qquad u(L,t) = 0. \tag{15.7}$$

The vibration response can be expressed as a summation of natural modes, each having a characteristic spatial shape and temporal oscillation:

$$u(x,t) = \sum_{n=1}^{\infty} \left( A_n \sin \frac{n\pi x}{L} \right) \left( B_n \cos(\omega_n t) + C_n \sin(\omega_n t) \right), \tag{15.8}$$

where $\omega_n$ is the natural frequency given by

$$\omega_n = n\pi \frac{c}{L}, \qquad n = 1, 2, 3, \ldots \tag{15.9}$$

### 15.2.3 Artificial neural networks

ANNs (Sahoo et al., 2025b) act as powerful function approximators constructed from multiple layers of interconnected neurons. Each neuron generates a weighted combination of its inputs, and then applies a nonlinear activation function, equipping the neural network to represent complex and highly nonlinear mappings. For a feedforward ANN within the $l$ layers, the input to output relationship is written as

$$\mathbf{y} = f_\theta(\mathbf{x}) = \sigma_l(W_l\, \sigma_{l-1}(W_{l-1} \cdots \sigma_1(W_1 \mathbf{x} + b_1) \cdots + b_{l-1}) + b_l), \tag{15.10}$$

where $W_i$ and $b_i$ denote the trainable weight matrices and bias vectors associated with the $i^{th}$ layer, $\sigma_i(\cdot)$ is the activation function, and $\theta = \{W_i, b_i\}_{i=1}^{l}$ denotes the complete set of parameters. Training the neural network involves minimizing an appropriate loss function that evaluates the discrepancy between the model predicted outputs and reference values. This minimization is typically performed using optimization strategies, including gradient descent, stochastic gradient descent, Adam, or the limited-memory Broyden–Fletcher–Goldfarb–Shanno (L-BFGS) algorithm.

### 15.2.4 Physics-informed neural networks

PINNs integrate the physical governing laws into the training of neural networks. Instead of basing the training process only on available data points, PINNs incorporate the underlying differential equation along with the associated ICs and BCs through automatic differentiation. This framework allows the neural network to approximate solutions that are consistent with both observed data and with the governing physics.

Consider a second-order PDE defined on a domain $\Omega \subset \mathbb{R}^N$ and time interval $[0, T]$:

$$\mathcal{F}\left(u,\ \nabla u,\ \partial_t u,\ \nabla^2 u,\ \partial_{tt} u,\ x,\ t\right) = 0, \qquad (x, t) \in \Omega \times [0, T], \tag{15.11}$$

subject to ICs and BCs:

$$u(x, t) = \mathcal{I}(x) \quad \text{for } t = 0, \qquad u(x, t) = \mathcal{B}(x, t) \quad \text{for } x \in \partial\Omega. \tag{15.12}$$

To approximate the $u(x, t)$, a neural network surrogate $\tilde{u}(x, t; \theta)$ with trainable parameters $\theta$ is constructed. During the training, collocation points are distributed across the PDE domain as well as on the boundaries where the ICs and BCs are applied.

The overall loss of the PINN is constructed as

$$L_{\text{PINN}}(\theta) = \lambda_f L_f + \lambda_b L_b + \lambda_i L_i + \lambda_d L_d, \tag{15.13}$$

where

$$L_f = \frac{1}{|\mathcal{T}_f|} \sum_{(x,t) \in \mathcal{T}_f} \left|\mathcal{F}(\tilde{u},\ \nabla \tilde{u},\ \partial_t \tilde{u},\ \nabla^2 \tilde{u},\ \partial_{tt} \tilde{u},\ x,\ t)\right|^2, \tag{15.14}$$

$$L_b = \frac{1}{|\mathcal{T}_b|} \sum_{(x,t) \in \mathcal{T}_b} \left|\tilde{u}(x, t; \theta) - \mathcal{B}(x, t)\right|^2, \tag{15.15}$$

$$L_i = \frac{1}{|\mathcal{T}_i|} \sum_{x \in \mathcal{T}_i} \left|\tilde{u}(x, 0; \theta) - \mathcal{I}(x)\right|^2, \tag{15.16}$$

$$L_d = \frac{1}{|\mathcal{T}_d|} \sum_{(x,t) \in \mathcal{T}_d} \left|\tilde{u}(x, t; \theta) - u(x, t)\right|^2. \tag{15.17}$$

The coefficients $\lambda_f$, $\lambda_b$, $\lambda_i$, and $\lambda_d$ in the total loss function Eq. (15.13) serve as weighting factors that balance the effects of the PDE residual, BCs, ICs, and data terms. These weights decide how much importance each physical constraint receives during training.

### 15.2.5 Gradient-enhanced physics-informed neural networks

PINNs incorporate the differential equation within their loss formulation, enabling the network to learn physics-consistent solutions. However, minimizing only the residual can be inadequate for problems exhibiting oscillatory behavior or higher-order modes. This can cause slow convergence, unwanted fluctuations in the solution, and lower accuracy in stiff or fast changing systems. To address these issues, G-PINNs improve the PINN by also penalizing the spatial and temporal gradients of the residual, making the solution more stable and physically consistent.

Let $\mathcal{R}(x,t)$ denote the PDE residual defined in Eq. (15.11). For an exact solution to be satisfied, we have $\mathcal{R}(x,t) = 0$, its spatial and temporal derivatives also must vanish:

$$\nabla_{x,t}\mathcal{R}(x,t) = \mathbf{0}, \qquad (x,t) \in \Omega \times [0,T]. \tag{15.18}$$

Therefore the total loss of the G-PINN is constructed as

$$L_{\text{G-PINN}}(\theta) = L_{\text{PINN}}(\theta) + \sum_{j=1}^{N} \lambda_{g_j} L_{g_j}, \tag{15.19}$$

where,

$$L_{g_j} = \frac{1}{|\mathcal{T}_{g_j}|} \sum_{(x,t)\in\mathcal{T}_{g_j}} \left| \frac{\partial \mathcal{R}(x,t)}{\partial x_j} \right|^2. \tag{15.20}$$

The coefficients $\lambda_{g_j}$ regulate the strength of the gradient constraints in the loss function. Increasing these values amplifies the penalty on residual fluctuations, thereby reducing nonphysical oscillations and promoting a smoother and more accurate learned the solution. When combined with the PINNs loss weights, these gradient penalties provide a flexible mechanism to balance the importance of the PDE residual and its derivatives, resulting in improved physical consistency and faster convergence.

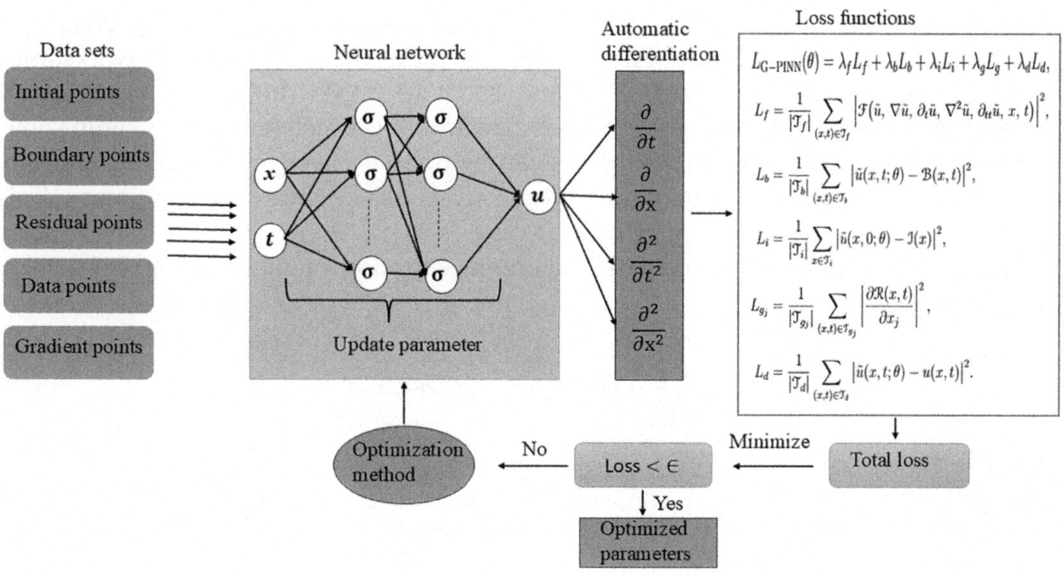


**FIGURE 15.1** Framework of the gradient-enhanced physics-informed neural network.

Overall, the incorporation of gradient-based residual terms substantially strengthens the enforcement of the underlying physics, making G-PINNs particularly effective for vibration problems governed by stiff operators or complex oscillatory dynamics. The complete framework of the G-PINN employed in this work is depicted in Fig. 15.1. The illustration of how gradient-enhanced residual terms are integrated into the PINN architecture.

## 15.3 Numerical simulations

In this section, three different vibration problems are examined to evaluate the performance of the PINN and G-PINN frameworks: (a) an undamped vibration mass-spring free system, (b) an undamped vibration mass-spring force system, and (c) longitudinal vibration of a rod. For each problem, solutions have been obtained using PINN and G-PINN, and then compared against the corresponding exact solutions. Graphical and tabular representations have been provided for comprehensive insight into the solution. These multiview comparisons allow a detailed assessment of the predictive capability of both approaches.

The accuracy of the predicted solutions is quantified using the absolute ($E_{\text{abs}}$) and relative error ($E_{\text{rel}}$) metrics, defined as

$$E_{\text{abs}} = \left|u_{\text{exact}} - u_{\text{pred}}\right|, \qquad E_{\text{rel}} = \frac{\left|u_{\text{exact}} - u_{\text{pred}}\right|}{\left|u_{\text{exact}}\right|}, \tag{15.21}$$

where $u_{\text{exact}}$ and $u_{\text{pred}}$ denote the exact solutions and predicted solutions, respectively.

All simulations in this study were carried out on an Acer Windows 11 workstation equipped with an AMD Ryzen 7 CPU and 8 GB of RAM. The neural models were implemented in Python 3.11.1 without relying on any GPU acceleration. This setup shows that both PINN and G-PINN architectures can be trained effectively using conventional desktop hardware, making the methodology practical and accessible for vibration analysis.

**Example 15.3.1.** Let us consider an undamped vibration of mass-spring free system described by

$$\frac{d^2u}{dt^2} + 64u = 0, \qquad t \in [0, 3], \tag{15.22}$$

with ICs,

$$u(0) = \frac{1}{4}, \qquad \frac{du}{dt}(0) = 1. \tag{15.23}$$

The exact solution of the system, as given in Sunday et al. (2016), is

$$u(t) = \frac{\sqrt{5}}{8}\cos\left(8t - 0.46\right). \tag{15.24}$$

Let $\tilde{u}(t;\theta)$ be the neural network approximation of the displacement. The residual and its temporal derivative are defined as

$$\mathcal{R}(t;\theta) = \frac{d^2\tilde{u}}{dt^2}(t;\theta) + 64\,\tilde{u}(t;\theta), \tag{15.25}$$

$$\mathcal{R}_t(t;\theta) = \frac{\partial f}{\partial t}(t;\theta). \tag{15.26}$$

The G-PINN loss components are given by

$$L_f = \frac{1}{|\mathcal{T}_f|}\sum_{t\in\mathcal{T}_f} \left|\mathcal{R}(t;\theta)\right|^2, \tag{15.27}$$

$$L_g = \frac{1}{|\mathcal{T}_g|}\sum_{t\in\mathcal{T}_g} \left|\mathcal{R}_t(t;\theta)\right|^2, \tag{15.28}$$

$$L_i = \frac{1}{|\mathcal{T}_i|}\sum_{t\in\mathcal{T}_i} \left(\left|\tilde{u}(0;\theta) - \frac{1}{4}\right|^2 + \left|\frac{d\tilde{u}}{dt}(0;\theta) - 1\right|^2\right). \tag{15.29}$$

Thus the total G-PINN loss is expressed as

$$L_{\text{G-PINN}} = \lambda_f L_f + \lambda_g L_g + \lambda_i L_i. \tag{15.30}$$

The PINN and G-PINN models were employed in a fully connected feedforward neural network comprising an input layer, 3 hidden layers of 20 neurons each, and an output layer. The `tanh` activation is used, and the parameters are assigned using the Glorot uniform to promote stable training. Both networks are trained for 10,000 epochs with the Adam optimizer using the learning rate of 0.0001, followed by fine-tuning with the help of L-BFGS algorithm to achieve higher accuracy and faster convergence. Since the architectural design and training setup are identical for the PINN and G-PINN, any improvement observed in the G-PINN results arises solely from the inclusion of gradient-enhanced residual terms.

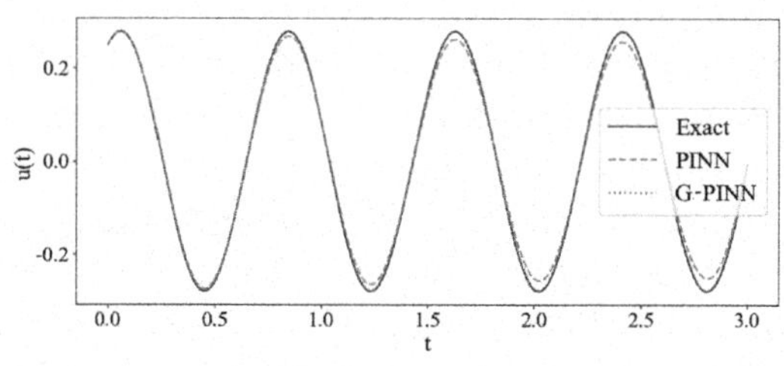


**FIGURE 15.2** Displacement of an undamped vibration mass-spring free system for Example 15.3.1.

Fig. 15.2 illustrates the displacement of an undamped mass–spring free vibration system, where the PINN and G-PINN are evaluated against the exact solution. All three curves exhibit periodic oscillations with the same frequency, but slight amplitude differences are noticeable between the PINN and exact solutions. The G-PINN prediction closely overlaps with the exact solution, accurately capturing both amplitude and phase throughout the vibration period. This confirms that the inclusion of gradient information improves the precision and physical consistency of the predicted displacement.

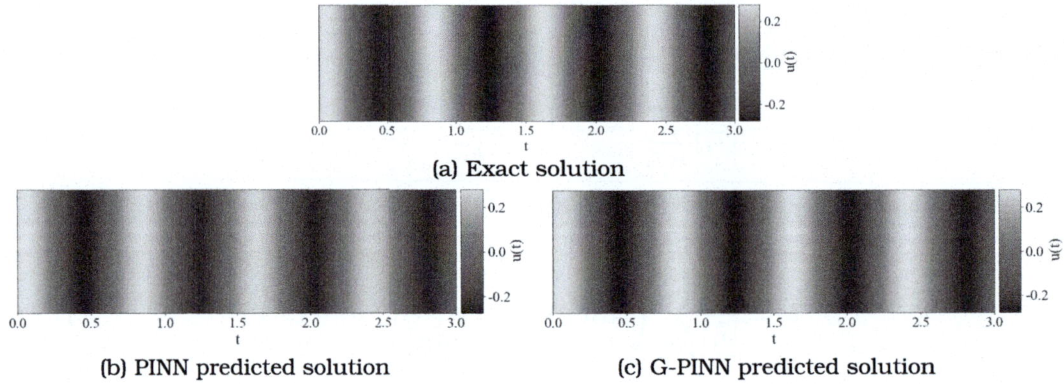


**FIGURE 15.3** Comparison of the exact solution with the PINN and G-PINN predictions for Example 15.3.1.

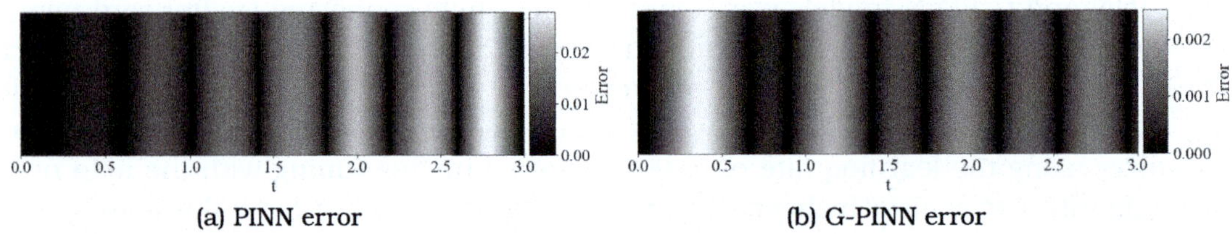


**FIGURE 15.4** Comparison of absolute error between PINN and G-PINN for Example 15.3.1.

Fig. 15.3 presents a comparison between the PINN and G-PINN predictions and the corresponding exact solution. Fig. 15.4 presents the heat maps of absolute error, where G-PINN clearly produces smaller error than PINN. The mean absolute error for the PINN model is $10^{-2}$, whereas for the G-PINN model it is significantly lower, around $10^{-3}$. This numerical comparison confirms that the G-PINN demonstrates a significant reduction in error, consistent with the smoother error observed in the color maps. Table 15.1 presents the exact solutions, PINN, and G-PINN predicted solutions along with their corresponding absolute and relative errors for $t \in [0, 3]$. The numerical results show that the G-PINN model consistently yields lower errors, with a maximum absolute error of approximately $10^{-3}$ and an average relative error below $10^{-2}$. In contrast, the PINN exhibits larger deviations, reaching a maximum absolute error $10^{-2}$ and a relative error exceeding $10^{-1}$ near the vibration peaks.

**Table 15.1** Comparison of solutions and errors at $t \in [0, 3]$ for Example 15.3.1.

| $t$ | Exact | PINN | G-PINN | $E_{\mathbf{abs}}$ | | $E_{\mathbf{rel}}$ | |
|---|---|---|---|---|---|---|---|
| | | | | PINN | G-PINN | PINN | G-PINN |
| 0.000000 | 0.250000 | 0.250000 | 0.250000 | 0.000e+00 | 0.000e+00 | 0.000e+00 | 0.000e+00 |
| 0.300000 | −0.100072 | −0.096847 | −0.097797 | 3.225e-03 | 2.275e-03 | 3.223e-02 | 2.273e-02 |
| 0.600000 | −0.102334 | −0.102315 | −0.101603 | 1.900e-05 | 7.310e-04 | 1.857e-04 | 7.146e-03 |
| 0.900000 | 0.251076 | 0.243073 | 0.251099 | 8.003e-03 | 2.300e-05 | 3.188e-02 | 9.162e-05 |
| 1.200000 | −0.268153 | −0.253909 | −0.266162 | 1.424e-02 | 1.991e-03 | 5.312e-02 | 7.423e-03 |
| 1.500000 | 0.143172 | 0.131838 | 0.142371 | 1.133e-02 | 8.010e-04 | 7.916e-02 | 5.598e-03 |
| 1.800000 | 0.056410 | 0.053494 | 0.057214 | 2.916e-03 | 8.040e-04 | 5.170e-02 | 1.426e-02 |
| 2.100000 | −0.226411 | −0.211357 | −0.225625 | 1.505e-02 | 7.860e-04 | 6.648e-02 | 3.472e-03 |
| 2.400000 | 0.277681 | 0.254712 | 0.276557 | 2.297e-02 | 1.124e-03 | 8.275e-02 | 4.047e-03 |
| 2.700000 | −0.183335 | −0.163107 | −0.181864 | 2.023e-02 | 1.471e-03 | 1.103e-01 | 8.025e-03 |
| 3.000000 | −0.007153 | −0.010008 | −0.007800 | 2.855e-03 | 6.470e-04 | 3.992e-01 | 9.049e-02 |

**Example 15.3.2.** Consider an undamped vibration of a mass-spring force system as

$$\frac{d^2u}{dt^2} + 4u = \sin(3t), \qquad t \in [0, 2\pi]. \tag{15.31}$$

The ICs are

$$u(0) = 0, \qquad \frac{du}{dt}(0) = 0. \tag{15.32}$$

The exact solution of the system, as given in Chatterjee et al. (2024), is

$$u(t) = -\frac{1}{5}\big(\sin(3t) - 1.5\,\sin(2t)\big). \tag{15.33}$$

Let $\tilde{u}(t;\theta)$ be the neural network approximation of $u(t)$. The residual and its temporal derivative are defined as

$$\mathcal{R}(t;\theta) = \frac{d^2\tilde{u}}{dt^2}(t;\theta) + 4\,\tilde{u}(t;\theta) - \sin(3t), \tag{15.34}$$

$$\mathcal{R}_t(t;\theta) = \frac{\partial \mathcal{R}}{\partial t}(t;\theta). \tag{15.35}$$

The G-PINN loss components are given by

$$L_f = \frac{1}{|\mathcal{T}_f|}\sum_{t\in\mathcal{T}_f} \left|\mathcal{R}(t;\theta)\right|^2, \tag{15.36}$$

$$L_g = \frac{1}{|\mathcal{T}_g|}\sum_{t\in\mathcal{T}_g} \left|\mathcal{R}_t(t;\theta)\right|^2, \tag{15.37}$$

$$L_i = \frac{1}{|\mathcal{T}_i|}\sum_{t\in\mathcal{T}_i} \left(\left|\tilde{u}(0;\theta) - 0\right|^2 + \left|\frac{d\tilde{u}}{dt}(0;\theta) - 0\right|^2\right). \tag{15.38}$$

Thus the total G-PINN loss is expressed as

$$L_{G-PINN} = \lambda_f L_f + \lambda_g L_g + \lambda_i L_i. \tag{15.39}$$

Both PINN and G-PINN use the fully connected architecture with 4 hidden layers of 60 neurons each and the `tanh` activation function. Glorot initialization and a fixed random seed are employed for reproducibility. Training starts with the Adam optimizer for 3000 epochs at a learning rate of 0.0001 and is then refined using the L-BFGS method.

Fig. 15.5 represents the displacement of an undamped vibration mass-spring force system. The results from the PINN and G-PINN models are compared with the exact solution. The G-PINN prediction closely matches the exact curve, whereas the PINN shows minor amplitude and phase differences.

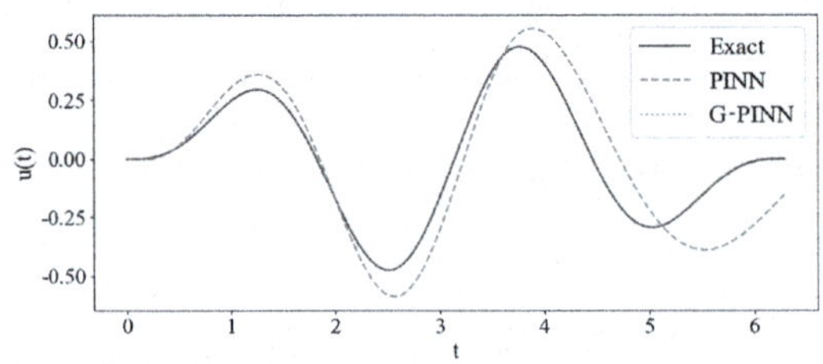


**FIGURE 15.5** Displacement of an undamped vibration mass-spring force system for Example 15.3.2.

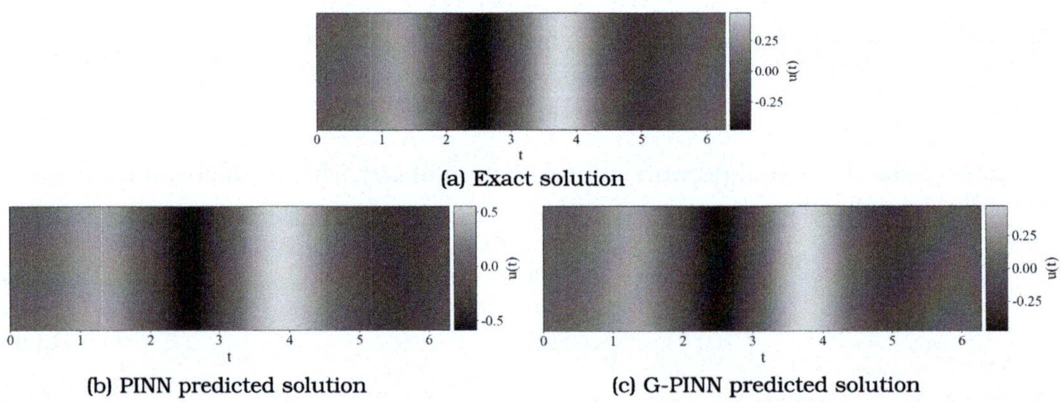


**FIGURE 15.6** Comparison of the exact solution with the PINN and G-PINN predictions for Example 15.3.2.

Fig. 15.6 compares the exact solution with the obtained solutions from the PINN and G-PINN models for the force vibration system. The G-PINN prediction exhibits a close match with the exact solution, accurately reproducing both the amplitude and phase of the oscillatory motion. In contrast, the PINN shows noticeable deviations near the peaks, indicating slower convergence and a phase lag in capturing the external excitation effects. Fig. 15.7 shows the corresponding heat maps of the absolute error for both models. The

G-PINN produces a more uniform and low-magnitude error around $10^{-3}$, whereas the PINN displays localized high-error regions up to $10^{-2}$, particularly where the displacement changes quickly. These observations confirm that incorporating gradient-based residual terms significantly enhances numerical accuracy and stability under forced vibration conditions.

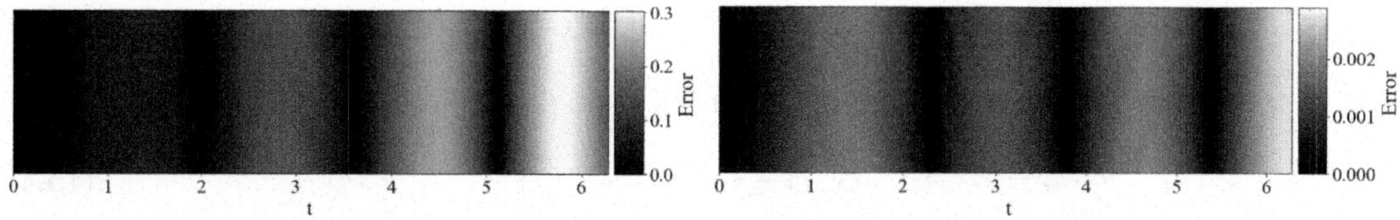


(a) PINN Error (b) G-PINN Error

**FIGURE 15.7** Comparison of absolute error between PINN and G-PINN for Example 15.3.2.

Table 15.2 provides the numerical results for the forced vibration problem using the PINN and G-PINN models. The tabulated values clearly show that G-PINN is closer to the exact solution, yielding consistently small errors throughout the entire time interval. The highest absolute error for G-PINN is about $10^{-3}$, whereas the PINN shows much larger deviations, reaching up to the order of $10^{-1}$ near the extreme vibration points. Similarly, the relative error in G-PINN remains below $10^{-2}$, while in PINN it becomes several times higher, and even exceeds unity at some points. These findings clearly show that the inclusion of gradient-based information in the G-PINN loss function enhances accuracy, reduces oscillatory deviations, and provides more stable and physically consistent results.

**Table 15.2** Comparison of solutions and errors at $t \in [0, 2\pi]$ for Example 15.3.2.

| $t$ | Exact | PINN | G-PINN | $E_{\text{abs}}$ | | $E_{\text{rel}}$ | |
|---|---|---|---|---|---|---|---|
| | | | | PINN | G-PINN | PINN | G-PINN |
| 0.000000 | 0.000000 | −0.000000 | −0.000000 | 0.000e+00 | 0.000e+00 | 0.000e+00 | 0.000e+00 |
| 0.628319 | 0.095164 | 0.114269 | 0.094404 | 1.911e-02 | 7.600e-04 | 2.009e-01 | 7.986e-03 |
| 1.256637 | 0.293893 | 0.357389 | 0.292128 | 6.349e-02 | 1.765e-03 | 2.161e-01 | 6.006e-03 |
| 1.884956 | −0.059236 | −0.038715 | −0.060462 | 2.052e-02 | 1.226e-03 | 3.463e-01 | 2.069e-02 |
| 2.513274 | −0.475528 | −0.584074 | −0.474918 | 1.085e-01 | 6.100e-04 | 2.282e-01 | 1.283e-03 |
| 3.141593 | −0.000943 | −0.123444 | 0.000485 | 1.225e-01 | 1.428e-03 | 1.299e+02 | 1.514e+00 |
| 3.769911 | 0.475528 | 0.531696 | 0.475597 | 5.617e-02 | 6.900e-05 | 1.181e-01 | 1.452e-04 |
| 4.398230 | 0.059236 | 0.290548 | 0.057516 | 2.313e-01 | 1.720e-03 | 3.906e+00 | 2.904e-02 |
| 5.026548 | −0.293893 | −0.235338 | −0.295282 | 5.855e-02 | 1.389e-03 | 1.992e-01 | 4.727e-03 |
| 5.654867 | −0.095164 | −0.380941 | −0.093986 | 2.858e-01 | 1.178e-03 | 3.004e+00 | 1.238e-02 |
| 6.283185 | 0.000000 | −0.155223 | 0.002743 | 1.552e-01 | 2.743e-03 | 0.000e+00 | 0.000e+00 |

**Example 15.3.3.** Let us examine the longitudinal vibration behavior of a rod subjected to

$$\frac{\partial^2 u}{\partial t^2} = \frac{\partial^2 u}{\partial x^2}, \qquad (x, t) \in [0, 1] \times [0, 1]. \tag{15.40}$$

The ICs are

$$u(x,0)=\sin(\pi x),\qquad \frac{\partial u}{\partial t}(x,0)=0, \tag{15.41}$$

and BCs are given by

$$u(0,t)=0,\qquad u(1,t)=0. \tag{15.42}$$

The exact solution of the system, as given in Alli et al. (2003), is

$$u(x,t)=\sin(\pi x)\cos(\pi t). \tag{15.43}$$

The neural network approximation to the exact solution $u(x,t)$ is represented by $\tilde{u}(x,t;\theta)$. The PDE residual and its derivatives are defined as

$$\mathcal{R}(x,t;\theta)=\frac{\partial^2\tilde{u}}{\partial t^2}(x,t;\theta)-\frac{\partial^2\tilde{u}}{\partial x^2}(x,t;\theta), \tag{15.44}$$

$$\mathcal{R}_t(x,t;\theta)=\frac{\partial\mathcal{R}}{\partial t}(x,t;\theta), \tag{15.45}$$

$$\mathcal{R}_x(x,t;\theta)=\frac{\partial\mathcal{R}}{\partial x}(x,t;\theta). \tag{15.46}$$

The G-PINN loss components are

$$L_f=\frac{1}{|\mathcal{T}_f|}\sum_{(x,t)\in\mathcal{T}_f}\left|\mathcal{R}(x,t;\theta)\right|^2, \tag{15.47}$$

$$L_g=\frac{1}{|\mathcal{T}_f|}\sum_{(x,t)\in\mathcal{T}_f}\left(\left|\mathcal{R}_t(x,t;\theta)\right|^2+\left|\mathcal{R}_x(x,t;\theta)\right|^2\right), \tag{15.48}$$

$$L_i=\frac{1}{|\mathcal{T}_i|}\sum_{x\in\mathcal{T}_i}\left(\left|\tilde{u}(x,0;\theta)-\sin(\pi x)\right|^2+\left|\frac{\partial\tilde{u}}{\partial t}(x,0;\theta)-0\right|^2\right), \tag{15.49}$$

$$L_b=\frac{1}{|\mathcal{T}_b|}\sum_{t\in\mathcal{T}_b}\left(\left|\tilde{u}(0,t;\theta)-0\right|^2+\left|\tilde{u}(1,t;\theta)-0\right|^2\right). \tag{15.50}$$

Thus the total G-PINN loss is expressed as

$$L_{\text{G-PINN}}=\lambda_f L_f+\lambda_g L_g+\lambda_i L_i+\lambda_b L_b. \tag{15.51}$$

The PINN and G-PINN models utilized in this study are implemented as fully connected feedforward neural networks composed of one input layer, 4 hidden layers consisting of 60 neurons each, and a single output layer that provides the approximate displacement $\tilde{u}(x,t)$ for the longitudinal vibration of a rod. The `tanh` is used as an activation function. All network parameters are initialized using the Glorot uniform scheme with a fixed random seed to ensure stable and reproducible training. Both models are optimized using the

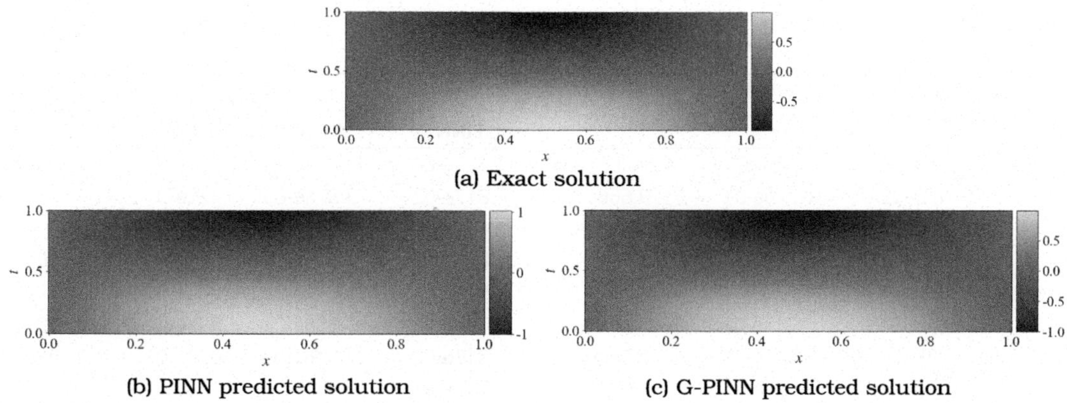


**FIGURE 15.8** Comparison of the exact solution with the PINN and G-PINN predictions for Example 15.3.3.

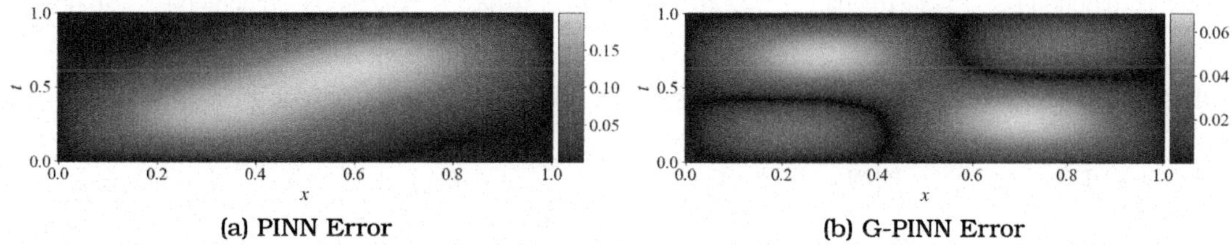


**FIGURE 15.9** Comparison of error between PINN and G-PINN for Example 15.3.3.

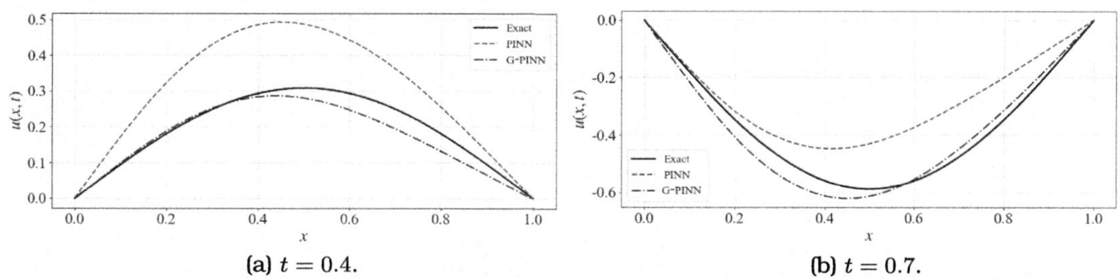


**FIGURE 15.10** Comparison of PINN and G-PINN predicted solutions with the exact solution (a) $t = 0.4$ s and (b) $t = 0.7$ s for $x \in [0, 1]$ for Example 15.3.3.

Adam algorithm for 20,000 epochs while using a learning rate of 0.0001. Because the architecture, hyperparameters, and training strategy are identical for the PINN and G-PINN, any improvement in the G-PINN results arises solely from the incorporation of gradient-enhanced residual terms in its loss formulation.

Fig. 15.8, the longitudinal vibration responses of the rod obtained using the exact, PINN, and G-PINN models are illustrated. The corresponding maps of absolute errors in Fig. 15.9 clearly show that the G-PINN achieves substantially lower error magnitudes across the entire spatial-temporal domain, reflecting its improved physical consistency and enhanced convergence stability. This comparison is further detailed in Figs. 15.10a

**Table 15.3** Comparison of solutions and errors at $t = 0.4$ s for $x \in [0, 1]$ for Example 15.3.3.

| $x$ | Exact | PINN | G-PINN | $E_{abs}$ | | $E_{rel}$ | |
|---|---|---|---|---|---|---|---|
| | | | | PINN | G-PINN | PINN | G-PINN |
| 0.00 | 0.000000 | 0.000202 | −0.001884 | 2.023e-04 | 1.884e-03 | 0.000e-00 | 0.000e-00 |
| 0.10 | 9.283e-02 | 1.710e-01 | 9.666e-02 | 7.817e-02 | 3.828e-03 | 8.421e-01 | 4.125e-02 |
| 0.20 | 1.765e-01 | 3.185e-01 | 1.825e-01 | 1.420e-01 | 6.054e-03 | 8.042e-01 | 3.429e-02 |
| 0.30 | 2.427e-01 | 4.243e-01 | 2.444e-01 | 1.816e-01 | 1.672e-03 | 7.486e-01 | 6.886e-03 |
| 0.40 | 2.849e-01 | 4.777e-01 | 2.744e-01 | 1.928e-01 | 1.049e-02 | 6.768e-01 | 3.682e-02 |
| 0.50 | 2.989e-01 | 4.783e-01 | 2.704e-01 | 1.794e-01 | 2.856e-02 | 5.999e-01 | 9.549e-02 |
| 0.60 | 2.849e-01 | 4.365e-01 | 2.394e-01 | 1.516e-01 | 4.546e-02 | 5.324e-01 | 1.596e-01 |
| 0.70 | 2.427e-01 | 3.573e-01 | 1.883e-01 | 1.146e-01 | 5.440e-02 | 4.725e-01 | 2.242e-01 |
| 0.80 | 1.765e-01 | 2.518e-01 | 1.279e-01 | 7.531e-02 | 4.863e-02 | 4.265e-01 | 2.755e-01 |
| 0.90 | 9.283e-02 | 1.296e-01 | 6.360e-02 | 3.672e-02 | 2.923e-02 | 3.955e-01 | 3.148e-01 |
| 1.00 | 0.000000 | −3.210e-04 | −2.465e-03 | 3.213e-04 | 2.465e-03 | 0.000e-00 | 0.000e-00 |

**Table 15.4** Comparison of solutions and errors at $t = 0.7$ s for $x \in [0, 1]$ for Example 15.3.3.

| $x$ | Exact | PINN | G-PINN | $E_{abs}$ | | $E_{rel}$ | |
|---|---|---|---|---|---|---|---|
| | | | | PINN | G-PINN | PINN | G-PINN |
| 0.00 | −0.000000 | 0.001139 | −0.000078 | 1.139e-03 | 7.773e-05 | 0.000e-00 | 0.000e-00 |
| 0.10 | −1.805e-01 | −1.707e-01 | −2.161e-01 | 9.843e-03 | 3.557e-02 | 5.454e-02 | 1.971e-01 |
| 0.20 | −3.432e-01 | −3.135e-01 | −4.036e-01 | 2.972e-02 | 6.037e-02 | 8.659e-02 | 1.759e-01 |
| 0.30 | −4.719e-01 | −4.056e-01 | −5.385e-01 | 6.632e-02 | 6.656e-02 | 1.406e-01 | 1.411e-01 |
| 0.40 | −5.540e-01 | −4.397e-01 | −6.065e-01 | 1.143e-01 | 5.251e-02 | 2.063e-01 | 9.478e-02 |
| 0.50 | −5.813e-01 | −4.212e-01 | −6.059e-01 | 1.602e-01 | 2.451e-02 | 2.757e-01 | 4.218e-02 |
| 0.60 | −5.540e-01 | −3.674e-01 | −5.494e-01 | 1.866e-01 | 4.618e-03 | 3.369e-01 | 8.336e-03 |
| 0.70 | −4.719e-01 | −2.880e-01 | −4.453e-01 | 1.839e-01 | 2.664e-02 | 3.899e-01 | 5.644e-02 |
| 0.80 | −3.432e-01 | −1.963e-01 | −3.112e-01 | 1.469e-01 | 3.196e-02 | 4.279e-01 | 9.313e-02 |
| 0.90 | −1.805e-01 | −9.867e-02 | −1.602e-01 | 8.185e-02 | 2.031e-02 | 4.536e-01 | 1.126e-01 |
| 1.00 | −0.000000 | 0.000833 | −0.000717 | 8.334e-04 | 7.167e-04 | 0.000e-00 | 0.000e-00 |

and 15.10b, where the PINN and G-PINN outputs at $t = 0.4$ s and $t = 0.7$ s are directly compared with the exact solution.

At both instances, the G-PINN solution follows the analytical solution curve almost perfectly, whereas the PINN shows underestimation near the boundaries and overshooting near the center of the rod. This demonstrates the improved accuracy of the gradient-enhanced formulation in capturing spatial oscillations. Tables 15.3 and 15.4 present a quantitative comparison of the exact, PINN, and G-PINN solutions along the rod length evaluated at $t = 0.4$ s and $t = 0.7$ s, respectively. The numerical results show that the G-PINN consistently yields smaller absolute and relative errors, typically of the order of $10^{-3}$ to $10^{-2}$, whereas the PINN errors reach up to $10^{-1}$ in some regions. Overall, the G-

PINN achieves more uniform accuracy across the spatial domain, confirming its superior performance and reliability for dynamic problems.

## 15.4 Conclusion

This chapter presents, the G-PINN framework for three classical vibration problems an undamped vibration mass-spring free and force system, and also the longitudinal vibration of a rod to demonstrate its improved capability over standard PINNs. By augmenting the loss function with spatial and temporal derivatives of the PDE residual, G-PINN effectively strengthens the enforcement of underlying physics, enabling the network to better capture sharp gradients and oscillatory patterns. Across all cases, G-PINN consistently delivered more accurate and physically consistent solutions by incorporating gradient-enhanced residual terms that reinforced the governing differential equations. The method captured oscillatory amplitudes, phase evolution, and transient behavior with higher fidelity in both free and forced vibration scenarios, while also producing smoother spatial-temporal displacement fields for the rod vibration problem. Despite these advantages, G-PINN requires additional computational cost due to higher-order derivative evaluations and may demand careful tuning of loss weights for different systems. Nevertheless, the results highlight its strong potential of this approach for solving complex dynamical systems. Future work may extend this framework to damped, nonlinear, or multidegree of freedom structures, develop adaptive weighting strategies, explore domain decomposition methods for long-time dynamics simulations, and incorporate uncertainty quantification to enhance robustness for real-world applications.

## CRediT authorship contribution statement

**S.S:** Conceptualization, visualization, software, experiment, writing original draft. **R.K.V:** Supervision, investigation, formal analysis, writing: review & editing. **A.K.S:** Software, validation, writing: review & editing.

## Funding

The authors confirm that they did not receive any funding to carry out this work.

## Declaration of competing interest

The authors confirm that they have no conflicts of interest related to this chapter.

## Data availability

No datasets were used or generated during the preparation of this article.

# References


Alli, H., Uçar, A., Demir, Y., 2003. The solutions of vibration control problems using artificial neural networks. Journal of the Franklin Institute 340 (5), 307–325.

Chakraverty, S., Sahoo, A.K., Mohapatra, D., 2025. Artificial Neural Networks and Type-2 Fuzzy Set: Elements of Soft Computing and Its Applications. Elsevier.

Chang, C.M., Lin, T.K., Chang, C.W., 2018. Applications of neural network models for structural health monitoring based on derived modal properties. Measurement 129, 457–470.

Chatterjee, T., Friswell, M.I., Adhikari, S., Khodaparast, H.H., 2024. Matlab implementation of physics informed deep neural networks for forward and inverse structural vibration problems. Aerospace Research Communications 2, 13194.

Chen, L., Zhang, H.Y., Liu, S.W., Chan, S.L., 2023. Second-order analysis of beam-columns by machine learning-based structural analysis through physics-informed neural networks. Advanced Steel Construction 19 (4), 411–420.

Cheng, C., Meng, H., Li, Y.Z., Zhang, G.T., 2021. Deep learning based on PINN for solving 2 DOF vortex induced vibration of cylinder. Ocean Engineering 240, 109932.

Delic, G., 2025. Numerical solution of ordinary differential equations. In: Guide to Numerical Algorithm Design and Development: Including Legacy Examples from Fortran and MathCAD in High Precision. Springer, pp. 197–232.

Hu, H., Qi, L., Chao, X., 2024. Physics-informed neural networks (PINN) for computational solid mechanics: numerical frameworks and applications. Thin-Walled Structures 205, 112495.

Kapoor, T., Wang, H., Núñez, A., Dollevoet, R., 2023. Physics-informed neural networks for solving forward and inverse problems in complex beam systems. IEEE Transactions on Neural Networks and Learning Systems 35 (5), 5981–5995.

Kumar, S., Sahoo, A.K., Chakraverty, S., 2023. Physics-informed machine learning framework for approximating the modified Degasperis-Procesi equation. In: 2023 International Conference on Ambient Intelligence Knowledge Informatics and Industrial Electronics (AIKIIE). IEEE, pp. 1–6.

Kumar, Y., Vats, R.K., 2025. A deep Kolmogorov–Arnold network framework for solving time–fractional partial differential equations. Applied Soft Computing, 114379.

Kumar, Y., Kumar Vats, R., Kumar Nain, A., Yadav, S., 2025. Robust deep learning framework using Hermite interpolation for time-dependent Caputo fractional differential equations. Mathematical Methods in the Applied Sciences, 1–30.

Lagaris, I.E., Likas, A., Fotiadis, D.I., 1998. Artificial neural networks for solving ordinary and partial differential equations. IEEE Transactions on Neural Networks 9 (5), 987–1000.

Lai, Z., Mylonas, C., Nagarajaiah, S., Chatzi, E., 2021. Structural identification with physics-informed neural ordinary differential equations. Journal of Sound and Vibration 508, 116196.

Li, J., Wu, X., Li, Z., 2025. Physics-informed neural network approach for solving vortex-induced vibration problems of cylinder. Applied Ocean Research 161, 104661.

Raissi, M., Perdikaris, P., Karniadakis, G.E., 2019. Physics-informed neural networks: a deep learning framework for solving forward and inverse problems involving nonlinear partial differential equations. Journal of Computational Physics 378, 686–707.

Rao, A., Vats, R.K., Yadav, S., 2024a. Numerical study of nonlinear time-fractional Caudrey–Dodd–Gibbon–Sawada–Kotera equation arising in propagation of waves. Chaos, Solitons and Fractals 184, 114941.

Rao, A., Vats, R.K., Yadav, S., 2024b. Analytical solution for time-fractional cold plasma equations via novel computational method. International Journal of Applied and Computational Mathematics 10 (1), 2.

Rao, S.S., 2019. Vibration of Continuous Systems. John Wiley & Sons.

Sahoo, A.K., Klein, I., 2025. MoRPI-PINN: a physics-informed framework for mobile robot pure inertial navigation. arXiv preprint. arXiv:2507.18206.

Sahoo, A.K., Klein, I., 2026. PiDR: physics-informed inertial dead reckoning for autonomous platforms. arXiv preprint. arXiv:2601.03040.

Sahoo, A.K., Kumar, S., Chakraverty, S., 2025a. Physics-informed neural network for vibration analysis of large membranes. Journal of Nonlinear, Complex and Data Science 25 (7–8), 505–521.

Sahoo, A.K., Sahoo, S.K., Chakraverty, S., 2025b. CL-PINN: a physics-informed neural networks framework for equatorial tsunami wave propagation in geophysical ocean. IEEE Access.

Saini, S., Vats, R.K., Sahoo, A.K., 2025. A-PINN: auxiliary physics-informed neural networks for structural vibration analysis in continuous Euler-Bernoulli beam. arXiv preprint. arXiv:2601.00866.
Sunday, J., James, A., Odekunle, M., Adesanya, A., 2016. Solutions to free undamped and free damped motion problems in mass-spring systems. American Journal of Computational and Applied Mathematics 6 (2), 82–91.
Yadav, S., Vats, R.K., Rao, A., 2025. Application of extended residual power series method for time-fractional Zakharov–Kuznetsov equations in ocean-based coastal wave. Pramāna 99 (3), 97.
Yang, L., Zhang, D., Karniadakis, G.E., 2020. Physics-informed generative adversarial networks for stochastic differential equations. SIAM Journal on Scientific Computing 42 (1), A292–A317.
Yu, J., Lu, L., Meng, X., Karniadakis, G.E., 2022. Gradient-enhanced physics-informed neural networks for forward and inverse PDE problems. Computer Methods in Applied Mechanics and Engineering 393, 114823.

# Index

Printed and bound by CPI Group (UK) Ltd, Croydon, CR0 4YY
14/09/2026
02216396-0003